Introduction to
Information
Theory
and
Data
Compression

Second Edition

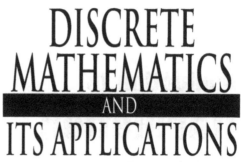

DISCRETE MATHEMATICS AND ITS APPLICATIONS

Series Editor

Kenneth H. Rosen, Ph.D.

AT&T Laboratories, Middletown, New Jersey

Abstract Algebra Applications with Maple,
Richard E. Klima, Ernest Stitzinger, and Neil P. Sigmon

Algebraic Number Theory, *Richard A. Mollin*

An Atlas of the Smaller Maps in Orientable and Nonorientable Surfaces,
David M. Jackson and Terry I. Visentin

An Introduction to Crytography, *Richard A. Mollin*

Combinatorial Algorithms: Generation Enumeration and Search,
Donald L. Kreher and Douglas R. Stinson

The CRC Handbook of Combinatorial Designs,
Charles J. Colbourn and Jeffrey H. Dinitz

Cryptography: Theory and Practice, Second Edition, *Douglas R. Stinson*

Design Theory, *Charles C. Lindner and Christopher A. Rodgers*

Frames and Resolvable Designs: Uses, Constructions, and Existence,
Steven Furino, Ying Miao, and Jianxing Yin

Fundamental Number Theory with Applications, *Richard A. Mollin*

Graph Theory and Its Applications, *Jonathan Gross and Jay Yellen*

Handbook of Applied Cryptography,
Alfred J. Menezes, Paul C. van Oorschot, and Scott A. Vanstone

Handbook of Constrained Optimization,
Herbert B. Shulman and Venkat Venkateswaran

Handbook of Discrete and Combinatorial Mathematics, *Kenneth H. Rosen*

Handbook of Discrete and Computational Geometry,
Jacob E. Goodman and Joseph O'Rourke

Introduction to Information Theory and Data Compression,
Darrel R. Hankerson, Greg A. Harris, and Peter D. Johnson

Network Reliability: Experiments with a Symbolic Algebra Environment,
Daryl D. Harms, Miroslav Kraetzl, Charles J. Colbourn, and John S. Devitt

RSA and Public-Key Cryptography
Richard A. Mollin

Quadratics, *Richard A. Mollin*

Verification of Computer Codes in Computational Science and Engineering,
Patrick Knupp and Kambiz Salari

Darrel Hankerson
Greg A. Harris
Peter D. Johnson, Jr.

Introduction to

Information Theory and Data Compression

Second Edition

CRC Press
Taylor & Francis Group
Boca Raton   London   New York

CRC Press is an imprint of the
Taylor & Francis Group, an **informa** business

A CHAPMAN & HALL BOOK

CRC Press
Taylor & Francis Group
6000 Broken Sound Parkway NW, Suite 300
Boca Raton, FL 33487-2742

First issued in paperback 2019

No claim to original U.S. Government works

ISBN-13: 978-1-58488-313-5 (hbk)
ISBN-13: 978-0-367-39543-8 (pbk)

Library of Congress Card Number   2002041506

Library of Congress Cataloging-in-Publication Data

Hankerson, Darrel R.
Introduction to information theory and data compression / Darrel R. Hankerson, Greg A.
Harris, Peter D. Johnson.--2nd ed.
    p. cm. (Discrete mathematics and its applications)
Includes bibliographical references and index.
ISBN 1-58488-313-8 (alk. paper)
1. Information theory. 2. Data compression (Computer science) I. Harris, Greg A. II.
Johnson, Peter D. (Peter Dexter), 1945- III. Title. IV. Series.
Q360.H35 2003
005.74′6—dc21

2002041506
CIP

Visit the Taylor & Francis Web site at
http://www.taylorandfrancis.com

and the CRC Press Web site at
http://www.crcpress.com

Preface

This textbook is aimed at graduate students and upper level undergraduates in mathematics, engineering, and computer science. The material and the approach of the text were developed over several years at Auburn University in two independent courses, Information Theory and Data Compression. Although the material in the two courses is related, we think it unwise for information theory to be a prerequisite for data compression, and have written the data compression section of the text so that it can be read by or presented to students with no prior knowledge of information theory. There are references in the data compression part to results and proofs in the information theory part of the text, and those who are interested may browse over those references, but it is not absolutely necessary to do so. In fact, perhaps the best pedagogical order of approach to these subjects is the reverse of the apparent logical order: students will come to information theory curious and better prepared for having seen some of the definitions and theorems of that subject playing a role in data compression.

Our main aim in the data compression part of the text, as well as in the course it grew from, is to acquaint the students with a number of significant lossless compression techniques, and to discuss two lossy compression methods. Our aim is for the students to emerge competent in and broadly conversant with a large range of techniques. We have striven for a "practical" style of presentation: here is what you do and here is what it is good for. Nonetheless, proofs are provided, sometimes in the text, sometimes in the exercises, so that the instructor can have the option of emphasizing the mathematics of data compression to some degree.

Information theory is of a more theoretical nature than data compression. It provides a vocabulary and a certain abstraction that can bring the power of simplification to many different situations. We thought it reasonable to treat it as a mathematical theory and to present the fundamental definitions and elementary results of that theory in utter abstraction from the particular problems of communication through noisy channels, which inspired the theory in the first place. We bring the theory to bear on noisy channels in Chapters 3 and 4.

The treatment of information theory given here is extremely elementary. The channels are memoryless and discrete, and the sources are all "zeroth-order," one-state sources (although more complicated source models are discussed in Chapter 7). We feel that this elementary approach is appropriate for the target audience, and that, by leaving more complicated sources and channels out of the picture, we more effectively impart the grasp of Information Theory that we hope our students will take with them.

The exercises range from the routine to somewhat lengthy problems that introduce additional material or establish more difficult results. An asterisk by an exercise or section indicates that the material is off the main road, so to speak, and might reasonably be skipped. In the case of exercises, it may also indicate that the problem is hard and/or unusual.

In the data compression portion of the book, a number of projects require the use of a computer. Appendix A documents Octave and Matlab scripts written by the authors that can be used on some of the exercises and projects involving transform methods and images, and that can also serve as building blocks for other explorations. The software can be obtained from the authors' site, listed in Appendix C. In addition, the site contains information about the book, an online version of Appendix A, and links to other sites of interest.

Organization

Here's a brief synopsis of each chapter and appendix.

Chapter 1 contains an introduction to the language and results of probability theory.

Chapter 2 presents the elementary definitions of information theory, a justification of the quantification of information on which the theory is based, and the fundamental relations among various sorts of information and entropy.

Chapter 3 is about information flow through discrete memoryless noisy channels.

Chapter 4 is about coding text from a discrete source, transmitting the encoded text through a discrete memoryless noisy channel, and decoding the output. The "classical" fundamental theorems of information theory, including the Noisy Channel Theorem, appear in this chapter.

Chapter 5 begins the material of the data compression portion of this book. Replacement schemes are discussed and the chapter concludes with the Noiseless Coding Theorem, proved here for a binary code alphabet. (It appears in Chapter 4 in more general form.)

Chapter 6 discusses arithmetic coding, which is of considerable interest since it is optimal in a certain way that the replacement schemes are not. Considerations for both an "ideal" scheme and for practical implementation on a computer are presented.

Chapter 7 focuses on the modeling aspects of Chapters 5 and 6 (Chapter 8 continues the discussion). Since coding methods such as those presented in Chapter 6 can (in theory) produce optimal-length output for a given model of the source, much of the interest in improving compression in statistical schemes lies in improving the model of the source. Higher-order models attempt to use larger contexts for predictions. In the second edition, a section on probabilistic finite state source automata has been added.

Chapter 8 considers another approach to modeling, using statistics that are updated as the source is read and encoded. These have the advantage that no statistical study needs to be done in advance and the scheme can also detect changes in the nature of the source.

Chapter 9 discusses popular dictionary methods. These have been widely used, in part due to their simplicity, speed, and relatively good compression. Applications such as Ross Williams' LZRW1 algorithm, Unix *compress*, and GNU zip (*gzip*) are examined.

Chapter 10 develops the Fourier, cosine, and wavelet transforms, and discusses their use in compression of signals or images. The lossy scheme in JPEG is presented as a widely-used standard that relies on transform techniques. The chapter concludes with an introduction to wavelet-based compression.

Appendix A documents the use of the "JPEGtool" collection of Octave and Matlab scripts in understanding JPEG-like image compression.

Appendix B contains the source listing for Ross Williams' LZRW1-A algorithm, which rather concisely illustrates a viable dictionary compression method.

Appendix C contains material that didn't fit elsewhere. The first section lists sources for information and code for many areas of data compression. The second section contains a few notes on patents affecting the field. The final section contains a semi-famous story illustrating some of the misunderstandings about compression.

Appendix D offers solutions and notes on the exercises.

Acknowledgments

We'd like to thank Ross Williams for permission to reprint his LZRW1-A algorithm, and for notes on his sources. Alistair Moffat provided preprints and alerted us to other information concerning arithmetic coding. Ian H. Witten was kind enough to respond to our questions concerning a detail in *Text Compression*. We especially wish to acknowledge the help of four reviewers: Jean-loup Gailly offered many important suggestions concerning Chapter 9 and Appendix C, and granted permission to use portions of the "Frequently Asked Questions" document that he authors; Tom Lane suggested a number of improvements and clarifications in Chapter 10; we are grateful to James R. Wall and to Isidore Fleischer for reviewing portions of Chapters 1–5.

There are many folks who have made it easier for the community to understand the subject; some of their names are in this book. Others, working on "GNU Project" and other freely distributable software, made this book possible. The list of major contributors to this software is lengthy, and includes those involved with AUC TeX, dvips[k], Emacs, Ghostview and Ghostscript, GNU/Linux, the Independent JPEG Group, Info-ZIP, LaTeX, Netpbm, PiCTeX, Portable Network Graphics, TeX, xdvi[k], xfig, xv, XY-pic, and many GNU utilities such as bash, gawk, gcc and gdb, gzip, and make. We wish to especially thank the principal developer of Octave, John Eaton.

Thanks are due to A. Scottedward Hodel, Alfred Menezes, Stan Reeves, and Greg Roelofs for some early advice on the data compression course. Our students were also generous with their advice. Douglas Leonard and Luc Teirlinck provided some insightful suggestions and clarifications. Alfred Menezes gets the credit (and the blame) for setting us on the road to a course in data compression and this book.

Some of the computer resources for the course and book were made possible by a grant from the National Science Foundation, for which we are grateful. Our direct contacts at CRC Press were Bob Stern, Nora Konopka, Suzanne Lassandro, Tim Pletscher, Jamie Sigal, Mimi Williams, and Sylvia Wood, and it was a pleasure working with them.

Contents

Preface   v

Part I: Information Theory

1   Elementary Probability   1
1.1   Introduction 1
1.2   Events 3
1.3   Conditional probability 7
1.4   Independence 11
1.5   Bernoulli trials 13
1.6   An elementary counting principle 15
1.7*   On drawing without replacement 17
1.8   Random variables and expected, or average, value 18
1.9   The Law of Large Numbers 22

2   Information and Entropy   25
2.1   How is information quantified? 25
    2.1.1   Naming the units 27
    2.1.2   Information connecting two events 29
    2.1.3   The inevitability of Shannon's quantification of information 30
2.2   Systems of events and mutual information 33
2.3   Entropy 40
2.4   Information and entropy 43

3   Channels and Channel Capacity   47
3.1   Discrete memoryless channels 47
3.2   Transition probabilities and binary symmetric channels 50
3.3   Input frequencies 52
3.4   Channel capacity 56
3.5*   Proof of Theorem 3.4.3, on the capacity equations 67

4   Coding Theory   71
4.1   Encoding and decoding 71
4.2   Prefix-condition codes and the Kraft-McMillan inequality 75
4.3   Average code word length and Huffman's algorithm 79
    4.3.1   The validity of Huffman's algorithm 86
4.4   Optimizing the input frequencies 90
4.5   Error correction, maximum likelihood decoding, nearest code word decoding, and reliability 95

4.6   Shannon's Noisy Channel Theorem   106
4.7   Error correction with binary symmetric channels and equal source
       frequencies   111
4.8   The information rate of a code   115

Part II: Data Compression

5   **Lossless Data Compression by Replacement Schemes**   **119**
    5.1   Replacement via encoding scheme 120
    5.2   Review of the prefix condition   123
    5.3   Choosing an encoding scheme   126
          5.3.1   Shannon's method   127
          5.3.2   Fano's method   130
          5.3.3   Huffman's algorithm 131
    5.4   The Noiseless Coding Theorem and Shannon's bound 134

6   **Arithmetic Coding**   **141**
    6.1   Pure zeroth-order arithmetic coding: dfwld   142
          6.1.1   Rescaling while encoding   146
          6.1.2   Decoding 150
    6.2   What's good about dfwld coding: the compression ratio 155
    6.3   What's bad about dfwld coding and some ways to fix it 160
          6.3.1   Supplying the source word length   161
          6.3.2   Computation   162
          6.3.3   Must decoding wait until encoding is completed? 164
    6.4   Implementing arithmetic coding   167
    6.5   Notes   179

7   **Higher-order Modeling**   **181**
    7.1   Higher-order Huffman encoding   182
    7.2   The Shannon bound for higher-order encoding 186
    7.3   Higher-order arithmetic coding 191
    7.4   Statistical models, statistics, and the possibly unknowable truth . 193
    7.5   Probabilistic finite state source automata 197

8   **Adaptive Methods**   **205**
    8.1   Adaptive Huffman encoding   206
          8.1.1   Compression and readjustment 209
          8.1.2   Higher-order adaptive Huffman encoding   210
    8.2   Maintaining the tree in adaptive Huffman encoding: the method
           of Knuth and Gallager   212
          8.2.1   Gallager's method   215
          8.2.2   Knuth's algorithm   216
    8.3   Adaptive arithmetic coding 219
    8.4   Interval and recency rank encoding 221

      8.4.1   Interval encoding 221
      8.4.2   Recency rank encoding 224

9 Dictionary Methods   **229**
   9.1   LZ77 (sliding window) schemes 230
      9.1.1   An LZ77 implementation 232
      9.1.2   Case study: GNU zip 235
   9.2   The LZ78 approach 237
      9.2.1   The LZW variant 240
      9.2.2   Case study: Unix *compress* 242
   9.3   Notes 244

10 Transform Methods and Image Compression   **245**
   10.1   Transforms 247
   10.2   Periodic signals and the Fourier transform 249
      10.2.1   The Fourier transform and compression: an example . . 256
   10.3   The cosine and sine transforms 267
      10.3.1   A general orthogonal transform 270
      10.3.2   Summary 271
   10.4   Two-dimensional transforms 273
      10.4.1   The 2D Fourier, cosine, and sine transforms 275
      10.4.2   Matrix expressions for 2D transforms 279
   10.5   An application: JPEG image compression 281
   10.6   A brief introduction to wavelets 291
      10.6.1   2D Haar wavelets 296
   10.7   Notes 300

Appendices   **303**

A JPEGtool User's Guide   **303**
   A.1   Using the tools 304
   A.2   Reference 314
   A.3   Obtaining Octave 319

B Source Listing for LZRW1-A   **321**

C Resources, Patents, and Illusions   **333**
   C.1   Resources 333
   C.2   Data compression and patents 335
   C.3   Illusions 338

D Notes on and Solutions to Some Exercises   **343**

Bibliography   **357**

Index   **361**

Chapter 1

Elementary Probability

1.1 Introduction

Definition A *finite probability space* is a pair $(\mathcal{S}, P)$, in which $\mathcal{S}$ is a finite non-empty set and $P : \mathcal{S} \to [0, 1]$ is a function satisfying $\sum_{s \in \mathcal{S}} P(s) = 1$.

When the space $(\mathcal{S}, P)$ is fixed in the discussion, we will call $\mathcal{S}$ the *set of outcomes* of the space (or of the experiment with which the space is associated—see the discussion below), and P the *probability assignment* to the (set of) outcomes.

The "real" situations we are concerned with consist of an *action*, or *experiment*, with a finite number of mutually exclusive possible outcomes. For a given action, there may be many different ways of listing the possible outcomes, but all acceptable lists of possible outcomes satisfy this test: whatever happens, it shall be the case that one and only one of the listed outcomes will have occurred.

For instance, suppose that the experiment consists of someone jumping out of a plane somewhere over Ohio (parachute optional). Assuming that there is some way of defining the "patch upon which the jumper lands," it is possible to view this experiment as having infinitely many possible outcomes, corresponding to the infinitely many patches that might be landed upon. But we can collect these infinitely many possibilities into a finite number of different categories which are, one would think, much more interesting and useful to the jumper and everyone else concerned than are the undifferentiated infinity of fundamental possible outcomes. For instance, our finite list of possibilities might look like: (1) the jumper lands on some power line(s); (2) the jumper lands in a tree; ... ; (n) none of the above (in case we overlooked a possibility).

Clearly there are infinitely many ways to make a finite list of outcomes of this experiment. How would you, in practice, choose a list? That depends on your concerns. If the jumper is a parachutist, items like "lands in water" should probably be on the list. If the jumper is a suicidal terrorist carrying an atom bomb, items like "lands within 15 miles of the center of Cincinnati" might well be on the list. There is some art in the science of parsing the outcomes to suit your interest. Never forget the constraint that one and only one of the listed outcomes will occur, whatever happens. For instance, it is unacceptable to have "lands in water" and "lands within 15 miles of the center of Cincinnati" on the same list, since it is possible for the jumper to land in water within 15 miles of the center of Cincinnati. (See Exercise 1 at the end of this section.)

Now consider the definition at the beginning of this section. The set S is interpretable as the finite list of outcomes, or of outcome categories, of whatever experiment we have at hand. The function P is, as the term *probability assignment* suggests, an assessment or measure of the likelihoods of the different outcomes.

There is nothing in the definition that tells you how to provide yourself with S and P, given some actual experiment. The jumper-from-the-airplane example is one of a multitude that show that there may be cause for debate and occasion for subtlety even in the task of listing the possible outcomes of the given experiment. And once the outcomes are listed to your satisfaction, how do you arrive at a satisfactory assessment of the likelihoods of occurrence of the various outcomes? That is a *long* story, only the beginnings of which will be told in this chapter. There are plenty of people—actuaries, pollsters, quality-control engineers, market analysts, and epidemiologists—who make their living partly by their sophistication in the matter of assigning probabilities; assigning probabilities in different situations is the problem at the center of applied statistics.

To the novice, it may be heartening to note that two great minds once got into a confused dispute over the analysis of a very simple experiment. Before the description of the experiment and the dispute, we interject a couple of comments that will be referred to throughout this chapter.

Experiments with outcomes of equal likelihood

If it is judged that the different outcomes are equally likely, then the condition $\sum_{s\in S} P(s) = 1$ forces the probability assignment $P(s) = \frac{1}{|S|}$ for all $s \in S$, where $|S|$ stands for the size of S, also known as the number of elements in S, also known as the cardinality of S.

Coins. In this chapter, each coin shall have two sides, designated "heads" and "tails," or H and T, for short. On each flip, toss, or throw of a coin, one of these two "comes up." Sometimes H will be an abbreviation of the phrase "heads comes up," and T similarly. Thus, in the experiment of tossing a coin once, the only reasonable set of outcomes is abbreviable $\{H, T\}$.

A *fair* coin is one for which the outcomes H and T of the one-toss experiment are equally likely—i.e., each has probability $1/2$.

The D'Alembert-Laplace controversy

D'Alembert and Laplace were great mathematicians of the 18th and 19th centuries. Here is the experiment about the analysis of which they disagreed: a fair coin is tossed twice.

The assumption that the coin is fair tells us all about the experiment of tossing it once. Tossing it twice is the next-simplest experiment we can perform with this coin. How can controversy arise? Consider the question: what is the probability that heads will come up on each toss? D'Alembert's answer: $1/3$. Laplace's answer: $1/4$.

D'Alembert and Laplace differed right off in their choices of sets of outcomes. D'Alembert took $\mathcal{S}_D = \{\text{both heads, both tails, one head and one tail}\}$, and Laplace favored $\mathcal{S}_L = \{HH, HT, TH, TT\}$ where, for instance, HT stands for "heads on the first toss, tails on the second." Both D'Alembert and Laplace asserted that the outcomes in their respective sets of outcomes are equally likely, from which assertions you can see how they got their answers. Neither provided a convincing justification of his assertion.

We will give a plausible justification of one of the two assertions above in Section 1.3. Whether or not the disagreement between D'Alembert and Laplace is settled by that justification will be left to your judgment.

Exercises 1.1

1. Someone jumps out of a plane over Ohio. You are concerned with whether or not the jumper lands in water, and whether or not the jumper lands within 15 miles of the center of Cincinnati, and with nothing else. [Perhaps the jumper carries a bomb that will not go off if the jumper lands in water, and you have relatives in Cincinnati.]

   Give an acceptable list of possible outcomes, as short as possible, that will permit discussion of your concerns. [Hint: the shortest possible list has length 4.]

2. Notice that we can get D'Alembert's set of outcomes from Laplace's by "amalgamating" a couple of Laplace's outcomes into a single outcome.

   More generally, given any set S of outcomes, you can make a new set of outcomes $\widehat{S}$ by *partitioning* S into non-empty sets $P_1, \ldots, P_m$ and setting $\widehat{S} = \{P_1, \ldots, P_m\}$. [To say that subsets $P_1, \ldots, P_m$ of S *partition* S is to say that $P_1, \ldots, P_m$ are pairwise disjoint, i.e., $\emptyset = P_i \cap P_j$, $1 \leq i < j \leq m$, and cover S, i.e., $S = \bigcup_{i=1}^{m} P_i$. Thus, "partitioning" is "dividing up into non-overlapping parts."]

   How may different sets of outcomes can be made, in this way, from a set of outcomes with four elements?

1.2 Events

Throughout, $(\mathcal{S}, P)$ will be some finite probability space. An *event* in this space is a subset of $\mathcal{S}$. If $E \subseteq \mathcal{S}$ is an event, the *probability of E*, denoted $P(E)$, is

$$P(E) = \sum_{s \in E} P(s).$$

Some elementary observations:

(i) if $s \in S$, then $P(\{s\}) = P(s)$;

(ii) $P(\emptyset) = 0$;

(iii) $P(S) = 1$;

(iv) if the outcomes in S are equally likely, then, for each $E \subseteq S$, $P(E) = |E|/|S|$.

Events are usually described in plain English, by a sentence or phrase indicating what happens when that event occurs. The *set* indicated by such a description consists of those outcomes that satisfy, or conform to, the description. For instance, suppose an urn contains red, green, and yellow balls; suppose that two are drawn, without replacement. We take $S = \{rr, rg, ry, gr, gg, gy, yr, yg, yy\}$, in which, for instance, rg is short for "a red ball was drawn on the first draw, and a green on the second." Let $E = $ "no red ball was drawn." Then $E = \{gg, gy, yg, yy\}$.

Notice that the verbal description of an event need not refer to the set S of outcomes, and thus may be represented differently as a set of outcomes for different choices of S. For instance, in the experiment of tossing a coin twice, the event "both heads and tails came up" is a set consisting of a single outcome in D'Alembert's way of looking at things, but consists of two outcomes according to Laplace. (Notice that the event "tails came up on first toss, heads on the second" is not an admissible event in D'Alembert's space; this does not mean that D'Alembert was *wrong*, only that his analysis is insufficiently fine to permit discussion of certain events associable with the experiment. Perhaps he would argue that distinguishing between the tosses, labeling one "the first" and the other "the second," makes a different experiment from the one he was concerned with.)

Skeptics can, and should, be alarmed by the "definition" above of the probability $P(E)$ of an event E. If an event E has a description that makes no reference to the set S of outcomes, then E should have a probability that does not vary as you consider different realizations of E as a subset of different outcome sets S. Yet the probability of E is "defined" to be $\sum_{s \in E} P(s)$, which clearly involves S and P. This "definition" hides an assertion that deserves our scrutiny. The assertion is that, however an event E is realized as a subset of a set S of outcomes, if the probability assignment P to S is "correct," then the number $\sum_{s \in E} P(s)$ will be "correct," the "correct" probability of E by any "correct" assessment.

Here is an argument that seems to justify the equation $P(E) = \sum_{s \in E} P(s)$ in all cases where we agree that P is a correct assignment of probabilities to the elements of S, and that there is a correct à priori probability $P(E)$ of the event E, realizable as a subset of S. Let $\widehat{S} = (S \setminus E) \cup \{E\}$. That is, we are forming a new set of outcomes by amalgamating the outcomes in E into a single outcome, which we will denote by E. What shall the correct probability assignment $\widehat{P}$ to $\widehat{S}$ be? $\widehat{P}(E)$ ought to be the sought-after $P(E)$, the correct probability of E. Meanwhile, the outcomes of $S \setminus E$ are indifferent to our changed view of the experiment; we should have $P(s) = \widehat{P}(s)$ for $s \in S \setminus E$. Then $1 = \sum_{s \in S} \widehat{P}(s) = \widehat{P}(E) + \sum_{s \in S \setminus E} P(s) = P(E) + (1 - \sum_{s \in E} P(s))$, which implies the desired equation.

Definition Events E_1 and E_2 in a probability space (S, P) are *mutually exclusive* in case $P(E_1 \cap E_2) = 0$.

In common parlance, to say that two events are mutually exclusive is to say that they cannot *both* happen. Thus, it might seem reasonable to define E_1 and E_2 to be mutually exclusive if and only if $E_1 \cap E_2 = \emptyset$, a stronger condition than $P(E_1 \cap E_2) = 0$. It will be convenient to allow outcomes of experiments that have zero probability just because ruling out such outcomes may require a lengthy verbal digression or may spoil the symmetry of some array. In the service of this convenience, we define mutual exclusivity as above.

Example Suppose an urn contains a number of red and green balls, and exactly one yellow ball. Suppose that two balls are drawn, *without* replacement. If, as above, we take $S = \{rr, rg, ry, gr, gg, gy, yr, yg, yy\}$, then the outcome yy is impossible. However we assign probabilities to S, the only reasonable probability assignment to yy is zero. Thus, if $E_1 = $ "a yellow ball was chosen on the first draw" and $E_2 = $ "a yellow ball was chosen on the second draw," then E_1 and E_2 are mutually exclusive, even though $E_1 \cap E_2 = \{yy\} \neq \emptyset$.

Why not simply omit the impossible outcome yy from S? We may, for some reason, be performing this experiment on different occasions with different urns, and, for most of these, yy may be a possible outcome. It is a great convenience to be able to refer to the same set S of outcomes in discussing these different experiments.

Some useful observations and results As heretofore, (S, P) will be a probability space, and E, F, E_1, E_2, etc., will stand for events in this space.

1.2.1 If $E_1 \subseteq E_2$, then $P(E_1) \leq P(E_2)$.

1.2.2 If E_1 and E_2 are mutually exclusive, and $F_1 \subseteq E_1$ and $F_2 \subseteq E_2$, then F_1 and F_2 are mutually exclusive.

1.2.3 If $E_1, \ldots, E_m$ are pairwise mutually exclusive (meaning E_i and E_j are mutually exclusive when $1 \leq i < j \leq m$), then

$$P(\bigcup_{i=1}^{m} E_i) = \sum_{i=1}^{m} P(E_i).$$

For a "clean" proof, go by induction on m. For an instructive proof, first consider the case in which $E_1, \ldots, E_m$ are pairwise disjoint.

1.2.4 If $E \subseteq F \subseteq S$, then $P(F \setminus E) = P(F) - P(E)$.

Proof: Apply 1.2.3 with $m = 2$, $E_1 = E$ and $E_2 = F \setminus E$. $\qquad\square$

1.2.5 $P(E \cup F) + P(E \cap F) = P(E) + P(F)$.

Proof: Observe that $E \cup F = (E \cap F) \cup (E \setminus (E \cap F)) \cup (F \setminus (E \cap F))$, a union of pairwise disjoint events. Apply 1.2.3 and 1.2.4. □

1.2.6 $P(E) + P(S \setminus E) = 1$.

Proof: This is a corollary of 1.2.4. □

[When propositions are stated without proof, or when the proof is merely sketched, as in 1.2.4, it is hoped that the student will supply the details. The order in which the propositions are stated is intended to facilitate the verification. For instance, proposition 1.2.2 follows smoothly from 1.2.1.]

Example Suppose that, in a certain population, 40% of the people have red hair, 25% tuberculosis, and 15% have both. What percentage has neither?

The experiment that can be associated to this question is: choose a person "at random" from the population. The set S of outcomes can be identified with the population; the outcomes are equally likely if the selection process is indeed "random."

Let R stand for the event "the person selected has red hair," and T for the event "the person selected has tuberculosis." As subsets of the set of outcomes, R and T are the sets of people in the population which have red hair and tuberculosis, respectively. We are given that $P(R) = 40/100$, $P(T) = 25/100$, and $P(R \cap T) = 15/100$. Then

$$
\begin{aligned}
P(S \setminus (R \cup T)) &= 1 - P(R \cup T) \qquad \text{[by 1.2.6]} \\
&= 1 - [P(R) + P(T) - P(R \cap T)] \qquad \text{[by 1.2.5]} \\
&= 1 - \left[\frac{40}{100} + \frac{25}{100} - \frac{15}{100} \right] = \frac{50}{100}.
\end{aligned}
$$

Answer: 50% have neither.

Exercises 1.2

1. In a certain population, 25% of the people are small and dirty, 35% are large and clean, and 60% are small. What percentage are dirty?

2. In the experiment of tossing a fair coin twice, let us consider Laplace's set of outcomes, $\{HH, HT, TH, TT\}$. We do not know how to assign probabilities to these outcomes as yet, but surely HT and TH ought to have the same probability, and the events "heads on the first toss" and "heads on the second toss" each ought to have probability $1/2$.

   Do these considerations determine a probability assignment?

3. In a certain population, 30% of the people have acne, 10% have bubonic plague, and 12% have cholera. In addition,

8% have acne and bubonic plague,
7% have acne and cholera,
4% have bubonic plague and cholera,
and 2% have all three diseases.

What percentage of the population has none of the three diseases?

1.3 Conditional probability

Definition Suppose that (S, P) is a finite probability space, $E_1, E_2 \subseteq S$, and $P(E_2) \neq 0$. The *conditional probability* of E_1, *given* E_2, is $P(E_1 \mid E_2) = P(E_1 \cap E_2)/P(E_2)$.

Interpretation. You may as well imagine that you were not present when the experiment or action took place, and you received an incomplete report on what happened. You learn that event E_2 occurred (meaning the outcome was one of those in E_2), and nothing else. How shall you adjust your estimate of the probabilities of the various outcomes and events, in light of what you now know? The definition above proposes such an adjustment. Why this? Is it valid?

Justification. Supposing that E_2 has occurred, let's make a new probability space, $(E_2, \widehat{P})$, taking E_2 to be the new set of outcomes. What about the new probability assignment, $\widehat{P}$? We assume that the new probabilities $\widehat{P}(s)$, $s \in E_2$, are *proportional* to the old probabilities $P(s)$; that is, for some number r, we have $\widehat{P}(s) = r P(s)$ for all $s \in E_2$.

This might seem a reasonable assumption in many specific instances, but is it universally valid? Might not the knowledge that E_2 has occurred change our assessment of the relative likelihoods of the outcomes in E_2? If some outcome in E_2 was judged to be twice as likely as some other outcome in E_2 before the experiment was performed, must it continue to be judged twice as likely as the other after we learn that E_2 has occurred? We see no way to convincingly demonstrate the validity of this "proportionality" assumption, nor do we have in mind an example in which the assumption is clearly violated. We shall accept this assumption with qualms, and forge on.

Since $\widehat{P}$ is to be a probability assignment, we have

$$1 = \sum_{s \in E_2} \widehat{P}(s) = r \sum_{s \in E_2} P(s) = r P(E_2),$$

so $r = 1/P(E_2)$. Therefore, the probability that E_1 has occurred, given that E_2 has occurred, *ought* to be given by

$$P(E_1|E_2) = \widehat{P}(E_1 \cap E_2) = \sum_{s \in E_1 \cap E_2} \widehat{P}(s) = r \sum_{s \in E_1 \cap E_2} P(s) = \frac{P(E_1 \cap E_2)}{P(E_2)}.$$

End of justification.

Application to multi-stage experiments

Suppose that we have in mind an experiment with two stages, or sub-experiments. (Examples: tossing a coin twice, drawing two balls from an urn.) Let $x_1, \ldots, x_n$ denote the possible outcomes at the first stage, and $y_1, \ldots, y_m$ at the second. Suppose that the probabilities of the first-stage outcomes are known: say

$$P(\text{``}x_i \text{ occurs at the first stage''}) = p_i, \quad i = 1, \ldots, n.$$

Suppose that the probabilities of the y_j occurring are known, whenever it is known what happened at the first stage. Let us say that

$$P(\text{``}y_j \text{ occurs at the 2nd stage, supposing } x_i \text{ occurred at the first''}) = q_{ij}.$$

Now we will consider the full experiment. We take as the set of outcomes the set of ordered pairs

$$
\begin{aligned}
S &= \{x_1, \ldots, x_n\} \times \{y_1, \ldots, y_m\} \\
&= \{(x_i, y_j); i \in \{1, \ldots, n\}, j \in \{1, \ldots, m\}\},
\end{aligned}
$$

in which (x_i, y_j) is short for the statement "x_i occurred at the first stage, and y_j at the second." What shall the probability assignment P be?

Let $E_i = \text{``}x_i \text{ occurred at the first stage''} = \{(x_i, y_1), \ldots, (x_i, y_m)\}$, for each $i \in \{1, \ldots, n\}$, and let $F_j = \text{``}y_j \text{ occurred at the second stage''} = \{(x_1, y_j), \ldots, (x_n, y_j)\}$, for each $j \in \{1, \ldots, m\}$. Even though we do not yet know what P is, we supposedly know something about the probabilities of these events. What we know is that $P(E_i) = p_i$, and $P(F_j \mid E_i) = q_{ij}$, for each i and j. Therefore,

$$q_{ij} = P(F_j \mid E_i) = \frac{P(E_i \cap F_j)}{P(E_i)} = \frac{P(\{(x_i, y_j)\})}{p_i}$$

which implies

$$P((x_i, y_j)) = P(\{(x_i, y_j)\}) = p_i q_{ij}.$$

We now know how to assign probabilities to outcomes of two stage experiments, given certain information about the experiment that might, plausibly, be obtainable a priori, from the description of the experiment. To put it simply, you *multiply*. (But *what* do you multiply?) Something similar applies to experiments of three or more stages. It is left to you to formulate what you do to assign probabilities to the outcomes of such experiments.

Examples

1.3.1 An urn contains 5 red, 12 green, and 8 yellow balls. Three are drawn without replacement.

    (a) What is the probability that a red, a green, and a yellow ball will be drawn?

    (b) What is the probability that the last ball to be drawn will be green?

Solution and discussion. If we identify the set of outcomes with the set of all sequences of length three of the letters r, g, and y, in the obvious way (e.g., ryy stands for a red ball was drawn first, then two yellows), then we will have 27 outcomes; no need to list them all. The event described in part (a) is

$$E = \{rgy, ryg, gry, gyr, yrg, ygr\}.$$

These are assigned probabilities, thus:

$$P(rgy) = \frac{5}{25} \cdot \frac{12}{24} \cdot \frac{8}{23}, \quad P(ryg) = \frac{5}{25} \cdot \frac{8}{24} \cdot \frac{12}{23},$$

etc. How are these obtained? Well, for instance, to see that $P(rgy) = \frac{5}{25} \cdot \frac{12}{24} \cdot \frac{8}{23}$, you reason thus: on the first draw, we have 25 balls, all equally likely to be drawn (it is presumed from the description of the experiment), of which 5 are red, hence a probability of $\frac{5}{25}$ of a red on the first draw; *having drawn* that red, there is then a $\frac{12}{24}$ probability of a green on the second draw, and *having drawn* first a red, then a green, there is probability $\frac{8}{23}$ of a yellow on the third draw. Multiply.

In this case, we observe that all six outcomes in E have the same probability assignment. Thus the answer to (a) is $6 \cdot \frac{5}{25} \cdot \frac{12}{24} \cdot \frac{8}{23} = \frac{24}{115}$.

For (b), we cleverly take a different set of outcomes, and ponder the event

$$F = \{NNg, Ngg, gNg, ggg\},$$

where N stands for "not green." We have

$$\begin{aligned} P(F) &= \frac{13}{25} \cdot \frac{12}{24} \cdot \frac{12}{23} + \frac{13}{25} \cdot \frac{12}{24} \cdot \frac{11}{23} + \frac{12}{25} \cdot \frac{13}{24} \cdot \frac{11}{23} + \frac{12}{25} \cdot \frac{11}{24} \cdot \frac{10}{23} \\ &= \frac{12}{25} \Big[\frac{1}{23 \cdot 24} (13 \cdot (12 + 11) + (13 + 10) \cdot 11) \Big] \\ &= \frac{12}{25} \Big[\frac{1}{23 \cdot 24} 23 \cdot 24 \Big] = \frac{12}{25}, \end{aligned}$$

which is, interestingly, the probability of drawing a green on the first draw. Could we have foreseen the outcome of this calculation, and saved ourselves some trouble? It is left to you to decide whether or not the probability of drawing a green on draw number k, $1 \le k \le 25$, when we are drawing without replacement, might depend on k.

1.3.2 A room contains two urns, A and B. A contains nine red balls and one green ball; B contains four red balls and four green balls. The room is darkened, a man stumbles into it, gropes about for an urn, draws two balls without replacement, and leaves the room.

(a) What is the probability that both balls will be red?

(b) Suppose that one ball is red and one is green: what is the probability that urn A now contains only eight balls?

Solutions. (a) Using obvious and self-explanatory abbreviations,

$$P(\text{``both red''}) = P(Arr) + P(Brr) = \frac{1}{2}\frac{9}{10}\frac{8}{9} + \frac{1}{2}\frac{4}{8}\frac{3}{7}$$

(b) We calculate

$P(\text{``Urn } A \text{ was the urn chosen''} \mid \text{``One ball is red and one green''})$

$$= \frac{P(\{Arg, Agr\})}{P(\{Arg, Agr, Brg, Bgr\})} = \frac{\frac{1}{2}\frac{9}{10}\frac{1}{9} + \frac{1}{2}\frac{1}{10}\frac{9}{9}}{\frac{1}{2}\frac{9}{10}\frac{1}{9} + \frac{1}{2}\frac{1}{10}\frac{9}{9} + \frac{1}{2}\frac{4}{8}\frac{4}{7} + \frac{1}{2}\frac{4}{8}\frac{4}{7}} = \frac{7}{27}.$$

Notice that this result is satisfyingly less than $1/2$, the a priori probability that urn A was chosen.

Exercises 1.3

1. An urn contains six red balls, five green balls, and three yellow balls. Two are drawn without replacement. What is the probability that at least one is yellow?

2. Same question as in 1, except that *three* balls are drawn without replacement.

3. Same question as in 1, except that the drawing is *with* replacement.

4. An actuary figures that for a plane of a certain type, there is a 1 in 100,000 chance of a crash somewhere during a flight from New York to Chicago, and a 1 in 150,000 chance of a crash somewhere during a flight from Chicago to Los Angeles.

   A plane of that type is to attempt to fly from New York to Chicago and then from Chicago to Los Angeles.

   (a) What is the probability of a crash somewhere along the way? [Please do *not* use your calculator to convert to a decimal approximation.]

   (b) Suppose that you know that the plane crashed, but you know nothing else. What is the probability that the crash occurred during the Chicago-L.A. leg of the journey?

5. Urn A contains 11 red and seven green balls, urn B contains four red and one green, and urn C contains two red and six green balls. The three urns are placed in a dark room. Someone stumbles into the room, gropes around, finds an urn, draws a ball from it, lurches from the room, and looks at the ball. It is green. What is the probability that it was drawn from urn A?

   In this experiment, what is the probability that a red ball will be chosen? What is the proportion of red balls in the room?

6. Who was right, D'Alembert or Laplace? Or neither?

7. What is the probability of heads coming up exactly twice in three flips of a fair coin?

8. Let p_i and q_{ij} be as in the text preceding these exercises. What is the value of $\sum_{i=1}^{n}\sum_{j=1}^{m} p_i q_{ij}$?

9. On each of three different occasions, a fair coin is flipped twice.

   (a) On the first occasion, you witness the second flip, but not the first. You see that heads comes up. What is the probability that heads came up on both flips?

   (b) On the second occasion, you are not present, but are shown a video of one of the flips, you know not which; either is as likely as the other. On the video, heads comes up. What is the probability that heads came up on both flips?

   (c) On the third occasion, you are not present; a so-called friend teases you with the following information, that heads came up at least once in the two flips. What is the probability that heads came up on both flips?

*10. For planes of a certain type, the actuarial estimate of the probability of a crash during a flight (including take-off and landing) from New York to Chicago is p_1; from Chicago to L.A., p_2.

   In experiment #1, a plane of that type is trying to fly from New York to L.A., with a stop in Chicago. Let a denote the conditional probability that, if there is a crash, it occurs on the Chicago-L.A. leg of the journey.

   In experiment #2, two different planes of the fatal type are involved; one is to fly from N.Y. to Chicago, the other from Chicago to L.A. Let b denote the conditional probability that the Chicago-L.A. plane crashed, if it is known that at least one of the two crashed, and let c denote the conditional probability that the Chicago-L.A. plane crashed, if it is known that exactly one of the two crashed.

   Express a, b, and c in terms of p_1 and p_2, and show that $a \leq c \leq b$ for all possible p_1, p_2.

1.4 Independence

Definition Suppose that (S, P) is a finite probability space, and $E_1, E_2 \subseteq S$. The events E_1, E_2 are *independent* if and only if

$$P(E_1 \cap E_2) = P(E_1)P(E_2).$$

1.4.1 *Suppose that both $P(E_1)$ and $P(E_2)$ are non-zero. The following are equivalent:*

   (a) E_1 *and* E_2 *are independent;*

   (b) $P(E_1 \mid E_2) = P(E_1)$;

(c) $P(E_2 \mid E_1) = P(E_2)$.

Proof: Left to you.   □

The intuitive meaning of independence should be fairly clear from the proposition; if the events have non-zero probability, then two events are independent if and only if the occurrence of either has no influence on the likelihood of the occurrence of the other. Besides saying that two events are independent, we will also say that one event is independent of another.

We shall say that two *stages*, say the ith and jth, $i \neq j$, of a multi-stage experiment, are independent if and only if for any outcome x at the ith stage, and any outcome y at the jth stage, the *events* "x occurred at the ith stage" and "y occurred at the jth stage" are independent. This means that no outcome at either stage will influence the outcome at the other stage. It is intuitively evident, and can be proven, that when two stages are independent, any two events whose descriptions involve only those stages, respectively, will be independent. See Exercises 2.2.5 and 2.2.6.

Exercises 1.4

1. Suppose that E and F are independent events in a finite probability space $(\mathcal{S}, P)$. Show that

    (a) E and $\mathcal{S} \setminus F$ are independent;
    (b) $\mathcal{S} \setminus E$ and F are independent;
    (c) $\mathcal{S} \setminus E$ and $\mathcal{S} \setminus F$ are independent.

2. Show that each event with probability 0 or 1 is independent of every event in its space.

3. Suppose that $\mathcal{S} = \{a, b, c, d\}$, all outcomes equally likely, and $E = \{a, b\}$. List all the events in this space that are independent of E.

4. Suppose that $\mathcal{S} = \{a, b, c, d, e\}$, all outcomes equally likely, and $E = \{a, b\}$. List all the events in this space that are independent of E.

5. Urn A contains 3 red balls and 1 green ball, and urn B contains no red balls and 75 green balls. The action will be: select one of urns A, B, or C at random (meaning they are equally likely to be selected), and draw one ball from it.

    How many balls of each color should be in urn C, if the event "a green ball is selected" is to be independent of "urn C is chosen," and urn C is to contain as few balls as possible?

6. Is it possible for two events to be both independent and mutually exclusive? If so, in what circumstances does this happen?

7. Suppose that $\mathcal{S} = \{a, b, c, d\}$, and these outcomes are equally likely. Suppose that $E = \{a, b\}$, $F = \{a, c\}$, and $G = \{b, c\}$. Verify that E, F, and G

are pairwise independent. Verify that

$$P(E \cap F \cap G) \neq P(E)P(F)P(G).$$

Draw a moral by completing this sentence: just because $E_1, \ldots, E_k$ are pairwise independent, it does not follow that _____. What if $E_1, \ldots, E_k$ belong to the distinct, pairwise independent *stages* of a multistage experiment?

1.5 Bernoulli trials

Definition Suppose that $n \geq 0$ and k are integers; $\binom{n}{k}$, read as "n-choose-k" or "the binomial coefficient n, k," is the number of different k-subsets of an n-set (a set with n elements).

It is hoped that the reader is well-versed in the fundamentals of the binomial coefficients $\binom{n}{k}$, and that the following few facts constitute a mere review.

Note that $\binom{n}{k} = 0$ for $k > n$ and for $k < 0$, by the definition above, and that $\binom{n}{0} = \binom{n}{n} = 1$ and $\binom{n}{1} = n$ for all non-negative integers n. Some of those facts force us to adopt the convention that $0! = 1$, in what follows and forever after.

1.5.1 For $0 \leq k \leq n$, $\binom{n}{k} = \frac{n!}{k!(n-k)!}$. If $1 \leq k$, $\binom{n}{k} = \frac{n(n-1)\cdots(n-k+1)}{k!}$.

1.5.2 $\binom{n}{k} = \binom{n}{n-k}$.

1.5.3 $\binom{n}{k} + \binom{n}{k+1} = \binom{n+1}{k+1}$.

Definition An *alphabet* is just a non-empty finite set, and a *word of length n* over an alphabet A is a sequence, of length n, of elements of A, written without using parentheses and commas; for instance, 101 is a word of length 3 over $\{0, 1\}$.

1.5.4 *Suppose α and β are distinct symbols. Then $\binom{n}{k} = |\{w; w \text{ is a word of length } n \text{ over } \{\alpha, \beta\} \text{ and } \alpha \text{ appears exactly } k \text{ times in } w\}|$.*

1.5.5 *For any numbers α, β,*

$$(\alpha + \beta)^n = \sum_{k=0}^{n} \binom{n}{k} \alpha^k \beta^{n-k}.$$

1.5.6 $2^n = \sum_{k=0}^{n} \binom{n}{k}$.

Of the propositions above, 1.5.1 and 1.5.4 are the most fundamental, in that each of the others can be seen to follow from these two. However, 1.5.2 and 1.5.3 also have "combinatorial" proofs that appeal directly to the definition of $\binom{n}{k}$.

Proposition 1.5.5 is the famous binomial theorem of Pascal; it is from the role of the $\binom{n}{k}$ in this proposition that the term "binomial coefficient" arises.

Definitions A *Bernoulli trial* is an experiment with exactly two possible outcomes. A *sequence of independent Bernoulli trials* is a multi-stage experiment in which the stages are the same Bernoulli trial, and the stages are independent.

Tossing a coin is an example of a Bernoulli trial. So is drawing a ball from an urn, if we are distinguishing between only two types of balls. If the drawing is *with* replacement (and, it is understood, with mixing of the balls after replacement), a sequence of such drawings from an urn is a sequence of independent Bernoulli trials.

When speaking of some unspecified Bernoulli trial, we will call one possible outcome Success, or S, and the other Failure, or F. The distinction is arbitrary. For instance, in the Bernoulli trial consisting of a commercial DC-10 flight from New York to Chicago, you can let the outcome "plane crashes" correspond to the word Success in the theory, and "plane doesn't crash" to Failure, or the other way around.

Another way of saying that a sequence of Bernoulli trials (of the same type) is independent is: the probability of Success does not vary from trial to trial. Notice that if the probability of Success is p, then the probability of Failure is $1 - p$.

1.5.7 Theorem *Suppose the probability of Success in a particular Bernoulli trial is p. Then the probability of exactly k successes in a sequence of n independent such trials is*

$$\binom{n}{k} p^k (1 - p)^{n-k}.$$

Proof: Let the set of outcomes of the experiment consisting of the sequence of n independent Bernoulli trials be identified with $\{S, F\}^n$, the set of all sequences of length n of the symbols S and F. If u is such a sequence in which S appears exactly k times, then, by what we know about assigning probabilities to the outcomes of multi-stage experiments, we have

$$P(u) = p^k (1 - p)^{n-k}.$$

Therefore,

$$P(\text{"exactly } k \text{ successes"}) = \sum_{\substack{S \text{ occurs exactly} \\ k \text{ times in } u}} P(u)$$

$$= |\{u; S \text{ occurs exactly } k \text{ times in } u\}| p^k (1 - p)^{n-k}$$

$$= \binom{n}{k} p^k (1 - p)^{n-k},$$

by 1.5.4

In case $k < 0$ or $k > n$, when there are no such u, the truth of the theorem follows from the fact that $\binom{n}{k} = 0$.   □

Example Suppose an urn contains 7 red and 10 green balls, and 20 balls are drawn with replacement (and mixing) after each draw. What is the probability that (a) exactly 4, or (b) at least 4, of the balls drawn will be red?
Answers:

(a) $\binom{20}{4}(\frac{7}{17})^4(\frac{10}{17})^{16} = \frac{20 \cdot 19 \cdot 18 \cdot 17}{4 \cdot 3 \cdot 2}(\frac{7}{17})^4(\frac{10}{17})^{16}.$

(b)

$$\sum_{k=4}^{20}\binom{20}{k}(\frac{7}{17})^k(\frac{10}{17})^{20-k} = 1 - \sum_{k=0}^{3}\binom{20}{k}(\frac{7}{17})^k(\frac{10}{17})^{20-k}$$

$$= 1 - [(\frac{10}{17})^{20} + 20(\frac{7}{17})(\frac{10}{17})^{19} + \frac{20 \cdot 19}{2}(\frac{7}{17})^2(\frac{10}{17})^{18}$$

$$+ \frac{20 \cdot 19 \cdot 18}{3 \cdot 2}(\frac{7}{17})^3(\frac{10}{17})^{17}].$$

Observe that, in (b), the second expression for the probability is much more economical and evaluable than the first.

Exercises 1.5

1. An urn contains five red, seven green, and three yellow balls. Nine are drawn, *with* replacement. Find the probability that

    (a) exactly six of the balls drawn are green;
    (b) at least two of the balls drawn are yellow;
    (c) at most four of the balls drawn are red.

2. In eight tosses of a fair coin, find the probability that heads will come up

    (a) exactly three times;
    (b) at least three times;
    (c) at most three times.

3. Show that the probability of heads coming up exactly n times in $2n$ flips of a fair coin decreases with n.

*4. Find a simple representation of the polynomial $\sum_{k=0}^{n}\binom{n}{k}x^k(1-x)^{n-k}$.

1.6 An elementary counting principle

1.6.1 *Suppose that, in a k-stage experiment, for each i, $1 \le i \le k$, whatever may have happened in stages preceding, there are exactly n_i outcomes possible at the i th stage. Then there are $\prod_{i=1}^{k} n_i$ possible sequences of outcomes in the k stages.*

The proof can be done by induction on k. The word "preceding" in the statement above may seem to some to be too imprecise, and to others to be too precise, introducing an assumption about the order of the stages that need not be introduced. I shall leave the statement as it is, with "preceding" to be construed, in applications, as the applier deems wise.

The idea behind the wording is that the possible outcomes at the different stages may depend on the outcomes at other stages (those "preceding"), but whatever has happened at those other stages upon which the list of possible outcomes at the ith stage depends, the list of possible outcomes at the ith stage will always be of the same length, n_i.

For instance, with $k = 2$, suppose stage one is flipping a coin, and stage two is drawing a ball from one of two urns, RWB, which contains only red, white, and blue balls, and BGY, which contains only blue, green, and yellow balls. If the outcome at stage 1 is H, the ball is drawn from RWB; otherwise, from BGY. In this case, there are five possible outcomes at stage 2; yet $n_3 = 3$, and there are $2 \cdot 3 = 6$ possible sequences of outcomes of the experiment, namely HR, HW, HB, TB, TG, and TY, in abbreviated form. If, say, the second urn contained balls of only two different colors, then this two-stage experiment would not satisfy the hypothesis of 1.6.1.

In applying this counting principle, you think of a way to make, or construct, the objects you are trying to count. If you are lucky and clever, you will come up with a k-stage construction process satisfying the hypothesis of 1.6.1, and each object you are trying to count will result from exactly one sequence of outcomes or choices in the construction process. [But beware of situations in which the objects you want to count each arise from more than one construction sequence. See Exercise 4, below.]

For instance, to see that $|A_1 \times \cdots \times A_k| = \prod_{i=1}^{k} |A_i|$, when $A_1, \dots, A_k$ are sets, you think of making sequences $(a_1, \dots, a_k)$, with $a_i \in A_i$, $i = 1, \dots, k$, by the obvious process of first choosing a_1 from A_1, then a_2 from A_2, etc. For another instance, the number of different five-card hands dealable from a standard 52-card deck that are full houses is $13\binom{4}{3}12\binom{4}{2}$. [Why?]

Exercises 1.6

1. In the example above involving an urn, if the first urn contains balls of three different colors, and the second contains balls of two different colors, how many different possible sequences of outcomes are there in the two stages of the experiment?

2. How many different words of length ℓ are there over an alphabet with n letters? What does this have to do with 1.6.1?

3. In a certain collection of 23 balls, 15 are red and 8 are blue. How many different 7-subsets of the 23 balls are there, with 5 red and 2 blue balls?

4. How many different five-card hands dealable from a standard deck are "two-pair" hands?

1.7* On drawing without replacement

Suppose that an urn contains x red balls and y green balls. We draw n without replacement. What is the probability that exactly k will be red?

The set of outcomes of interest is identifiable with the set of all sequences, of length n, of the symbols r and g. We are interested in the event consisting of all such sequences in which r appears exactly k times.

We are not in the happy circumstances of Section 1.5, in which the probability of a red would be unchanging from draw to draw, but we do enjoy one piece of good fortune similar to something in that section: the different outcomes in the event "exactly k reds" all have the same probability. (Verify!) For $0 < k < n \le x+y$, that probability is

$$\frac{x(x-1)\cdots(x-k+1)y(y-1)\cdots(y-(n-k)+1)}{(x+y)(x+y-1)\cdots(x+y-n+1)}$$

$$= \frac{x!\,y!\,(x+y-n)!}{(x-k)!\,(y-(n-k))!\,(x+y)!};$$

this last expression is only valid when $k \le x$ and $n-k \le y$, which are necessary conditions for the probability to be non-zero. Under those conditions, the last expression is valid for $k = 0$ and $k = n$, as well.

By the same reasoning as in the proof of the independent Bernoulli trial theorem, 1.5.7, invoking 1.5.4, we have that

$$P(\text{"exactly } k \text{ reds"})$$

$$= \binom{n}{k}\frac{x(x-1)\cdots(x-k+1)y(y-1)\cdots(y-(n-k)+1)}{(x+y)\cdots(x+y-n+1)}$$

$$= \binom{n}{k}\frac{x!\,y!\,(x+y-n)!}{(x-k)!\,(y-(n-k))!\,(x+y)!}$$

for $0 < k < n$, provided we understand this last expression to be zero when $k > x$ or when $n-k > y$.

There is another way of looking at this experiment. Instead of drawing n balls one after the other, suppose you just reach into the urn and scoop up n balls all at once. Is this really different from drawing the balls one at a time, when you look at the final result? Supposing the two to be the same, we can take as the set of outcomes all n-subsets of the $x+y$ balls in the urn. Surely no n-subset is more likely than any other to be scooped up, so the outcomes are equiprobable, each with probability $1/\binom{x+y}{n}$ (provided $n \le x+y$). How many of these outcomes are in the event "exactly k reds are among the n balls selected?" Here is a two-stage method for forming such outcomes: first take a k-subset of the x red balls in the urn, then an $(n-k)$-subset of the y green balls (and put them together to make an n-set). Observe that different outcomes at either stage result in different n-sets, and that every n-set of these balls with exactly k reds is the result of one run of the two stages. By 1.6.1, it follows that

$$| \text{``exactly } k \text{ reds''} | = \binom{x}{k}\binom{y}{n-k}.$$

Thus $P(\text{``exactly } k \text{ reds''}) = \binom{x}{k}\binom{y}{n-k}/\binom{x+y}{n}$, provided $n \le x+y$.

Exercises 1.7

1. Verify that the two different ways of looking at drawing without replacement give the same probability for the event "exactly k balls are red."

2. Suppose an urn contains 10 red, 13 green, and 4 yellow balls. Nine are drawn, without replacement.

   (a) What is the probability that exactly three of the nine balls drawn will be yellow?

   (b) Find the probability that there are four red, four green, and one yellow among the nine balls drawn.

3. Find the probability of being dealt a flush (all cards of the same suit) in a five-card poker game.

1.8 Random variables and expected, or average, value

Suppose that (S, P) is a finite probability space. A *random variable* on this space is a function from S into the real numbers, $\mathbb{R}$. If $X : S \to \mathbb{R}$ is a random variable on (S, P), the *expected* or *average value* of X is

$$E(X) = \sum_{u \in S} X(u) P(u).$$

Random variables are commonly denoted X, Y, Z, or with subscripts: X_1, $X_2, \cdots$. It is sometimes useful, as in 1.8.2, to think of a random variable as a measurement on the outcomes, and the average value as the average of the measurements. The average value of a random variable X is sometimes denoted $\overline{X}$; you may recall that a bar over a letter connotes the arithmetic average, in elementary statistics.

The word "expected" in "expected value" requires interpretation. As numerous examples show (see 1.8.1, below), a random variable which takes only integer values can have a non-integer expected value. The point is that the expected value is not necessarily a possible value of the random variable, and thus is not really necessarily to be "expected" in any running of the experiment to which (S, P) is associated.

Examples

1.8.1 The experiment consists of n independent Bernoulli trials, with probability p of success on each trial. Let $X =$ "number of successes." Finding $E(X)$ in these circumstances is one of our goals; for now we'll have to be content with a special case.

Sub-example: $n = 3, p = 1/2$; let the experiment be flipping a fair coin three times, and let "success" be "heads." With $X =$ "number of heads," we have

$$E(X) = X(HHH)P(HHH) + X(HHT)P(HHT)$$
$$+ \cdots + X(TTT)P(TTT)$$
$$= \frac{1}{8}[3 + 2 + 2 + 1 + 2 + 1 + 1 + 0] = \frac{12}{8} = \frac{3}{2}.$$

1.8.2 We have some finite population (voters in a city, chickens on a farm) and some measurement M that can be applied to each individual of the population (blood pressure, weight, hat size, length of femur, ...). Once units are decided upon, $M(s)$ is a pure real number for each individual s. We can regard M as a random variable on the probability space associated with the experiment of choosing a member of the population "at random." Here S is the population, the outcomes are equiprobable (so P is the constant assignment $1/|S|$), and M assigns to each outcome (individual of the population) whatever real number measures that individual's M-measure. If $S = \{s_1, \ldots, s_n\}$, then $E(M) = (M(s_1) + \cdots + M(s_n))/n$, the arithmetic average, or mean, of the measurements $M(s), s \in S$.

1.8.3 Let an urn contain 8 red and 11 green balls. Four are drawn *without* replacement. Let X be the number of green balls drawn. What is $E(X)$?

Note that we have at our disposal two different views of this experiment. In one view, the set of outcomes is the set of ordered quadruples of the symbols r and g; i.e., $S = \{rrrr, rrrg, \ldots, gggg\}$. In the other view, an *outcome* is a four-subset of the 19 balls in the urn.

Taking the first view, we have

$$E(X) = 0 \cdot \frac{8 \cdot 7 \cdot 6 \cdot 5}{19 \cdot 18 \cdot 17 \cdot 16} + 1 \cdot 4 \cdot \frac{8 \cdot 7 \cdot 6 \cdot 11}{19 \cdot 18 \cdot 17 \cdot 16}$$
$$+ 2 \cdot \binom{4}{2} \frac{8 \cdot 7 \cdot 11 \cdot 10}{19 \cdot 18 \cdot 17 \cdot 16} + 3 \cdot 4 \frac{8 \cdot 11 \cdot 10 \cdot 9}{19 \cdot 18 \cdot 17 \cdot 16}$$
$$+ 4 \cdot \frac{11 \cdot 10 \cdot 9 \cdot 8}{19 \cdot 18 \cdot 17 \cdot 16}$$

Verify: $E(X) = 4 \cdot \frac{11}{19}$. Verify: the same answer is obtained if the other view of this experiment is taken. [Use 1.8.4, below, to express $E(X)$.]

1.8.4 Theorem *Suppose X is a random variable on (S, P). Then $E(X) = \sum_{x \in \mathbb{R}} x P(X = x)$.*

Here, "$X = x$" is the event $\{u \in \mathcal{S}; X(u) = x\} = X^{-1}(\{x\})$. Note that $X^{-1}(\{x\}) = \emptyset$ unless $x \in \text{ran}(X)$. Since $\mathcal{S}$ is finite, so is $\text{ran}(X)$. If $\text{ran}(X) \subseteq \{x_1, \ldots, x_r\}$, and $x_1, \ldots, x_r$ are distinct, then

$$\sum_{x \in \mathbb{R}} x P(X = x) = \sum_{k=1}^{r} x_k P(X = x_k).$$

Proof: Let $x_1, \ldots, x_r$ be the possible values that X might take; i.e., $\text{ran}(X) \subseteq \{x_1, \ldots, x_r\}$. Let, for $k \in \{1, \ldots, r\}$, $E_k = $ "$X = x_k$" $= X^{-1}(\{x_k\})$. Then $E_1, \ldots, E_r$ partition $\mathcal{S}$. Therefore,

$$E(X) = \sum_{u \in \mathcal{S}} X(u) P(u) = \sum_{k=1}^{r} \sum_{u \in E_k} X(u) P(u)$$

$$= \sum_{k=1}^{r} x_k \sum_{u \in E_k} P(u) = \sum_{k=1}^{r} x_k P(E_k). \qquad \square$$

1.8.5 Corollary *The expected or average number of successes, in a run of n independent Bernoulli trials with probability p of success in each, is*

$$\sum_{k=1}^{n} k \binom{n}{k} p^k (1-p)^{n-k}.$$

Proof: The possible values that $X = $ "number of successes" might take are $0, 1, \ldots, n$, and $P(X = k) = \binom{n}{k} p^k (1-p)^{n-k}$, for $k \in \{0, 1, \ldots, n\}$, by 1.5.7. $\qquad \square$

1.8.6 Theorem *If $X_1, \ldots, X_n$ are random variables on $(\mathcal{S}, P)$, and $a_1, \ldots, a_n$ are real numbers, then $E(\sum_{k=1}^{n} a_k X_k) = \sum_{k=1}^{n} a_k E(X_k)$.*

Proof:

$$E\left(\sum_{k=1}^{n} a_k X_k\right) = \sum_{u \in \mathcal{S}} \left(\sum_{k=1}^{n} a_k X_k\right)(u) P(u)$$

$$= \sum_{u \in \mathcal{S}} \sum_{k=1}^{n} a_k X_k(u) P(u)$$

$$= \sum_{k=1}^{n} a_k \sum_{u \in \mathcal{S}} X_k(u) P(u) = \sum_{k=1}^{n} a_k E(X_k). \qquad \square$$

1.8.7 Corollary *The expected, or average, number of successes, in a run of n independent Bernoulli trials with probability p of success in each, is np.*

Proof: First, let $n = 1$. Let X be the "number of successes." By definition and by the hypothesis,

$$E(X) = 0 \cdot (1-p) + 1 \cdot p = 1 \cdot p.$$

Now suppose that $n > 1$. Let $X =$ "number of successes." For $k \in \{1, \dots, n\}$, let $X_k =$ "number of successes on the kth trial." By the case already done, we have that $E(X_k) = p, k = 1, \dots, n$. Therefore,

$$E(X) = E(\sum_{k=1}^{n} X_k) = \sum_{k=1}^{n} E(X_k) = \sum_{k=1}^{n} p = np. \qquad \square$$

1.8.8 Corollary For $0 \leq p \leq 1$, and any positive integer n,

$$\sum_{k=1}^{n} k \binom{n}{k} p^k (1-p)^{n-k} = np.$$

Proof: This follows from 1.8.5 and 1.8.7, provided there exists, for each $p \in [0, 1]$, a Bernoulli trial with probability p of success. You are invited to ponder the existence of such trials (problem 1 at the end of this section).

Alternatively, using a sharper argument based on an elementary theorem about polynomials, the conclusion here (and the result called for in problem 2 at the end of this section) follows from 1.8.5 and 1.8.7 and the existence of such trials for only $n + 1$ distinct values of p. Therefore, the result follows from the existence of such trials for rational numbers p between 0 and 1. If $p = s/b$, where s and b are positive integers and $s < b$, let the Bernoulli trial with probability p of Success be: draw a ball from an urn containing s Success balls and $b - s$ Failure balls. $\qquad \square$

1.8.9 Corollary Suppose k balls are drawn, without replacement, from an urn containing x red balls and y green balls, where $1 \leq k \leq x + y$. The expected number of red balls to be drawn is $kx/(x+y)$.

Proof: It is left to the reader to see that the probability of a red being drawn on the ith draw, $1 \leq i \leq k$, is the same as the probability of a red being drawn on the first draw, namely $x/(x+y)$. [Before the drawing starts, the various $x + y$ balls are equally likely to be drawn on the ith draw, and x of them are red.] Once this is agreed to, the proof follows the lines of that of 1.8.7. $\qquad \square$

Exercises 1.8

1. Suppose $0 \leq p \leq 1$. Describe a Bernoulli trial with probability p of Success. You may suppose that there is a way of "picking a number at random" from a given interval.

2. Suppose that n is a positive integer. Find a simple representation of the polynomial $\sum_{k=1}^{n} k \binom{n}{k} x^k (1-x)^{n-k}$.

*3. State a result that is to 1.8.9 as 1.8.8 is to 1.8.7.

4. An urn contains 4 red and 13 green balls. Five are drawn. What is the expected number of reds to be drawn if the drawing is

    (a) with replacement?

(b) without replacement?

5. For any radioactive substance, if the material is not tightly packed, it is thought that whether or not any one atom decays is independent of whether or not any other atom decays. [We are all aware, I hope, that this assumption fails, in a big way, when the atoms are tightly packed together.]

For a certain mystery substance, let p denote the probability that any particular atom of the substance will decay during any particular 6-hour period at the start of which the atom is undecayed. A handful of this substance is left in a laboratory, in spread-out, unpacked condition, for 24 hours. At the end of the 24 hours it is found that approximately $1/10$ of the substance has decayed. Find p, approximately. [Hint: let n be the number of atoms, and think of the process of leaving the substance for 24 hours as n independent Bernoulli trials. Note that p here is not the probability of Success, whichever of two possibilities you choose to be Success, but that probability is expressible in terms of p. Assume that the amount of substance that decayed was approximately the "expected" amount.]

6. An actuary reckons that for any given year that you start alive, you have a 1 in 6,000 chance of dying during that year.

You are going to buy $100,000 worth of five-year term life insurance. However you pay for the insurance, let us define the payment to be *fair* if the life insurance company's *expected gain* from the transaction is *zero*.

(a) In payment plan number one, you pay a single premium at the beginning of the five-year-coverage period. Assuming the actuary's estimate is correct, what is the fair value of this premium?

(b) In payment plan number two, you pay five equal premiums, one at the beginning of each year of the coverage, provided you are alive to make the payment. Assuming the actuary's estimate is correct, what is the fair value of this premium?

7. (a) Find the expected total showing on the upper faces of two fair dice after a throw.

(b) Same question as in (a), except use backgammon scoring, in which doubles count quadruple. For instance, double threes count 12.

1.9 The Law of Large Numbers

1.9.1 Theorem *Suppose that, for a certain Bernoulli trial, the probability of Success is p. For a sequence of n independent such trials, let X_n be the random variable "number of Successes in the n trials." Suppose $\epsilon > 0$. Then*

$$P\left(\left|\frac{X_n}{n} - p\right| < \epsilon\right) \to 1 \text{ as } n \to \infty.$$

For instance, if you toss a fair coin 10 times, the probability that $|\frac{X_{10}}{10} - \frac{1}{2}| < \frac{1}{100}$, i.e., that the number of heads, say, is between 4.9 and 5.1, is just the probability that heads came up exactly 5 times, which is $\binom{10}{5}\frac{1}{2^{10}} = \frac{63}{256}$, not very close to 1. Suppose the same coin is tossed a thousand times; the probability that $|\frac{X_{1000}}{1000} - \frac{1}{2}| < \frac{1}{100}$ is the probability that between 490 and 510 heads appear in the thousand flips. By approximation methods that we will not go into here, this probability can be shown to be approximately 0.45. The probability of heads coming up between 4900 and 5100 times in 10,000 flips is around 0.95.

For a large number of independent Bernoulli trials of the same species, it is plausible that the *proportion* of Successes "ought" to be near p, the probability of Success on each trial. The Law of Large Numbers, 1.9.1, gives precise form to this plausibility. This theorem has a purely mathematical proof that will not be given here.

Theorem 1.9.1 is not the only way of stating that the proportion of Successes will tend, with high probability, to be around p, the probability of Success on each trial. Indeed, Feller [18] maligns this theorem as the weakest and least interesting of the various laws of large numbers available. Still, 1.9.1 is the best-known law of large numbers, and 'twill serve our purpose.

Exercises 1.9

*1. With n, ϵ, p, and X_n as in 1.9.1, express $P(|\frac{X_n}{n} - p| \le \epsilon)$ explicitly as a sum of terms of the form $\binom{n}{k}p^k(1-p)^{n-k}$. You will need the symbols $\lceil \cdot \rceil$ and $\lfloor \cdot \rfloor$, which stand for "round up" and "round down," respectively.

2. An urn contains three red and seven green balls. Twenty are drawn, with replacement. What is the probability of exactly six reds being drawn? Of five, six, or seven reds being drawn?

3. An urn contains an unknown number of red balls, and 10 balls total. You draw 100 balls, with replacement; 42 are red. What is your best guess as to the number of red balls in the urn?

*4. A pollster asks some question of 100 people, and finds that 42 of the 100 give "favorable" responses. The pollster estimates from this result that (probably) between 40 and 45 percent of the total population would be "favorable" on this question, if asked.

Ponder the similarities and differences between this inference and that in problem 3. Which is more closely analogous to the inference you used in doing Exercise 1.8.5?

Chapter *2*

Information and Entropy

2.1 How is information quantified?

Information theory leaped fully clothed from the forehead of Claude Shannon in 1948 [63]. The foundation of the theory is a quantification of information, a quantification that a few researchers had been floundering toward for 20 or 30 years (see [29] and [56]). The definition will appear strange and unnatural at first glance. The purpose of this first section of Chapter 2 is to acquaint the reader with certain issues regarding this definition, and finally to present the brilliant proof of its inevitability due, as far as we know, to Aczél and Daroczy in 1975 [1].

To begin to make sense of what follows, think of the familiar quantities *area* and *volume*. These are associated with certain kinds of objects—planar regions or regions on surfaces, in the case of area; bodies in space, in the case of volume. The assignment of these quantities to appropriate objects is *defined*, and the definitions can be quite involved; in fact, the final chapter on the question of mere definitions of area and volume was perhaps not written until the twentieth century, with the introduction of Lebesgue measure.

These definitions are not simply masterpieces of arbitrary mathematical cleverness—they have to respond to certain human agreements on the nature of these quantities. The more elaborate definitions have to agree with simpler ways of computing area on simple geometric figures, and a planar region composed of two other non-overlapping planar regions should have area equal to the sum of the areas of the two.

The class of objects to which the quantity *information* will be attached are occurrences of events associated with probabilistic experiments; another name for this class is *random phenomena*. It is supposed[1] that every such event or phenomenon E has a pre-assigned, a priori probability $P(E)$ of occurrence. Here is Shannon's definition of the "self-information" $I(E)$ of an event E:

$$I(E) = \log 1/P(E) = -\log P(E).$$

---

[1]Take care! Paradoxes and absurdities are known to be obtainable by loose manipulation of this assumption. These are avoidable by staying within the strict framework of a well-specified probabilistic experiment, in each situation.

If $P(E) = 0$, $I(E) = \infty$.

Feinstein [17] used other terminology that many would find more explana-
tory than "self-information": $I(E)$ is the amount of information *disclosed* or
given off by the occurrence of E. This terminology coaxes our agreement to the
premise that information ought to be a quantity attached to random phenomena
with prior probabilities. And if that is agreed to, then it seems unavoidable that
the quantity of information must be some function of the prior probability, i.e.,
$I(E) = f(P(E))$ for some function f, just because prior probability is the only
quantity associated with all random phenomena, the only thing to work with.

Lest this seem a frightful simplification of the wild world of random phenomena, to compute information content as a function of prior probability alone,
let us observe that this sort of simplification happens with other quantities; indeed, such simplification is one of the charms of quantification. Planar regions
of radically different shapes and with very different topological properties can
have the same area. Just so; why shouldn't a stock market crash in Tokyo and
an Ebola virus outbreak in the Sudan possibly release the same amount of information? The quantification of information should take no account of the
"quality" or category of the random phenomenon whose occurrence releases
the information.

Suppose we agree that $I(E)$ ought to equal $f(P(E))$ for some function f
defined on $(0, 1]$, at least, for all probabilistic events E. Then why did Shannon
take $f(x) = \log(1/x)$? (And which log are we talking about? But we will
deal with that question in the next subsection.) We will take up the question
of Shannon's inspiration, and Aczél and Daroczy's final word on the matter, in
Section 2.1.3. But to get acclimated, let's notice some properties of $f(x) =
\log(1/x)$, with log to any base > 1: f is a decreasing, non-negative function on
$(0, 1]$, and $f(1) = 0$. These seem to be necessary properties for a function to be
used to quantify information, via the equation $I(E) = f(P(E))$. Since $I(E)$
is to be a quantity, it should be non-negative. The smaller the prior probability
of an event the greater the quantity of information released when it occurs, so
f should be decreasing. And an event of prior probability 1 should release no
information at all when it occurs, so $f(1)$ should be 0.

Even among functions definable by elementary formulas, there are an infinite number of functions on $(0, 1]$ satisfying the requirements noted above; for
instance, $1 - x^q$ and $(1/x)^q - 1$ satisfy those requirements, for any $q > 0$. One
advantage that $\log(1/x)$ has over these functions is that it converts products to
sums, and a lot of products occur in the calculation of probabilities. As we shall
see in section 2.1.3, this facile, shallow observation in favor of $\log(1/x)$ as the
choice of function to be used to quantify information is remarkably close to the
reason why $\log(1/x)$ is the only possible choice for that purpose.

2.1.1 Naming the units

For any $a, b > 0$, $a \neq 1 \neq b$, and $x > 0$, $\log_a x = (\log_a b) \log_b x$; that is, the functions $\log x$ to different bases are just constant multiples of each other. So, in Shannon's use of log in the quantification of information, changing bases is like changing units. Choosing a base amounts to choosing a unit of information. What requires discussion is the name of the unit that Shannon chose when the base is 2: Shannon chose to call that unit a *bit*.

Yes, the unit name when $\log = \log_2$ is the very same abbreviation of "binary digit" widely reported to have been invented by J. W. Tukey, who was at Bell Labs with Shannon in the several years before [63] appeared. (In [76] we read that the word "bit", with the meaning of "binary digit", first appeared *in print* in "A mathematical theory of communication.")

Now, we do not normally pay much attention to unit names in other contexts. For example, "square meter" as a unit of area seems rather self-explanatory. But in this case the connection between "bit" as a unit of information, an arbitrarily divisible quantifiable substance, like a liquid, and "bit" meaning a binary digit, either 0 or 1, is not immediately self-evident to human intuition; yet Shannon uses the two meanings interchangeably, as has virtually every other information theorist since Shannon (although Solomon Golomb, in [24], is careful to distinguish between the two). We shall attempt to justify the unit name, and, in the process, to throw light on the meaning of the information unit when the base of the logarithm is a positive integer greater than 2.

Think of one square meter of area as the greatest amount of area that can be squeezed into a square of side length one meter. (You may object that when one has a square of side length 1 meter, one already has a maximum area "squeezed" into it. Fine; just humor us on this point.) Reciprocally, the square meter measure of the area of a planar region is the side length, in meters, of the smallest square into which the region can hypothetically be squeezed, by deformation without shrinking or expanding (don't ask for a rigorous definition here!).

With this in mind, let us take, in analogy to a planar region, an entire probabilistic experiment, initially unanalyzed as to its possible outcomes; and now let it be analyzed, the possible outcomes broken into a list $E_1, \ldots, E_m$ of pairwise mutually exclusive events which exhaust the possibilities: $P(\cup_{i=1}^{m} E_i) = \sum_{i=1}^{m} P(E_i) = 1$ (recall 1.2.3). If you wish, think of each E_i as a single outcome, in a set of outcomes. Assume $P(E_i) > 0$, $i = 1, \ldots, m$.

It may be objected that rather than analogizing a planar region by an entire probabilistic experiment, a planar region to which the quantity area is assigned should be analogous to the kind of thing to which the quantity information is assigned, namely a single event. This is a valid objection.

In what follows, rather than squeezing the information contained in a single event into a "box" of agreed size, we will be squeezing the information contained in the ensemble of the events $E_1, \ldots, E_m$ into a very special box, the set of binary words of a certain length. We will compare the average information content of the events $E_1, \ldots, E_m$ with that length. This comparison will be taken to indicate what the maximum average (over a list like $E_1, \ldots, E_m$) number of units of information can be represented by the typical (aren't they all?) binary word of that length.

We admit that this is all rather tortuous, as a justification for terminology. Until someone thinks of something better, we seem to be forced to this approach by the circumstance that we are trying to squeeze the information content of events into binary words, whereas, in the case of area, we deform a region to fit into another region of standard shape. If we considered only one event, extracted without reference to the probabilistic experiment to which it is associated, we could let it be named with a single bit, 0 or 1, and this does not seem to be telling us anything. Considering a non-exhaustive ensemble of events associated with the same probabilistic experiment (pairwise mutually exclusive so that their information contents are separate) we have a generalization of the situation with a single event; we can store a lot of information by encoding with relatively short binary words, just because we are ignoring the full universe of possibilities. Again, this does not seem to lead to a satisfactory conclusion about the relation between information and the length of binary words required to store it. What about looking at ensembles of events from possibly different probabilistic experiments? Again, unless there is some constraint on the number of these events and their probabilities, it does not seem that encoding these as binary words of fixed length tells us anything about units of information, any more than in the case when the events are associated with the same probabilistic experiment.

We realize that this discussion is not wholly convincing; perhaps someone will develop a more compelling way of justifying our setup in the future. For now, let us return to $E_1, \ldots, E_m$, pairwise mutually exclusive events with $\sum_{i=1}^{m} P(E_i) = 1$. If we agree that $I(E) = -\log P(E)$ for any event E, with log to some base > 1, then the average information content of an event in the list $E_1, \ldots, E_m$ is

$$H(E_1, \ldots, E_m) = \sum_{i=1}^{m} P(E_i) I(E_i) = -\sum_{i=1}^{m} P(E_i) \log P(E_i).$$

(Recall Section 1.8.)

As is conventional, let $\ln = \log_e$, the natural logarithm.

2.1.1 Lemma For $x > 0$, $\ln x \leq x - 1$, with equality when and only when $x = 1$.

Indication of proof Apply elementary calculus to $f(x) = x - 1 - \ln x$ to see that $f(x) \geq 0$ on $(0, \infty)$, with equality only when $x = 1$.

2.1.2 Theorem If $p_1, \ldots, p_m$ are positive numbers summing to 1, then $-\sum_{i=1}^{m} p_i \log p_i \leq \log m$, with equality if and only if $p_i = 1/m$, $i = 1, \ldots, m$.

Proof: Let $c = \log e > 0$. Since $\sum p_i = 1$,

$$(-\sum_{i=1}^{m} p_i \log p_i) - \log m = \sum_{i=1}^{m} p_i (\log(1/p_i) - \log m)$$

$$= \sum_{i=1}^{m} p_i \log(1/(mp_i))$$

$$= c \sum_{i=1}^{m} p_i \ln(1/(mp_i)) \le c \sum_{i=1}^{m} p_i (\frac{1}{mp_i} - 1)$$

$$= c\left(\sum_{i=1}^{m}(1/m) - \sum_{i=1}^{m} p_i\right) = c(1-1) = 0,$$

by Lemma 2.1.1, with equality if and only if $1/(mp_i) = 1$ for each $i = 1, \ldots, m$.

□

Now back to considering $E_1, \ldots, E_m$. Let k be an integer such that $m \le 2^k$ and let us put the E_i in one-to-one correspondence with m of the binary words of length k. That is, the E_i have been encoded, or named, by members of $\{0, 1\}^k$, and thereby we consider the ensemble $E_1, \ldots, E_m$ to be stored in $\{0, 1\}^k$.

By Theorem 2.1.2, the average information content among the E_i satisfies $H(E_1, \ldots, E_m) = -\sum_{i=1}^{m} P(E_i) \log P(E_i) \le \log m \le \log 2^k = k$, if $\log = \log_2$; and equality can be achieved if $m = 2^k$ and $P(E_i) = 1/m, i = 1, \ldots, m$. That is, the greatest average number of information units per event contained in a "system of events", as we shall call them, which can be stored as k-bit binary words, is k, if the unit corresponds to $\log = \log_2$. And that, ladies and gentlemen, is why we call the unit of information a *bit* when $\log = \log_2$.

In case $\log = \log_n$, for an integer $n > 2$, we would like to call the unit of information a nit, but the term probably won't catch on. Whatever it is called, the discussion preceding can be adapted to justify the equivalence of the information unit, when $\log = \log_n$, and a single letter of an n-element alphabet.

2.1.2 Information connecting two events

Let $\log = \log_b$ for some $b > 1$, and suppose that E and F are events in the same probability space (i.e., associated with the same probabilistic experiment).

If $P(F) > 0$, the conditional information contained in E, conditional upon F, denoted $I(E \mid F)$, is

$$I(E \mid F) = -\log P(E \mid F) = -\log \frac{P(E \cap F)}{P(F)}.$$

If $P(E \cap F) = 0$ we declare $I(E \mid F) = \infty$.

The *mutual information* of (or between) E and F, denoted $I(E, F)$, is

$$I(E, F) = \log \frac{P(E \cap F)}{P(E)P(F)}, \text{ if } P(E)P(F) > 0,$$

and $I(E, F) = 0$ otherwise, i.e., if either $P(E) = 0$ or $P(F) = 0$.

If Shannon's quantification of information is agreed to, and if account is taken of the justification of the formula for conditional probability given in Section 1.3, then there should be no perplexity regarding the definition of $I(E \mid F)$. But it is a different story with $I(E, F)$. For one thing, $I(E, F)$ can be positive or negative. Indeed, if $P(E)P(F) > 0$ and $P(E \cap F) = 0$, we have no choice but to set $I(E, F) = -\infty$; also, $I(E, F)$ can take finite negative values, as well as positive ones.

We do not know of a neat justification for the term "mutual information", applied to $I(E, F)$, but quite a strong case for the terminology can be built on circumstantial evidence, so to speak. The mutual information function and the important index based on it, the mutual information between two systems of events, to be introduced in Section 2.2, behave as one would hope that indices so named would behave. As a first instance of this behavior, consider the following, the verification of which is left as an exercise.

2.1.3 Proposition $I(E, F) = 0$ *if and only if E and F are independent events.*

2.1.3 The inevitability of Shannon's quantification of information

Shannon himself provided a demonstration (in [63]) that information must be quantified as he proposed, given that it is to be a quantity attached to random phenomena, and supposing certain other fundamental premises about its behavior. His demonstration was mathematically intriguing, and certainly contributed to the shocked awe with which "A mathematical theory of communication" was received. However, after the initial astonishment at Shannon's virtuosity wears off, one notices a certain infelicity in this demonstration, arising from the abstruseness of those certain other fundamental premises referred to above. These premises are not about information directly, but about something called *entropy*, defined in Section 2.3 as the average information content of events in a system of events. [Yes, we have already seen this average in Section 2.1.1.] Defined thus, entropy can also be regarded as a function on the space of all finite probability vectors, and it is as such that certain premises—we could call them axioms—about entropy were posed by Shannon. He then showed that if entropy, defined with respect to information, is to satisfy these axioms, then information must be defined as it is.

The problem with the demonstration has to do with our assent to the axioms. This assent is supposed to arise from a prior acquaintance with the word "entropy," connoting disorder or unpredictability, in thermodynamics or the kinetic theory of gases. Even supposing an acquaintance with entropy in those contexts, there are a couple of intellectual leaps required to assent to Shannon's axioms for entropy: why should this newly defined, information-theoretic entropy carry the connotation of the older entropy, and how does this connotation translate into the specific axioms set by Shannon?

The demonstration of Feinstein [17] is of the same sort as Shannon's, with a somewhat more agreeable set of axioms, lessening the vertigo associated with the second of the intellectual leaps mentioned above. The first leap remains. Why should we assent to requirements on something called entropy just because we are calling it entropy, a word that occurs in other contexts?

Here are some requirements directly on the function f appearing in $I(E) = f(P(E))$ enunciated by Aczél and Daroczy [1]:

(i) $f(x) \geq 0$ for all $x \in (0, 1]$;

(ii) $f(x) > 0$ for all $x \in (0, 1)$; and

(iii) $f(pq) = f(p) + f(q)$ for all $p, q \in (0, 1]$.

Requirements (i) and (ii) have been discussed in section 2.1.1. Notice that there is no requirement that f be decreasing here.

Obviously requirement (iii) above is a strong requirement, and deserves considerable comment. Suppose that $p, q \in (0, 1]$. Suppose that we can find events E and F, possibly associated with different probabilistic experiments, such that $P(E) = p$ and $P(F) = q$. (We pass over the question of whether or not probabilistic experiments providing events of arbitrary prior probability can be found.) Now imagine the two-stage experiment consisting of performing copies of the experiments associated with E and F independently. Let G be the event "E occurred in the one experiment and F occurred in the other". Then, as we know from section 1.3, $P(G) = pq$, so $I(G) = f(pq)$.

On the other hand, the independence of the performance of the probabilistic experiments means that the information given off by the occurrence of E in one, and F in the other, ought to be the sum of the information quantities disclosed by the occurrence of each separately. This is like saying that the area of a region made up of two non-overlapping regions ought to be the sum of the areas of the constituent regions. Thus we should have

$$f(pq) = I(G) = I(E) + I(F) = f(p) + f(q).$$

We leave it to the reader to scrutinize the heart of the matter, the contention that because the two probabilistic experiments are performed with indifference, or obliviousness, to each other, the information that an observer will obtain from the occurrences of E and F, in the different experiments, ought to be $I(E) + I(F)$. We make the obvious remark that if you receive something—say, money—from one source, and then some more money from a totally different source, then the total amount of money received will be the sum of the two amounts received.

We achieve the purpose of this subsection by proving a slightly stronger version of Aczél and Daroczy's result, that if f satisfies (i), (ii), and (iii), above, then $f(x) = -\log_b x$ for some $b > 1$ for all $x \in (0, 1]$.

2.1.4 Theorem Suppose that f is a real-valued function on $[0, 1)$ satisfying

(a) $f(x) \geq 0$ for all $x \in (0, 1]$;

(b) $f(\alpha) > 0$ for some $\alpha \in (0, 1]$; and

(c) $f(pq) = f(p) + f(q)$ for all $p, q \in (0, 1]$.

Then for some $b > 1$, $f(x) = -\log_b x$ for all $x \in (0, 1]$.

Proof: First we show that f is monotone non-increasing on $(0, 1]$. Suppose that $0 < x < y \le 1$. Then $f(x) = f(y\frac{x}{y}) = f(y) + f(x/y) \ge f(y)$, by (a) and (c).

Now we use a standard argument using (c) alone to show that $f(x^r) = rf(x)$ for any $x \in (0, 1]$ and any positive rational r. First, using (c) repeatedly, or, if you prefer, proceeding by induction on m, it is straightforward to see that for each $x \in (0, 1]$ and each positive integer m, $f(x^m) = mf(x)$. Now suppose that $x \in (0, 1]$ and that m and n are positive integers. Then $x^m, x^{m/n} \in (0, 1]$, and

$$mf(x) = f(x^m) = f((x^{m/n})^n) = nf(x^{m/n}),$$

so $f(x^{m/n}) = \frac{m}{n} f(x)$.

By (c), $f(1) = f(1) + f(1)$, so $f(1) = 0$. Therefore the α mentioned in (b) is not 1. As b ranges over $(1, \infty)$, $-\log_b \alpha = -\frac{\ln \alpha}{\ln b}$ ranges over $(0, \infty)$. Therefore, for some $b > 1$, $-\log_b \alpha = f(\alpha)$. By the result of the paragraph preceding and the properties of $\log_b$, the functions f and $-\log_b$ agree at each point α^r, r a positive rational.

The set of such points is dense in $(0, 1]$. An easy way to see this is to note that $\ln \alpha^r = r \ln \alpha$, so $\{\ln \alpha^r ; r$ is a positive rational$\}$ is dense in $(-\infty, 0)$, by the well-known density of the rationals in the real numbers; and the inverse of $\ln$, the exponential function, being continuous and increasing, will map a dense set in $(-\infty, 0)$ onto a dense set in $(0, 1]$.

We have that f and $-\log_b$ are both non-increasing, they agree on a dense subset of $(0, 1]$, and $-\log_b$ is continuous. We conclude that $f = -\log_b$ on $(0, 1]$. The argument forcing this conclusion is left as an exercise.   □

Exercises 2.1

1. Regarding the experiment described in Exercise 1.3.5, let $E_A = $ "urn A was chosen" and $F_g = $ "a green ball was drawn". Write explicitly: (a) $I(E_A, F_g)$; (b) $I(E_A \mid F_g)$; (c) $I(F_g \mid E_A)$.

2. Suppose that E is an event in some probability space, and $P(E) > 0$. Show that $I(E, E) = I(E)$, and that $I(E \mid E) = 0$.

3. Suppose that E and F are events in some probability space. Show that $I(E, F) = 0$ if and only if E and F are independent. Show that $I(E, F) = I(E) - I(E \mid F)$, if $P(E)P(F) > 0$. Show that $I(E, F) \le \min(I(E), I(F))$. Show that $I(E, F) = I(F)$ if and only if E is essentially contained in F, meaning, $P(E \setminus F) = 0$.

4. Fill in the proof of Lemma 2.1.1.

5. Suppose that $p_1, \ldots, p_n, q_1, \ldots, q_n$ are positive numbers and $\sum_i p_i = 1 = \sum_i q_i$. Show that $\sum_{i=1}^{m} p_i \log(1/q_i) \leq \sum_{i=1}^{m} p_i \log(1/p_i)$ with equality if and only if $p_i = q_i, i = 1, \ldots, n$. [Hint: look at the proof of Theorem 2.1.2.]

6. Show that if f and g are monotone non-increasing real-valued functions on a real interval I which agree on a dense subset of I, and g is continuous, then $f = g$ on I. Give an example to show that the conclusion is not valid if the assumption that g is continuous is omitted.

2.2 Systems of events and mutual information

Suppose that (S, P) is a finite probability space. A *system of events* in (S, P) is a finite indexed collection $\mathcal{E} = [E_i; i \in I]$ of pairwise mutually exclusive events satisfying $1 = P(\bigcup_{i \in I} E_i)$.

Remarks

2.2.1 When $1 = P(\bigcup_{i \in I} E_i)$, it is common to say that the E_i *exhaust* S.

2.2.2 Note that if the E_i are pairwise mutually exclusive, then $P(\bigcup_{i \in I} E_i) = \sum_{i \in I} P(E_i)$, by 1.2.3.

2.2.3 Any *partition* of S is a system of events in (S, P) (see exercise 1.1.2), and partitioning is the most obvious way of obtaining systems of events. For instance, in the case of n Bernoulli trials, with $S = \{S, F\}^n$, if we take $E_k =$ "exactly k successes," then $E_0, \ldots, E_n$ partition S.

It is possible to have a system of events in (S, P) which does not partition S only when S contains outcomes with zero probability. Just as it is convenient to allow outcomes of zero probability to be elements of sets of outcomes, it is convenient to allow events of zero probability in systems of events. One aspect of this convenience is that when we derive new systems from old, as we shall, we do not have to stop and weed out the events of probability zero in the resultant system.

In deriving or describing systems of events we may have repeated events in the system, $E_i = E_j$ for some indices $i \neq j$. In this case, $P(E_i) = 0$. For better or for worse, the formality of defining a system of events as an indexed collection, rather than as a set or list of events, permits such repetition.

2.2.4 If $\mathcal{E} = [E_i; i \in I]$ is a system of events in (S, P), we can take $\mathcal{E}$ as a new set of outcomes. The technical niceties are satisfied since $1 = P(\bigcup_{i \in I} E_i) = \sum_{i \in I} P(E_i)$; in viewing $\mathcal{E}$ as a *set* we regard the E_i as distinct, even when they are not.

Taking $\mathcal{E}$ as a new set of outcomes involves a certain change of view. The old outcomes are merged into "larger" conglomerate outcomes, the events E_i. The changes in point of view achievable in this way are constrained by the choice of the original set of outcomes, S. Notice that S itself can be thought of as a system of events, if we identify each $s \in S$ with the event $\{s\}$.

If $\mathcal{F}$ is a system of events in (S, P), and we choose to view $\mathcal{F}$ as a set of outcomes, then we can form systems of events in the "new" space $(\mathcal{F}, P)$ just as $\mathcal{F}$ is formed from S. Systems of events in $(\mathcal{F}, P)$ are really just systems in (S, P) that bear a certain relation to $\mathcal{F}$.

2.2.5 Definition Suppose $\mathcal{E} = [E_i; i \in I]$ and $\mathcal{F} = [F_j; j \in J]$ are systems of events in a finite probability space (S, P); we say that $\mathcal{E}$ is an *amalgamation* of $\mathcal{F}$ in case for each $j \in J$ there is some $i \in I$ such that $P(E_i \cap F_j) = P(F_j)$.

2.2.6 Definition If (S, P) is a finite probability space and $E, F \subseteq S$, we will say that $F \subseteq E$ *essentially* if $P(F \setminus E) = 0$, and $F = E$ essentially if $F \subseteq E$ essentially and $E \subseteq F$ essentially.

To make sense of Definition 2.2.5, notice that $P(E_i \cap F_j) = P(F_j)$ is equivalent to $P(F_j \setminus (E_i \cap F_j)) = 0$, which means that F_j is contained in E_i, except, possibly, for outcomes of zero probability. So, the condition in the definition says that each F_j is essentially contained in some E_i. By Corollary 2.2.8, below, $P(F_j) = \sum_{i \in I} P(E_i \cap F_j)$ for each $j \in J$, so if $0 < P(F_j) = P(E_{i_0} \cap F_j)$ for some $i_0 \in I$, then i_0 is unique and $P(E_i \cap F_j) = 0$ for all $i \in I$, $i \neq i_0$. This says that, when $\mathcal{E}$ is an amalgamation of $\mathcal{F}$, each F_j of positive probability is essentially contained in exactly one E_i. Since the F_j are mutually exclusive and essentially cover S, it also follows that each E_i is essentially (neglecting outcomes of zero probability) the union of the F_j it essentially contains; i.e., the E_i are obtained by "amalgamating" the F_j somehow. (Recall exercise 1.1.2.)

Indeed, the most straightforward way to obtain amalgamations of $\mathcal{F}$ is as follows: partition J into non-empty subsets $J_1, \ldots, J_k$, and set $\mathcal{E} = [E_1, \ldots, E_k]$, with $E_i = \bigcup_{j \in J_i} F_j$. It is left to you to verify that $\mathcal{E}$ thus obtained is an amalgamation of $\mathcal{F}$. We shall prove the insinuations of the preceding paragraph, and see that every amalgamation is essentially (neglecting outcomes of probability zero) obtained in this way. This formality will also justify the interpretation of an amalgamation of $\mathcal{F}$ as a system of events in the new space $(\mathcal{F}, P)$, treating $\mathcal{F}$ as a set of outcomes. Readers who already see this interpretation and abhor formalities may skip to 2.2.10.

2.2.7 Lemma *If* (S, P) *is a finite probability space,* $E, F \subseteq S$, *and* $P(F) = 1$, *then* $P(E) = P(E \cap F)$.

Proof: $P(E) = P(E \cap F) + P(E \cap (S \setminus F)) = P(E \cap F) + 0$, since $E \cap (S \setminus F) \subseteq S \setminus F$ and $P(S \setminus F) = 1 - P(F) = 1 - 1 = 0$. $\square$

2.2.8 Corollary *If* $\mathcal{F} = [F_j; j \in J]$ *is a system of events in* (S, P), *and* $E \subseteq S$, *then* $P(E) = \sum_{j \in J} P(E \cap F_j)$.

Proof: Suppose $j_1, j_2 \in J$ and $j_1 \neq j_2$. We have $(E \cap F_{j_1}) \cap (E \cap F_{j_2}) \subseteq F_{j_1} \cap F_{j_2}$, so $0 \leq P((E \cap F_{j_1}) \cap (E \cap F_{j_2})) \leq P(F_{j_1} \cap F_{j_2}) = 0$ since F_{j_1} and F_{j_2} are mutually exclusive. Thus $E \cap F_{j_1}$ and $E \cap F_{j_2}$ are mutually exclusive. It follows that $\sum_{j \in J} P(E \cap F_j) = P(\bigcup_{j \in J}(E \cap F_j)) = P(E \cap (\bigcup_{j \in J} F_j)) = P(E)$ by 2.2.7, taking $F = \bigcup_{j \in J} F_j$. □

2.2.9 Theorem *Suppose that $\mathcal{F} = [F_j; j \in J]$ is a system of events in a finite probability space $(\mathcal{S}, P)$, and I is a finite non-empty set. An indexed collection $\mathcal{E} = [E_i; i \in I]$ of subsets of S is an amalgamation of $\mathcal{F}$ if and only if there is a partition $[J_i; i \in I]$ of J (into not necessarily non-empty sets) such that, for each $i \in I$, $E_i = \bigcup_{j \in J_i} F_j$ essentially.*

Proof: The proof of the "if" assertion is left to the reader. Note that as part of this proof it should be verified that $\mathcal{E}$ is a system of events.

Suppose that $\mathcal{E}$ is an amalgamation of $\mathcal{F}$. If $i \in I$ and $P(E_i) > 0$, set $J_i = \{j \in J; P(E_i \cap F_j) > 0\}$. If $j \in J$ and $P(F_j) > 0$, then there is a unique $i_0 \in I$ such that $j \in J_{i_0}$, by the argument in the paragraph following Definition 2.2.6. Note that $P(F_j) = P(F_j \cap E_i)$ if $j \in J_i$, by that argument.

Thus we have pairwise disjoint sets $J_i \subseteq J$ containing every $j \in J$ such that $P(F_j) > 0$. If $P(F_j) = 0$, put j into one of the J_i, it doesn't matter which. If $P(E_i) = 0$, set $J_i = \emptyset$. The J_i, $i \in I$, now partition J. It remains to be seen that $E_i = \bigcup_{j \in J_i} F_j$ essentially for each $i \in I$. This is clear if $P(E_i) = 0$. If $P(E_i) > 0$, then, since $P(E_i \cap F_j) > 0$ only for $j \in J_i$, we have

$$P(E_i) = \sum_{j \in J} P(E_i \cap F_j) \quad \text{[by Corollary 2.2.8]}$$

$$= \sum_{j \in J_i} P(E_i \cap F_j) = P(\bigcup_{j \in J_i} (E_i \cap F_j)), \text{ so}$$

$$P(E_i \setminus \bigcup_{j \in J_i} F_j) = P(E_i \setminus \bigcup_{j \in J_i} (E_i \cap F_j))$$

$$= P(E_i) - P(\bigcup_{j \in J_i} (E_i \cap F_j)) = 0.$$

On the other hand,

$$P(E_i) = \sum_{j \in J_i} P(E_i \cap F_j) = \sum_{j \in J_i} P(F_j)$$

implies that

$$P(\bigcup_{j \in J_i} F_j \setminus E_i) = P(\bigcup_{j \in J_i} (F_j \setminus (E_i \cap F_j)))$$

$$= \sum_{j \in J_i} [P(F_j) - P(E_i \cap F_j)] = 0.$$

Thus $E_i = \bigcup_{j \in J_i} F_j$, essentially. □

When $\mathcal{E}$ is an amalgamation of $\mathcal{F}$, we say that $\mathcal{E}$ is *coarser* than $\mathcal{F}$, and $\mathcal{F}$ is *finer* than $\mathcal{E}$. Given $\mathcal{E}$ and $\mathcal{F}$, neither necessarily coarser than the other, there is an obvious way to obtain a coarsest system of events which is finer than each of $\mathcal{E}$, $\mathcal{F}$. (See Exercise 2.2.2.)

2.2.10 Definition Suppose that $\mathcal{E} = [E_i; i \in I]$ and $\mathcal{F} = [F_j; j \in J]$ are systems of events in a finite probability space $(\mathcal{S}, P)$. The *joint system* associated with $\mathcal{E}$ and $\mathcal{F}$, denoted $\mathcal{E} \wedge \mathcal{F}$, is

$$\mathcal{E} \wedge \mathcal{F} = [E_i \cap F_j; (i, j) \in I \times J].$$

$\mathcal{E} \wedge \mathcal{F}$ is also called the *join* of $\mathcal{E}$ and $\mathcal{F}$.

2.2.11 Theorem *If $\mathcal{E}$ and $\mathcal{F}$ are systems of events in $(\mathcal{S}, P)$ then $\mathcal{E} \wedge \mathcal{F}$ is a system of events in $(\mathcal{S}, P)$.*

Proof: If $(i, j), (i', j') \in I \times J$ and $(i, j) \neq (i', j')$, then either $i \neq i'$ or $j \neq j'$. Suppose that $i \neq i'$. Since $(E_i \cap F_j) \cap (E_{i'}' \cap F_j') \subseteq E_i \cap E_{i'}'$, we have

$$0 \leq P((E_i \cap F_j) \cap (E_{i'}' \cap F_j')) \leq P(E_i \cap E_{i'}') = 0,$$

so $E_i \cap F_j$ and $E_{i'}' \cap F_j'$ are mutually exclusive. The case $i = i'$ but $j \neq j'$ is handled symmetrically.

Next, since mutual exclusivity has already been established,

$$P\left(\bigcup_{(i,j) \in I \times J} (E_i \cap F_j) \right) = \sum_{(i,j) \in I \times J} P(E_i \cap F_j)$$

$$= \sum_{i \in I} \sum_{j \in J} P(E_i \cap F_j)$$

$$= \sum_{i \in I} P(E_i) \qquad \text{[by Corollary 2.2.8]}$$

$$= 1. \qquad\qquad \square$$

Definition Suppose that $\mathcal{E} = [E_i; i \in I]$ and $\mathcal{F} = [F_j; j \in J]$ are systems of events in some finite probability space $(\mathcal{S}, P)$; $\mathcal{E}$ and $\mathcal{F}$ are *statistically independent* if and only if E_i and F_j are independent events, for each $i \in I$ and $j \in J$.

Statistically independent systems of events occur quite commonly in association with multistage experiments in which two of the stages are "independent" – i.e., outcomes at one of the two stages do not influence the probabilities of the outcomes at the other. For instance, think of an experiment consisting of flipping a coin, and then drawing a ball from an urn, and two systems, one consisting of the two events "heads came up", "tails came up", with reference to the coin flip, and the other consisting of the events associated with the colors of the balls that might be drawn. For a more formal discussion of stage, or component, systems associated with multistage experiments, see Exercise 2.2.5.

2.2.12 Definition Suppose that $\mathcal{E} = [E_i; i \in I]$ and $\mathcal{F} = [F_j; j \in J]$ are systems of events in a finite probability space $(\mathcal{S}, P)$. The *mutual information between* $\mathcal{E}$ and $\mathcal{F}$ is

$$I(\mathcal{E}, \mathcal{F}) = \sum_{i \in I} \sum_{j \in J} P(E_i \cap F_j) I(E_i, F_j).$$

In the expression for $I(\mathcal{E}, \mathcal{F})$ above, $I(E_i, F_j)$ is the mutual information between events E_i and F_j as defined in Section 2.1.2: $I(E_i, F_j) = \log \frac{P(E_i \cap F_j)}{P(E_i) P(F_j)}$ if $P(E_i) P(F_j) > 0$, and $I(E_i, F_j) = 0$ otherwise. If we adopt the convention that $0 \log(\text{anything}) = 0$, then we are permitted to write

$$I(\mathcal{E}, \mathcal{F}) = \sum_{i \in I} \sum_{j \in J} P(E_i \cap F_j) \log \frac{P(E_i \cap F_j)}{P(E_i) P(F_j)};$$

this rewriting will turn out to be a great convenience.

The mutual information between two systems of events will be an extremely important parameter in the assessment of the performance of communication channels, in Chapters 3 and 4, so it behooves us to seek some justification of the term "mutual information." Given $\mathcal{E}$ and $\mathcal{F}$, we can think of the mapping $(i, j) \to I(E_i, F_j)$ as a random variable on the system $\mathcal{E} \wedge \mathcal{F}$, and we then see that $I(\mathcal{E}, \mathcal{F})$ is the average value of this random variable, the "mutual information between events" random variable. This observation would justify the naming of $I(\mathcal{E}, \mathcal{F})$, if only we were quite sure that the mutual-information-between-events function is well named. It seems that the naming of both mutual informations, between events and between systems of events, will have to be justified by the behavior of these quantities. We have seen one such justification in Proposition 2.1.3, and there is another in the next Theorem. This theorem, by the way, is quite surprising in view of the fact that the terms in the sum defining $I(\mathcal{E}, \mathcal{F})$ can be negative.

2.2.13 Theorem *Suppose that $\mathcal{E}$ and $\mathcal{F}$ are systems of events in a finite probability space. Then $I(\mathcal{E}, \mathcal{F}) \geq 0$ with equality if and only if $\mathcal{E}$ and $\mathcal{F}$ are statistically independent.*

Proof: Let $c = \log e$, so that $\log x = c \ln x$ for all $x > 0$. If $i \in I$, $j \in J$, and $P(E_i \cap F_j) > 0$, we have, by Lemma 2.1.1,

$$P(E_i \cap F_j) \log \frac{P(E_i) P(F_j)}{P(E_i \cap F_j)} \leq c P(E_i \cap F_j) \left[\frac{P(E_i) P(F_j)}{P(E_i \cap F_j)} - 1 \right]$$
$$= c \left[P(E_i) P(F_j) - P(E_i \cap F_j) \right]$$

with equality if and only if $\frac{P(E_i) P(F_j)}{P(E_i \cap F_j)} = 1$, i.e., if and only if E_i and F_j are independent events. If $P(E_i \cap F_j) = 0$ then, using the convention that $0 \log(\text{anything}) = 0$, we have

$$0 = P(E_i \cap F_j) \log \frac{P(E_i) P(F_j)}{P(E_i \cap F_j)} \leq c [P(E_i) P(F_j) - P(E_i \cap F_j)],$$

again, with equality if and only if $P(E_i) P(F_j) = 0 = P(E_i \cap F_j)$. Thus

$$-I(\mathcal{E},\mathcal{F}) = \sum_{i \in I} \sum_{j \in J} P(E_i \cap F_j) \log \frac{P(E_i)P(F_j)}{P(E_i \cap F_j)}$$

$$\leq c \sum_{i \in I} \sum_{j \in J} [P(E_i)P(F_j) - P(E_i \cap F_j)]$$

$$= c \left[\sum_{i \in I} P(E_i) \sum_{j \in J} P(F_j) - \sum_{i \in I} \sum_{j \in J} P(E_i \cap F_j) \right]$$

$$= c[1 - 1] = 0 \qquad \text{[note Theorem 2.2.11]}$$

with equality if and only if E_i and F_j are independent, for each $i \in I$, $j \in J$. $\square$

Exercises 2.2

1. You have two fair dice, one red, one green. You roll them once. We can make a probability space referring to this experiment in a number of different ways. Let

   $\mathcal{S}_1 = \{$"i appeared on the red die, j on the green"; $i, j \in \{1, \ldots, 6\}\}$,

   $\mathcal{S}_2 = \{$"i appeared on one of the dice, and j on the other";

   $\quad i, j \in \{1, \ldots, 6\}\}$,

   $\mathcal{S}_3 = \{$"the sum of the numbers appearing on the dice was k";

   $\quad k \in \{2, \ldots, 12\}\}$, and

   $\mathcal{S}_4 = \{$"even numbers on both dice", "even on the red, odd on the green",

   $\quad$ "even on the green, odd on the red", "odd numbers on both dice"$\}$.

   Which pairs of these sets of outcomes have the property that neither is an amalgamation of the other?

2. Suppose that $\mathcal{E}$ and $\mathcal{F}$ are systems of events in a finite probability space.

   (a) Prove that each of $\mathcal{E}, \mathcal{F}$ is an amalgamation of $\mathcal{E} \wedge \mathcal{F}$. [Thus, $\mathcal{E} \wedge \mathcal{F}$ is finer than each of $\mathcal{E}, \mathcal{F}$.]

   (b) Suppose that each of $\mathcal{E}, \mathcal{F}$ is an amalgamation of a system of events $\mathcal{G}$. Show that $\mathcal{E} \wedge \mathcal{F}$ is an amalgamation of $\mathcal{G}$. [So $\mathcal{E} \wedge \mathcal{F}$ is the coarsest system of events, among those that are finer than each of $\mathcal{E}, \mathcal{F}$.] Here is a hint for (b): Suppose that E, F, and G are events in $\mathcal{E}, \mathcal{F}$ and $\mathcal{G}$, respectively, and $P(G \cap E \cap F) > 0$. Then $P(G \cap E), P(G \cap F) > 0$. By the assumption that $\mathcal{E}, \mathcal{F}$ are amalgamations of $\mathcal{G}$ and an argument in the proof of Theorem 2.2.9, it follows that G is essentially contained in E, and in F. So ...

3. Two fair dice, one red, one green, are rolled once. Let $\mathcal{E} = [E_1, \ldots, E_6]$, where $E_i = $ "i came up on the red die", and $\mathcal{F} = [F_2, \ldots, F_{12}]$, where $F_j = $ "the sum of the numbers that came up on the two dice was j". Write $I(\mathcal{E}, \mathcal{F})$ explicitly, in a form that permits calculation once a base for "log" is specified.

4. Regarding the experiment described in Exercise 1.3.5, let E_U = "urn U was chosen", $U \in \{A, B, C\}$, F_r = "a red ball was drawn", F_g = "a green ball was drawn", $\mathcal{E} = [E_A, E_B, E_C]$, and $\mathcal{F} = [F_r, F_g]$. Write $I(\mathcal{E}, \mathcal{F})$ explicitly, in a form that permits calculation, given a base for "log".

5. Suppose we have a k-stage experiment, with possible outcomes $x_1^{(i)}, \ldots, x_{n_i}^{(i)}$ at the ith stage, $i = 1, \ldots, k$. Let us take, as we often do, $\mathcal{S} = \{(x_{j_1}^{(i)}, \ldots, x_{j_k}^{(k)}), 1 \le j_i \le n_i, i = 1, \ldots, k\}$, the set of all sequences of possible outcomes at the different stages.

   For $1 \le i \le k$, $1 \le j \le n_i$, let $E_j^{(i)}$ = "$x_j^{(i)}$ occurred at the ith stage", and $\mathcal{E}^{(i)} = [E_j^{(i)}; j = 1, \ldots, n_i]$. We will call $\mathcal{E}^{(i)}$ the ith *stage*, or ith *component*, system of events in $(\mathcal{S}, P)$. Two stages will be called *independent* if and only if their corresponding component systems are statistically independent.

   In the case $k = 2$, let us simplify notation by letting the possible outcomes at the first stage be $x_1, \ldots, x_n$ and at the second stage $y_1, \ldots, y_m$. Let E_i = "x_i occurred at the first stage", $i = 1, \ldots, n$, F_j = "y_j occurred at the second stage", $j = 1, \ldots, m$, be the events comprising the component systems $\mathcal{E}$ and $\mathcal{F}$, respectively.

   (a) Let $p_i = P(E_i)$ and $q_{ij} = P(F_j \mid E_i)$, $i = 1, \ldots, n$, $j = 1, \ldots, m$, as in Section 1.3. Assume that each p_i is positive. Show that $\mathcal{E}$ and $\mathcal{F}$ are statistically independent if and only if the q_{ij} depend only on j, not i. (That is, for any $i_1, i_2 \in \{1, \ldots, n\}$ and $j \in \{1, \ldots, m\}$, $q_{i_1 j} = q_{i_2 j}$.)

   (b) Verify that if $\mathcal{S}$ is regarded as a system of events in the space $(\mathcal{S}, P)$ (i.e., identify each pair (x_i, y_j) with the event $\{(x_i, y_j)\}$), then $\mathcal{S} = \mathcal{E} \wedge \mathcal{F}$.

   *(c) Suppose that three component systems of a multistage experiment, say $\mathcal{E}^{(1)}, \mathcal{E}^{(2)}$, and $\mathcal{E}^{(3)}$, are pairwise statistically independent. Does it follow that they are *jointly* statistically independent? This would mean that for all $1 \le i \le n_1$, $1 \le j \le n_2$, and $1 \le k \le n_3$, $P(E_i^{(1)} \cap E_j^{(2)} \cap E_k^{(3)}) = P(E_i^{(1)}) P(E_j^{(2)}) P(E_k^{(3)})$. Take note of Exercise 1.4.7.

6. (a) Suppose that $E_1, \ldots, E_k, F_1, \ldots, F_r$ are events in a finite probability space $(\mathcal{S}, P)$ satisfying

      (i) $E_1, \ldots, E_k$ are pairwise mutually exclusive;
      (ii) $F_1, \ldots, F_r$ are pairwise mutually exclusive; and
      (iii) for each $i \in \{1, \ldots, k\}$, $j \in \{1, \ldots, r\}$, E_i and F_j are independent.

      Show that $\bigcup_{i=1}^{k} E_i$ and $\bigcup_{j=1}^{r} F_j$ are independent.

   (b) Suppose that $\mathcal{E}, \widehat{\mathcal{E}}, \mathcal{F}$, and $\widehat{\mathcal{F}}$ are systems of events in some finite probability space, and $\widehat{\mathcal{E}}$ and $\widehat{\mathcal{F}}$ are amalgamations of $\mathcal{E}$ and $\mathcal{F}$, respectively. Show that if $\mathcal{E}$ and $\mathcal{F}$ are statistically independent, then so are $\widehat{\mathcal{E}}$ and $\widehat{\mathcal{F}}$. [You did the hard work in part (a).]

(c) It is asserted at the end of Section 1.4 that "when two stages [of a multi-stage experiment] are independent, any two events whose descriptions involve only those stages, respectively, will be independent." Explain how this assertion is a special case of the result of 6(a), above.

7. Succinctly characterize the systems of events $\mathcal{E}$ such that $I(\mathcal{E}, \mathcal{E}) = 0$.

8. Suppose that $\mathcal{E}$ and $\mathcal{F}$ are systems of events in some finite probability space, and that $\widehat{\mathcal{E}}$ is an amalgamation of $\mathcal{E}$. Show that $I(\widehat{\mathcal{E}}, \mathcal{F}) \leq I(\mathcal{E}, \mathcal{F})$. [You may have to use Lemma 2.1.1.]

2.3 Entropy

Suppose that $\mathcal{E} = [E_i; i \in I]$ and $\mathcal{F} = [F_j; j \in J]$ are systems of events in some finite probability space $(\mathcal{S}, P)$. The *entropy* of $\mathcal{E}$, denoted $H(\mathcal{E})$, is

$$H(\mathcal{E}) = -\sum_{i \in I} P(E_i) \log P(E_i).$$

The *joint entropy of the systems* $\mathcal{E}$ and $\mathcal{F}$ is the entropy of the joint system,

$$H(\mathcal{E} \wedge \mathcal{F}) = -\sum_{i \in I} \sum_{j \in J} P(E_i \cap F_j) \log P(E_i \cap F_j).$$

The *conditional entropy of* $\mathcal{E}$, *conditional upon* $\mathcal{F}$, is

$$H(\mathcal{E} \mid \mathcal{F}) = \sum_{i \in I} \sum_{j \in J} P(E_i \cap F_j) I(E_i \mid F_j)$$

$$= -\sum_{i \in I} \sum_{j \in J} P(E_i \cap F_j) \log \frac{P(E_i \cap F_j)}{P(F_j)}.$$

Remarks

2.3.1 In the definitions above, we continue to observe the convention that $0 \log(\text{anything}) = 0$. As in the preceding section, the base of the logarithm is unspecified; any base greater than 1 may be used.

2.3.2 Taking $\mathcal{E}$ as a new set of outcomes for the probability space, we see that $H(\mathcal{E})$ is the average value of the self-information of the events (now outcomes) in $\mathcal{E}$. Similarly, the joint and conditional entropies are average values of certain kinds of information.

2.3.3 The word *entropy* connotes disorder, or uncertainty, and the number $H(\mathcal{E})$ is to be taken as a measure of the disorder or uncertainty inherent in the system $\mathcal{E}$. What sort of disorder or uncertainty is associable with a system $\mathcal{E}$, and why is $H(\mathcal{E})$, as defined here, a good measure of it?

A system of events represents a way of looking at an experiment. The pesky little individual outcomes are grouped into events according to some organizing principle. We have the vague intuition that the more finely we divide the set of outcomes into events, the greater the "complexity" of our point of view, and the less order and simplicity we have brought to the analysis of the experiment. Thus, there is an intuitive feeling that some systems of events are more complex, less simple, than others, and it is not too great a stretch to make complexity a synonym, in this context, of disorder or uncertainty.

As to why H is a suitable measure of this felt complexity: as with mutual information, the justification resides in the behavior of the quantity. We refer to the theorem below, and to the result of Exercise 4 at the end of this section. The theorem says that $H(\mathcal{E})$ is a minimum (zero) when and only when $\mathcal{E}$ consists of one big event that is certain to occur, together, possibly, with massless events (events with probability zero). Surely this is the situation of maximum simplicity, minimum disorder. (A colleague has facetiously suggested that such systems of events be called "Mussolini systems". The reference is to the level of order in the system; in correlation to this terminology, an event of probability one may be called a Mussolini.) The theorem also says that, for a fixed value of $|\mathcal{E}| = |I|$, the greatest value $H(\mathcal{E})$ can take is achieved when and only when the events in $\mathcal{E}$ are equally likely. This taxes the intuition a bit, but it does seem that having a particular number of equiprobable events is a more "uncertain" or "complex" situation than having the same number of events, but with some events more likely than others.

The result of Exercise 2.3.4 is that if $\widehat{\mathcal{E}}$ is obtained from $\mathcal{E}$ by amalgamation, then $H(\widehat{\mathcal{E}}) \leq H(\mathcal{E})$. To put this the other way around, if you obtain a new system from an old system by dividing the old events into smaller events, the entropy goes up, as it should.

Shannon ([63] and [65]) introduced an axiom system for entropy, a series of statements that the symbol H *ought* to satisfy to be worthy of the name *entropy*, and showed that the definition of entropy given here is the only one compatible with these requirements. As previously mentioned, Feinstein [17] did something similar, with (perhaps) a more congenial axiom system. For an excellent explanation of these axioms and further references on the matter, see the book by Dominic Welsh [81] or that of D. S. Jones [37]. We shall not pursue further the question of the validity of the definition of H, nor its uniqueness.

2.3.4 Theorem *Suppose that $\mathcal{E} = [E_i; i \in I]$ is a system of events in a finite probability space. Then $0 \leq H(\mathcal{E}) \leq \log |I|$. Equality at the lower extreme occurs if and only if all but one of the events in $\mathcal{E}$ have probability zero. [That one event would then be forced to have probability 1, since $\sum_{i \in I} P(E_i) = P(\cup_{i \in I} E_i) = 1$.] Equality occurs at the upper extreme if and only if the events in $\mathcal{E}$ are equally likely. [In this case, each event in $\mathcal{E}$ would have probability $1/|I|$.]*

Proof: It is straightforward to see that the given conditions for equality at the two extremes are sufficient.

Since $0 \leq P(E_i) \leq 1$ for each $i \in I$, $-P(E_i) \log P(E_i) \geq 0$ with equality if and only if either $P(E_i) = 0$ (by convention) or $P(E_i) = 1$. Thus $H(\mathcal{E}) \geq 0$, and equality forces $P(E_i) = 0$ or 1 for each i. Since the E_i are pairwise mutually exclusive, and $\sum_i P(E_i) = 1$, $H(\mathcal{E}) = 0$ implies that exactly one of the E_i has probability 1 and the rest have probability zero.

Let $c = \log e$. We have

$$H(\mathcal{E}) - \log|I| = \sum_{i \in I} P(E_i) \log \frac{1}{P(E_i)} - \sum_{i \in I} P(E_i) \log|I|$$

$$= c \sum_{i \in I} P(E_i) \ln(P(E_i)|I|)^{-1}$$

$$\leq c \sum_{i \in I} P(E_i)[(P(E_i)|I|)^{-1} - 1] \quad \text{(by Lemma 2.1.1)}$$

$$= c \left[\sum_{i \in I} |I|^{-1} - \sum_{i \in I} P(E_i) \right] = c[1 - 1] = 0,$$

with equality if and only if $P(E_i)|I| = 1$ for each $i \in I$. $\qquad \square$

The following theorem gives a useful connection between conditional entropy and the set-wise relation between two systems. Notice that if $\mathcal{E}$ is an amalgamation of $\mathcal{F}$, then whenever you know which event in $\mathcal{F}$ occurred, you also know which event in $\mathcal{E}$ occurred; i.e., there is no uncertainty regarding $\mathcal{E}$.

2.3.5 Theorem *Suppose that $\mathcal{E} = [E_i; i \in I]$ and $\mathcal{F} = [F_j; j \in J]$ are systems of events in some finite probability space. Then $H(\mathcal{E}|\mathcal{F}) = 0$ if and only if $\mathcal{E}$ is an amalgamation of $\mathcal{F}$.*

Proof: $H(\mathcal{E}|\mathcal{F}) = \sum_{i \in I} \sum_{j \in J} P(E_i \cap F_j) \log \frac{P(F_j)}{P(E_i \cap F_j)} = 0 \Leftrightarrow$ for each $i \in I, j \in J$, $P(E_i \cap F_j) \log \frac{P(F_j)}{P(E_i \cap F_j)} = 0$, since the terms of the sum above are all non-negative.

If $P(F_j) = 0$ then $P(E_i \cap F_j) = P(F_j)$ for any choice of $i \in I$. Since $P(F_j) = \sum_{i \in I} P(E_i \cap F_j)$ by Corollary 2.2.8, if $P(F_j) > 0$ then $P(E_i \cap F_j) > 0$ for some $i \in I$, and then $P(E_i \cap F_j) \log \frac{P(F_j)}{P(E_i \cap F_j)} = 0$ implies $P(E_i \cap F_j) = P(F_j)$. Thus $H(\mathcal{E}|\mathcal{F}) = 0$ implies that $\mathcal{E}$ is an amalgamation of $\mathcal{F}$, and the converse is straightforward to see. $\qquad \square$

Exercises 2.3

1. Treating the sets of outcomes as systems of events, write out the entropies of each of $\mathcal{S}_1, \ldots, \mathcal{S}_4$ in Exercise 2.2.1.

2. In the experiment of n independent Bernoulli trials with probability p of success on each trial, let $E_k =$ "exactly k successes," and $\mathcal{E} = [E_0, \ldots, E_n]$. Let $\mathcal{S} = \{S, F\}^n$, and treat $\mathcal{S}$ as a system of events (i.e., each element of $\mathcal{S}$,

regarded as an outcome of the experiment, is also to be thought of as an event). Write out both $H(\mathcal{E})$ and $H(\mathcal{S})$.

3. For a system $\mathcal{E}$ of events, show that $I(\mathcal{E},\mathcal{E}) = H(\mathcal{E})$.

4. (a) Show that, if $x_1,\ldots,x_n \geq 0$, then

$$\left(\sum_{i=1}^{n} x_i\right) \log\left(\sum_{i=1}^{n} x_i\right) \geq \sum_{i=1}^{n} x_i \log x_i.$$

   (b) Show that if $\widehat{\mathcal{E}}$ is obtained from $\mathcal{E}$ by amalgamation, then $H(\widehat{\mathcal{E}}) \leq H(\mathcal{E})$.

5. Suppose that $\mathcal{E} = [E_i; i \in I]$ and $\mathcal{F} = [F_j; j \in J]$ are systems of events in some finite probability space. Under what conditions on $\mathcal{E}$ and $\mathcal{F}$ will it be the case that $H(\mathcal{E} \mid \mathcal{F}) = H(\mathcal{E})$? [See Theorem 2.4.4, next section.]

6. Suppose that $\mathcal{E}$ and $\mathcal{F}$ are systems of events in probability spaces associated with two (different) experiments. Suppose that the two experiments are performed *independently*, and the set of outcomes of the compound experiment is identified with $\mathcal{S}_1 \times \mathcal{S}_2$, where $\mathcal{S}_1$ and $\mathcal{S}_2$ are the sets of outcomes for the two experiments separately. Let

$$\mathcal{E}\cdot\mathcal{F} = [E \times F; E \in \mathcal{E}, F \in \mathcal{F}].$$

Verify that $\mathcal{E}\cdot\mathcal{F}$ is a system of events in the space of the compound experiment.

Show that $H(\mathcal{E}\cdot\mathcal{F}) = H(\mathcal{E}) + H(\mathcal{F})$. Will this result hold (necessarily) if the two experiments are not independent?

2.4 Information and entropy

Throughout this section, $\mathcal{E}$ and $\mathcal{F}$ will be systems of events in some finite probability space.

2.4.1 Theorem $I(\mathcal{E},\mathcal{F}) = H(\mathcal{E}) + H(\mathcal{F}) - H(\mathcal{E}\wedge\mathcal{F}).$

Proof:

$$I(\mathcal{E},\mathcal{F}) = \sum_i \sum_j P(E_i \cap F_j) \log \frac{P(E_i \cap F_j)}{P(E_i)P(F_j)}$$

$$= \sum_i \sum_j P(E_i \cap F_j) \log P(E_i \cap F_j)$$

$$- \sum_i \sum_j P(E_i \cap F_j) \log P(E_i) - \sum_i \sum_j P(E_i \cap F_j) \log P(F_j)$$

$$= -H(\mathcal{E} \wedge \mathcal{F}) - \sum_i P(E_i) \log P(E_i)$$

$$- \sum_j P(F_j) \log P(F_j) \quad \text{[using Corollary 2.2.8]}$$

$$= -H(\mathcal{E} \wedge \mathcal{F}) + H(\mathcal{E}) + H(\mathcal{F}). \qquad \Box$$

2.4.2 Corollary $I(\mathcal{E}, \mathcal{F}) \le H(\mathcal{E}) + H(\mathcal{F})$.

2.4.3 Corollary $H(\mathcal{E} \wedge \mathcal{F}) \le H(\mathcal{E}) + H(\mathcal{F})$, with equality if and only if $\mathcal{E}$ and $\mathcal{F}$ are statistically independent.

2.4.4 Theorem $H(\mathcal{E} \mid \mathcal{F}) = H(\mathcal{E} \wedge \mathcal{F}) - H(\mathcal{F}) = H(\mathcal{E}) - I(\mathcal{E}, \mathcal{F})$.

2.4.5 Corollary $H(\mathcal{E} \mid \mathcal{F}) \le H(\mathcal{E})$, with equality if and only if $\mathcal{E}$ and $\mathcal{F}$ are statistically independent.

2.4.6 Corollary $I(\mathcal{E}, \mathcal{F}) \le \min(H(\mathcal{E}), H(\mathcal{F}))$.

Proof: It will suffice to see that $I(\mathcal{E}, \mathcal{F}) \le H(\mathcal{E})$. This follows from Theorem 2.4.4 and the observation that $H(\mathcal{E} \mid \mathcal{F}) \ge 0$. $\qquad \Box$

Notice that Corollary 2.4.6 is much stronger than Corollary 2.4.2.

Exercises 2.4

1–4. Prove 2.4.2, 2.4.3, 2.4.4, and 2.4.5, above.

5. From 2.4.1 and 2.3.4 deduce necessary and sufficient conditions on $\mathcal{E}$ and $\mathcal{F}$ for $I(\mathcal{E}, \mathcal{F}) = H(\mathcal{E}) + H(\mathcal{F})$.

6. Express $H(\mathcal{E} \wedge \mathcal{E})$ and $H(\mathcal{E} \mid \mathcal{E})$ as simply as possible.

7. Three urns, A, B, and C, contain colored balls, as follows:

> A contains three red and five green balls,
> B contains one red and two green balls, and
> C contains seven red and six green balls.

An urn is chosen, at random, and then a ball is drawn from that urn. Let the urn names also stand for the event that that urn was chosen, and let $R =$ "a red ball was chosen," and $G =$ "a green ball was chosen." Let $\mathcal{E} = \{A, B, C\}$ and $\mathcal{F} = \{R, G\}$. Write out $I(\mathcal{E}, \mathcal{F})$, $H(\mathcal{E})$, $H(\mathcal{F})$, $H(\mathcal{E} \wedge \mathcal{F})$, $H(\mathcal{E} \mid \mathcal{F})$, and $H(\mathcal{F} \mid \mathcal{E})$. If, at any stage, you can express whatever you are trying to express in terms of items already written out, do so.

8. Regarding Corollary 2.4.6: under what conditions on $\mathcal{E}$ and $\mathcal{F}$ is it in the case that $I(\mathcal{E}, \mathcal{F}) = H(\mathcal{E})$?

9. With the urns of problem 7 above, we play a new game. First draw a ball from urn A; if it is red, draw a ball from urn B; if the ball from urn A is green, draw a ball from urn C. Let $\mathcal{E}$ and $\mathcal{F}$ be the first and second

stage systems of events for this two-stage experiment; i.e., $\mathcal{E} = [R, G]$ and $\mathcal{F} = [\tilde{R}, \tilde{G}]$, where, for instance, $R =$ "the first ball drawn was red" and $\tilde{R} =$ "the second ball drawn was red."

(a) Write out $I(\mathcal{E}, \mathcal{F})$, $H(\mathcal{E})$, $H(\mathcal{F})$, $H(\mathcal{E} \wedge \mathcal{F})$, $H(\mathcal{E} \mid \mathcal{F})$, and $H(\mathcal{F} \mid \mathcal{E})$ in this new situation.

(b) Suppose now that you are allowed to transfer balls between urns B and C. How would you rearrange the balls in those urns to maximize $I(\mathcal{E}, \mathcal{F})$? What is that maximum value?

(c) How would you rearrange the balls in urns B and C to minimize $I(\mathcal{E}, \mathcal{F})$? What is that minimum value?

(d) Answer the same questions in (b) and (c) with $I(\mathcal{E}, \mathcal{F})$ replaced by $H(\mathcal{E} \mid \mathcal{F})$.

(e) Under which of the rearrangements you produced in (b), (c), and (d) is $\mathcal{E}$ an amalgamation of $\mathcal{F}$? Under which is $\mathcal{F}$ an amalgamation of $\mathcal{E}$? Under which are $\mathcal{E}$ and $\mathcal{F}$ statistically independent?

Chapter 3

Channels and Channel Capacity

3.1 Discrete memoryless channels

A *channel* is a communication device with two ends, an input end, or *transmitter*, and an output end, or *receiver*. A *discrete* channel accepts for transmission the characters of some finite alphabet $A = \{a_1, \ldots, a_n\}$, the *input alphabet*, and delivers characters of an *output alphabet* $B = \{b_1, \ldots, b_k\}$ to the receiver. Every time an input character is accepted for transmission, an output character subsequently arrives at the receiver. That is, we do not encompass situations in which the channel responds to an input character by delivering several output characters, or no output. Such situations may be defined out of existence: once the input alphabet and the channel are fixed, the output alphabet is defined to consist of all possible outputs that may result from an input. For instance, suppose $A = \{0, 1\}$, the *binary alphabet*, and suppose that it is known that the channel is rickety, and may *fuzz* the input digit so that the receiver cannot tell which digit, 0 or 1, is being received, or may *stutter* and deliver two digits, either of which might be fuzzed, upon the transmission of a single digit. Then, with $*$ standing for "fuzzy digit," we are forced to take $B = \{0, 1, *, 00, 01, 10, 11, 0*, 1*, *0, *1, **\}$.

For a finite alphabet A, we let, as convention dictates,

$$A^\ell = \text{the Cartesian product of } A \text{ with itself } \ell \text{ times}$$
$$= \text{the set of words of length } \ell, \text{ over } A.$$

Further, let

$$A^+ = \bigcup_{\ell=1}^{\infty} A^\ell = \text{the set of all (non-empty) words over } A.$$

Note that if A is the input alphabet of a channel, then any finite non-empty subset of A^+ could be taken as the input alphabet of the same channel. Changing the input alphabet in this way will necessitate a change in the output alphabet. For instance, if $A = \{0, 1\}$, and the corresponding output alphabet is $B = \{0, 1, *\}$, then if we take $\widehat{A} = \{00, 11\}$, the new output alphabet will be

$$\widehat{B} = \{00, 01, 0*, 10, 11, 1*, *0, *1, **\}.$$

It is possible to vary the output alphabet by merging or amalgamating letters; for instance, if $B = \{0, 1, *, x\}$, we could take $\widetilde{B} = \{0, 1, \alpha\}$, with α meaning "either $*$ or x." This might be a shrewd simplification if, for instance, the original letters $*$ and x are different sorts of error indicators, and the distinction is of no importance.

Another common method of simplifying the output alphabet involves modifying the channel by "adding a coin flip." For instance, if $B = \{0, 1, *\}$ and you really do not want to bother with $*$, you can flip a coin whenever $*$ is received to decide if it will be read as 0 or 1. The coin need not be fair. The same idea can be used to shrink B from any finite size down to any smaller size of 2 or more. The details of the process depend on the particular situation; they are left to the ingenuity of the engineer. See Exercise 3.2.6.

It may be that there are *fundamental* input and output alphabets forced upon us by the physical nature of the channel; or, as in the case of the telegraph, for which the time-hallowed input and output alphabet is {dot, dash} (or, as Shannon [65] has it, {dot, dash, short pause (between letters), long pause (between words)}), it may be that some fundamental input alphabet is strongly recommended, although not forced, by the physical nature of the channel. In the most widespread class of examples, the *binary* channels, the input alphabet has size 2, and we usually identify the input characters with 0 and 1. Note that, for any channel that accepts at least two input characters, we can always confine ourselves to two input characters, and thus make the channel binary.

The telegraph provides a historically fundamental example of a channel; it is a somewhat uninteresting, or misleading, example for the student of information theory, because it is so reliable. Over the telegraph, if a "dot" is transmitted, then a "dot" is received (unless the lines are down), and the same goes for "dash." What makes life interesting in modern times is "channel noise"; you cannot be dead certain what the output will be for a given input. Modern channels run from outer space, to the ocean floor, to downtown Cleveland—a lot can go wrong. Specks of dust momentarily lodge in the receiver, birds fly up in front of the transmitter, a storm briefly disrupts the local electromagnetic environment—it's a wonder that successful communication ever takes place.

We take account of the uncertainty of communication by regarding the attempt to transmit a single digit as a probabilistic experiment. Before we become thoroughly engaged in working out the consequences of this view, it is time to announce a blanket assumption that will be in force from here on in this text: our channels will all be *memoryless*. This means that the likelihood of b_j being the output when a_i is the input does not vary with local conditions, nor with recent history, for each i and j. These unvarying likelihoods are called transition probabilities and will be discussed in the next section.

Please note that this assumption may well be invalid in a real situation. For instance, when you hear a "skip" from a record on a turntable,[1] your estimate

---

[1] If you are unacquainted with "records" and "turntables," ask the nearest elderly person about them.

of the probability of a skip in the near future changes drastically. You now estimate that there is a great likelihood of another skip soon, because experience tells you that these skips occur for an underlying *reason*, usually a piece of fluff or lint caught on the phonograph needle. Just so, in a great many situations wobbles and glitches in the communication occur for some underlying reason that will not go away for a while, and the assumption of memorylessness is rendered invalid. What can you do in such situations? There is a good deal of theory and practice available on the subject of correcting *burst* errors, as they are called in some parts. This theory and practice will not, however, be part of this course. We are calling your attention to the phenomenon of burst errors, and to the indefensibility of our blanket assumption of memorylessness in certain situations, just because one of the worst things you can do with mathematics is to misapply it to situations outside the umbrella of assumption. Probabilistic assumptions about randomness and independence are very tricky, and the assumption of memorylessness of a channel is one such.

This is not the place to go into detail, but let us assure you that you can misapply a result about randomly occurring phenomena (such as the glitches, skips, and wobbles in transmissions over our memoryless channels are assumed to be) to "show," in a dignified, sincere manner, that the probability that the sun will *not* rise tomorrow is a little greater than $1/3$. The moral is that you should stare and ponder a bit, to see if your mathematics applies to the situation at hand, and if it doesn't, don't try to force it.

Exercises 3.1

1. (a) Suppose $A = \{0, 1\}$ and $B = \{0, 1, *\}$; suppose we decide to use $\widehat{A} = \{0000, 1111\}$ as the new input alphabet, for some reason. How large will the new output alphabet be?

   (b) In general, for any input alphabet A and output alphabet B, with $|B| = k$, if we take a new input alphabet $\widehat{A} \subset A^{\ell}$, how many elements will the new output alphabet have? What will the new output alphabet be?

2. A certain binary channel has the binary alphabet as its output alphabet, as well: $A = B = \{0, 1\}$. This channel has a memory, albeit a very short one. At the start of a transmission, or right after the successful transmission of a digit, the probability of a correct transmission is p (regardless of which digit, 0 or 1, is being transmitted); right after an error (0 input, 1 output, or 1 input, 0 output), the probability of a correct transmission is q. (If this situation were real, we would plausibly have $1/2 < q < p < 1$.) In terms of p and q, find

   (a) the probability that the string 10001 is received, if 11101 was sent;

   (b) the probability that 10111 received, if 11101 was sent;

   (c) the probability of exactly two errors in transmitting a binary word of length 5;

(d) the probability of two or fewer errors, in transmitting a binary word of length n.

3. Another binary channel has $A = B = \{0, 1\}$, and no memory; the probability of a correct transmission is p, for each digit transmitted. Find the probabilities in problem 2, above, for this channel.

3.2 Transition probabilities and binary symmetric channels

Now we shall begin to work out the consequences of the assumption of memorylessness. Let the input alphabet be $A = \{a_1, \ldots, a_n\}$, and the output alphabet be $B = \{b_1, \ldots, b_k\}$. By the assumption of memorylessness, the probability that b_j will be received, *if a_i is the input character*, depends only on i, j, and the nature of the channel, not on the weather nor the recent history of the channel. We denote this probability by q_{ij}.

The q_{ij} are called the *transition probabilities* of the channel, and the $n \times k$ matrix $Q = [q_{ij}]$ is the matrix of transition probabilities. After the input and output alphabets have been agreed upon, Q depends on the hardware, the channel itself; or, we could say that Q is a property of the channel. In principle, q_{ij} could be estimated by testing the channel: send a_i many times and record how often b_j is received. In practice, such testing may be difficult or impossible, and the q_{ij} are either estimated through theoretical considerations, or remain hypothetical. Note that $\sum_{j=1}^{k} q_{ij} = 1$, for each i; that is, the row sums of Q are all 1.

A *binary symmetric channel* (BSC, for short) is a memoryless channel with $A = B = \{0, 1\}$ like that described in Exercise 3.1.3; whichever digit, 0 or 1, is being transmitted, the probability p that it will get through correctly is called the *reliability* of the channel. Usually, $1/2 < p < 1$, and we hope p is close to 1. Letting 0 and 1 index the transition probabilities in the obvious way, the matrix of transition probabilities for a binary symmetric channel with reliability p is

$$Q = \begin{bmatrix} q_{00} & q_{01} \\ q_{10} & q_{11} \end{bmatrix} = \begin{bmatrix} p & 1-p \\ 1-p & p \end{bmatrix}.$$

The word "symmetric" in "binary symmetric channel" refers to the symmetry of Q, or to the fact that the channel treats the digits 0 and 1 symmetrically.

Observe that sending any particular binary word of length n through a binary symmetric channel with reliability p is an instance of n independent Bernoulli trials, with probability p of Success on each trial (if you count a correct transmission of a digit as a Success). Thus, the probability of exactly k errors ($n - k$ Successes) in such a transmission is $\binom{n}{k} p^{n-k}(1-p)^k$, and the average or expected number of errors is $n(1-p)$.

Exercises 3.2

1. For a particular memoryless channel we have $A = \{0, 1\}$, $B = \{0, 1, *\}$, and the channel treats the input digits symmetrically; each digit has probability p of being transmitted correctly, probability q of being switched to the other digit, and probability r of being fuzzed, so that the output is $*$. Note that $p + q + r = 1$.

   (a) Give the matrix of transition probabilities, in terms of p, q, and r.

   (b) In terms of n, p, and k, what is the probability of exactly k errors (where an *error* is either a fuzzed digit or a switched digit) in the transmission of a binary word of length n, over this channel?

   (c) Suppose that $*$ is eliminated from the output alphabet by means of coin flip, with a fair coin. Whenever $*$ is received, the coin is flipped; if heads comes up, the $*$ is read is 0, and if tails comes up, it is read as 1. What is the new matrix of transition probabilities? Is the channel now binary symmetric?

   (d) Suppose that $*$ is eliminated from the output alphabet by merging it with 1. That is, whenever $*$ is received, it is read as 1 (this amounts to a coin flip with a very unfair coin). What is the new matrix of transition probabilities? Is the channel now binary symmetric?

2. A binary symmetric channel has reliability p.

   (a) What is the minimum value of p allowable, if there is to be at least a 95% chance of no errors at all in the transmission of a binary word of length 15?

   (b) Give the inequality that p must satisfy if there is to be at least a 95% chance of no more than one error in the transmission of a binary word of length 15. For the numerically deft and/or curious: is the minimum value of p satisfying this requirement significantly less than the minimum p satisfying the more stringent requirement in part (a)?

   (c) What is the minimum value of p allowable if the average number of errors in transmitting binary words of length 15 is to be no greater than $1/2$?

3. $A = B = \{0, 1\}$, and the channel is memoryless, but is not a binary symmetric channel because it treats 0 and 1 differently. The probability is p_0 that 0 will be transmitted correctly, and p_1 that 1 will be transmitted correctly.

   In terms of p_0 and p_1, find the probabilities described in Exercise 3.1.2 (a) and (b). Also, if a binary word of length n has z zeros and $n - z$ ones, with $n \geq 2$, find, in terms of p_0, p_1, n, and z, the probability of two or fewer errors in the transmission of the word.

4. Suppose we decide to take A^2 as the new input alphabet. Then B^2 will be the new output alphabet. How will the new transition probabilities $q_{(i,i')(j,j')}$ be related to the old transition probabilities q_{ij}?

5. We have a binary symmetric channel with reliability p. We take $\widehat{A} = \{000,$ $111\}$ as the new input alphabet. Find the new output alphabet and the new transition probabilities.

6. Here is a quite general way of modifying the output alphabet of a discrete channel that includes the idea of "amalgamation" discussed in Section 3.1 and the idea of "amalgamation with a coin flip" broached in Exercise 3.2.1. We may as well call this method *probabilistic amalgamation*. Suppose that $A = \{a_1, \ldots, a_n\}$ and $B = \{b_1, \ldots, b_k\}$ are the input and output alphabets, respectively, of a discrete memoryless channel, with transition probabilities q_{ij}. Let $\widetilde{B} = \{\beta_1, \ldots, \beta_m\}$, $m \geq 2$, be a new (output) alphabet, and let u_{jt}, $j = 1, \ldots, k$, $t = 1, \ldots, m$, be probabilities satisfying $\sum_{t=1}^{m} u_{jt} = 1$ for each $j = 1, \ldots, k$. We make $\widetilde{B}$ into the new output alphabet of the channel by declaring that b_j *will be read as* β_t *with probability* u_{jt}. That is, whenever b_j is the output letter, a probabilistic experiment is performed with outcomes $\beta_1, \ldots, \beta_m$ and corresponding probabilities $u_{j1}, \ldots, u_{jm}$ to determine which of the new output letters will be the output.

(a) In each of Exercises 3.2.1 (c) and (d) identify $\widetilde{B}$ and give the matrix $U = [u_{tj}]$.

(b) In general, supposing that B has been replaced by $\widetilde{B}$ as described, express the new matrix of transition probabilities $\widetilde{Q} = [\widetilde{q}_{it}]$ for the new channel with input alphabet A and output alphabet $\widetilde{B}$ in terms of the old matrix of transition probabilities Q and the matrix of probabilities $U = [u_{jt}]$.

(c) Suppose that $A = \{0, 1\}$, $B = \{0, 1, *\}$, and

$$Q = \begin{bmatrix} q_{00} & q_{01} & q_{0*} \\ q_{10} & q_{11} & q_{1*} \end{bmatrix} = \begin{bmatrix} .9 & .02 & .08 \\ .05 & .88 & .07 \end{bmatrix}.$$

Find a way to probabilistically amalgamate B to $\widetilde{B} = A$, so that the resulting channel is binary symmetric, and $u_{00} = u_{11} = 1$. (That is, find a 3×2 matrix $U = [u_{jt}]$ that will do the job.) Is there any other way (i.e., possibly with $u_{00} \neq 1$ or $u_{11} \neq 1$) to probabilistically amalgamate B to A to give a BSC with a greater reliability?

3.3 Input frequencies

As before, we have a memoryless channel with input alphabet $A = \{a_1, \ldots, a_n\}$, output alphabet $B = \{b_1, \ldots, b_k\}$, and transition probabilities q_{ij}. For $i \in \{1, \ldots, n\}$, let p_i denote the *relative frequency of transmission*, or *input frequency*, of the input character a_i. In a large number of situations, it makes sense to think of p_i as the proportion of the occurrences of a_i in the text (written in input alphabetic characters) to be transmitted through the channel.

There is a bit of ambiguity here: by "the text to be transmitted" do we mean some particular segment of input text, or a "typical" segment of input text, or the totality of all possible input text that ever will or could be transmitted? This is one of those ambiguities that will never be satisfactorily resolved, we think; we shall just admit that "the input text" may mean different things on different occasions. Whatever it means, p_i is to be thought of as the (hypothetical) probability that a character selected at random from the input text will be a_i. This probability can sometimes be estimated by examining particular segments of text. For instance, if you count the number of characters, including punctuation marks and blanks, in this text, from the beginning of this section until the end of this sentence, and then tally the number of occurrences of the letter 'e' in the same stretch of text, you will find that 'e' accounts for a little less than 1/10 of all characters; its relative frequency is estimated, by this tally, to be around 0.096. You can take this as an estimate of the input frequency of the letter 'e' for any channel accepting the typographical characters of this text as input. This estimate is likely to be close to the "true" input frequency of 'e', if such there be, provided the text segments to be transmitted are not significantly different in kind from the sample segment from which 0.096 was derived. On the other hand, you might well doubt the validity of this estimate in case the text to be transmitted were the translation of "Romeo and Juliet" into Polish.

There are situations in which there is a way to estimate the input frequencies other than by inspecting a segment of input text. For instance, suppose we are trying to transmit data by means of a *binary code*; each datum is represented by a binary word, a member of $\{0, 1\}^+$, and the binary word is input to a binary channel. We take $A = \{0, 1\}$. Now, the input frequencies p_0 and p_1 of 0 and 1, respectively, will depend on the frequencies with which the various data emerge from the data source, and on how these are encoded as binary words. We know, in fact we control, the latter, but the former may well be beyond our powers of conjecture. If the probabilities of the various data emerging are known a priori, and the encoding scheme is agreed upon, then p_0 and p_1 can be calculated straightforwardly (see Exercise 1 at the end of this section).

Otherwise, when the relative frequencies of the source data are not known beforehand, it is a good working rule that different data are to be regarded as equally likely. The justification for this rule is ignorance; since probability in practice is an a priori assessment of likelihood, in case there is no prior knowledge you may as well assess the known possibilities as equally likely.

We now return to the general case, with $A = \{a_1, \ldots, a_n\}$ and a_i having input frequency p_i. Observe that $\sum_{i=1}^{n} p_i = 1$. Also, note that the probabilities p_i have nothing to do with the channel; they depend on how we use the input alphabet to form text. They are therefore manageable, in principle; we feel that if we know enough about what is to be transmitted, we can make arrangements (in the *encoding* of the messages to be sent) so that the input frequencies of $a_1, \ldots, a_n$ are as close as desired to any prescribed values $p_1, \ldots, p_n \geq 0$ satisfying $\sum_{i=1}^{n} p_i = 1$. The practical difficulties involved in approaching prescribed input frequencies are part of the next chapter's subject. For now, we will ignore those difficulties and consider the p_i to be *variables*; we pretend to be able to vary them, within the constraints that $p_i \geq 0$, $i = 1, \ldots, n$ and $\sum_i p_i = 1$. In this respect the p_i are quite different from the q_{ij}, about which we can do nothing; the transition probabilities are constant parameters, forced upon us by the choice of channel.

We now focus on the act of attempting to transmit a single input character. We regard this as a two-stage experiment. The first stage: selecting some a_i for transmission. The second stage: observing which b_j emerges at the receiving end of the channel. We take, as the set of outcomes,

$$\mathcal{S} = \{(a_i, b_j); i \in \{1, \ldots, n\}, j \in \{1, \ldots, k\}\},$$

in which (a_i, b_j) is short for "a_i was selected for transmission, and b_j was received." We commit further semantic atrocities in the interest of brevity: a_i will stand for the event

$$\{(a_i, b_1), \ldots, (a_i, b_k)\} = \text{``}a_i \text{ was selected for transmission,''}$$

as well as standing for the ith input character; similarly b_j will sometimes denote the event "b_j was received." Readers will have to be alert to the context, in order to divine what means what. For instance, in the sentence "$P(a_i) = p_i$," it is evident that a_i stands for an event, not a letter.

3.3.1 With P denoting the probability assignment to $\mathcal{S}$, and noting the abbreviations introduced above, it seems that we are given the following:

(i) $P(a_i) = p_i$, and

(ii) $P(b_j \mid a_i) = q_{ij}$, whence

(iii) $P(a_i, b_j) = P(a_i \cap b_j) = p_i q_{ij}$, and

(iv) $P(b_j) = \sum_{t=1}^{n} p_t q_{tj}$.

The probabilities $P(b_j)$ in (iv) are called the *output frequencies* of b_j, $j = 1$, $\ldots, k$. It is readily checked that $P(b_j) \geq 0$ and $\sum_{j=1}^{k} P(b_j) = 1$.

Now, the careful and skeptical reader will, we hope, experience a shiver of doubt in thinking all of this over. Putting aside qualms about memorylessness and the invariability of the q_{ij}, there is still an infelicity in the correspondence between the "model" and "reality" in this two-stage experiment view of transmission of a single character, and the problem is in the first stage. In order to assert (i), above, we must view the process of "selecting an input character" as similar to drawing a ball from an urn; we envision a large urn, containing balls colored $a_1, \ldots, a_n$, with proportion p_i of them colored a_i, $i = 1, \ldots, n$. Attempting to transmit a string of input symbols means successively drawing balls from this urn, with replacement and remixing after each draw; this is what our "model" says we are up to.

The problem is that this does not seem much like what we actually do when dealing with input text. If you are at some point in the text, it doesn't seem that the next character pops up at random from an urn; it seems that the probabilities of the various characters appearing next in the text ought to be affected by where we are in the text, by what has gone before. This is certainly the way it is with natural languages; for instance, in English, 'u' almost always follows 'q' and 'b' rarely follows 'z'. Thus, for the English-speaking reader, the "draw from an urn" model of "selecting the next character for transmission" breaks down badly, when the input text is in English.

Notice also the situation of Exercise 3.3.1. Because of the way the source messages are encoded, it seems intuitively obvious that whenever a 0 is input, the probability of the next letter for transmission being 0 is greater than p_0, the relative frequency of 0 in the input text. (And that is, in fact, the case. You might verify that, after a 0, assuming we know nothing else of what has been transmitted already, the probability that the next letter will be 0 is $17/24$, while $p_0 < 1/2$.)

Nevertheless, we shall hold to the simplifying assumption that p_i, the proportion of a_i's in the input text, is also the probability that the *next* letter is a_i, at any point in the input stream. This assumption is valid if we are ignorant of grammar and spelling in the input language; we are again, as with the transition probabilities, in the weird position of bringing a probability into existence by assuming ignorance. In the case of the transition probabilities, that assumption of ignorance is usually truthful; in the present case, it is more often for convenience, because it is difficult to take into account what we know of the input language. There are ways to analyze information transfer through discrete channels with account taken of grammar and/or spelling in the input language—see Shannon's paper [65], and the discussion in Chapter 7 of this text. We shall not burden the reader here with that more difficult analysis, but content ourselves with a crude but useful simplification, in this introduction to the subject.

Exercises 3.3

1. Suppose a data, or message, source gives off, from time to time, any one of three data, or messages, M_1, M_2, and M_3. M_1 accounts for 30% of all emanations from the source, M_2 for 50%, and M_3 for 20%.

    These messages are to be transmitted using a binary channel. To this end, M_1 is encoded as 11111, M_2 as 100001, and M_3 as 1100. Find p_0 and p_1, the input frequencies of 0 and 1, respectively, into the channel to be used for this task.

    [Hint: suppose that a large number N of messages are in line to be transmitted, with $3N/10$ of them instances of M_1, $N/2$ of them M_2, and $N/5$ of them M_3. Count up the number of 0's and the number of 1's in the corresponding input text.]

2. Same question as in Exercise 1 above, except that nothing is known about the relative frequencies of M_1, M_2, and M_3; apply the convention of assuming that M_1, M_2, and M_3 are equally likely.

3. A binary symmetric channel with reliability p is used in a particular communication task for which the input frequencies of 0 and 1 are $p_0 = 2/3$ and $p_1 = 1/3$. Find the output frequencies of 0 and 1 in terms of p. [Hint: apply 3.3.1(iv).]

4. Let $A = \{a_1, a_2, a_3\}$, $B = \{b_1, b_2, b_3\}$,

$$Q = \begin{bmatrix} .94 & .04 & .02 \\ .01 & .93 & .06 \\ .03 & .04 & .93 \end{bmatrix},$$

$p_1 = .4$, $p_2 = .5$, and $p_3 = .1$. Find the output frequencies, $P(b_1)$, $P(b_2)$, and $P(b_3)$.

5. Suppose, in using the channel of the preceding problem, there is a *cost* associated with each attempted transmission of a single input letter. Suppose the (i, j)-entry of the following matrix gives the cost, to the user, of b_j being received when a_i was sent, in some monetary units:

$$C = \begin{bmatrix} 0 & 5 & 9 \\ 10 & 0 & 2 \\ 4 & 2 & 0 \end{bmatrix}.$$

(a) Express, in terms of p_1, p_2, and p_3, the average cost per transmission-of-a-single-input-letter of using this channel. Evaluate when $p_1 = .4$, $p_2 = .5$, and $p_3 = .1$.

(b) What choice of p_1, p_2, p_3 minimizes the average cost-per-use of this channel? Would the user be wise to aim to minimize that average cost?

3.4 Channel capacity

A, B, q_{ij}, and the p_i will be as in the preceding section. With the a_i standing for events, not characters, $A = \{a_1, \ldots, a_n\}$ is a system of events in the probability space associated with the two-stage experiment of sending a single character through a memoryless channel with input alphabet A and output alphabet B. Observe that we have taken on yet another risk of misunderstanding; A will sometimes be an alphabet, sometimes a system of events, and you must infer which from the context. When a system of events, A, is the *input system* of events, for the channel with input alphabet A. Similarly, $B = \{b_1, \ldots, b_k\}$ will sometimes stand for a system of events called the *output system*.

We are interested in communication, the transfer of information; it is reasonable to suppose that we ought, therefore, to be interested in the mutual information between the input and output systems,

$$I(A, B) = \sum_{i=1}^{n} \sum_{j=1}^{k} P(a_i \cap b_j) \log \frac{P(a_i \cap b_j)}{P(a_i)P(b_j)}$$

$$= \sum_{i=1}^{n} \sum_{j=1}^{k} p_i q_{ij} \log \frac{p_i q_{ij}}{p_i \sum_{t=1}^{n} p_t q_{tj}}$$

$$= \sum_{i=1}^{n} p_i \sum_{j=1}^{k} q_{ij} \log \frac{q_{ij}}{\sum_{t=1}^{n} p_t q_{tj}}.$$

$I(A, B)$ is a function of the variables $p_1, \ldots, p_n$, the input frequencies. It would be interesting to know the maximum value that $I(A, B)$ can have. That maximum value is called the *capacity* of the channel, and any values of $p_1, \ldots, p_n$ for which that value is achieved are called *optimal input* (or *transmission*) *frequencies* for the channel. If you accept $I(A, B)$ as an index, or measure, of the potential effectiveness of communication attempts using this channel, then the capacity is the fragile acme of effectiveness. This peak is achieved by optimally adjusting the only quantities within our power to adjust, once the hardware has been established and the input alphabet has been agreed to, namely, the input frequencies. The main result of this section will show how to find the optimal input frequencies (in principle). But before launching into the technical details, let us muse a while on the meaning of what it is that we are optimizing.

3.4.1 *Shannon's interpretation of $I(A, B)$ as rate of information transfer or flow.* Suppose that input letters are arriving at the transmitter at the rate of r letters per second. The average information content of an input letter is $H(A)$; therefore, since the average of a sum is the sum of the averages, there are, on average, $rH(A)$ units of information per second arriving at the transmitter. The information flow is mussed up a bit by the channel; at what average rate is information "flowing" through the channel?

C. E. Shannon's answer [63, 65]: at the rate $rI(A, B) = r(H(A) - H(A \mid B))$. This answer becomes plausible if you bear down on the interpretation of $H(A \mid B)$ as a measure of the average *uncertainty* of the input letter, conditional upon knowing the output letter. Shannon calls $H(A \mid B)$ the "average ambiguity of the received signal," or "the equivocation," and this last terminology has taken root. Note that "the equivocation" is not dependent on the channel alone, but also on the input frequencies. In Shannon's interpretation, it is the amount of information removed, on average, by the channel from the input stream, per input letter.

The validity of this interpretation is bolstered by the role the equivocation plays in Shannon's Noisy Channel Theorem, which we will encounter later. For right now, here is an elementary example due to Shannon himself.

Let the base of log be 2, so the units of information are bits. Suppose we have a binary symmetric channel with reliability .99, and the input is streaming into the receiver at the rate of 1000 symbols (binary digits) per second, with input frequencies $p_0 = p_1 = 1/2$. These are, we shall see soon, optimal, and give

$I(A, B) = .99 \log 1.98 + .01 \log .02 \approx .919$. By the interpretation of $I(A, B)$ under consideration, this says that information is flowing to the receiver at the average rate of $1000(.919) = 919$ bits per second.

You might object that, on average, 990 of the 1000 digits arriving at the receiver each second are *correct* (i.e., equal to the digit transmitted), so perhaps 990 bits/second ought to be the average rate of information flow to the receiver. Shannon points out that by that reasoning, if the reliability of the channel were $1/2$, i.e., if the channel were perfectly useless, you would compute that information flow to the receiver at 500 bits/second, on average, whereas the true rate of information flow in this case ought to be zero. The problem is, whether $p = 1/2$ or $p = .99$, we do not know which of the $1000p$ correct digits (on average, each second) are correct; our uncertainty in this regard means that our estimate of the rate of information flow to the receiver ought to be revised downward from $1000p$ bits/sec. (Verify: $p \log_2 2p + (1 - p) \log_2 2(1 - p) < p$, $1/2 \leq p < 1$.) Why this particular revision, from 990 down to 919 bits/sec? This is where $H(A \mid B) = -(.01 \log .01 + .99 \log .99) \approx .081$ comes in; supposing you know which letter, 0 or 1, is received, $H(A \mid B)$ is the entropy, i.e., average uncertainty, of the *input letter* (system), so it is a good measure of the amount of information to be subtracted from one (the number of bits just received) due to uncertainty about what was sent. (Convinced? Feel uncertain about something? Well, that's entropy, and it's good for you, taken in moderation.)

It is preferable to speak of $I(A, B)$ as the average information *flow through* the channel, or *flow to* the receiver, per input letter, rather than as the average amount of information *arriving at* the receiver (per input letter). The latter might reasonably be taken to be $H(B)$, which is, indeed, the average amount of information contained in the set of outcomes of the probabilistic experiment of "choosing" an input letter and then attempting to transmit it, if we were to take B as the set of outcomes; and taking B as the set of outcomes does seem to respond to the question of how much information is *arriving at* the receiver, per input letter. But $H(B)$ as a measure of information has no connection with how well the channel is communicating the input stream. For instance, for a BSC with reliability $1/2$, $H(B) = \log 2$, while surely the level of communication ought to be $0 = I(A, B)$.

Use of the word "flow" in this context will aid in understanding the Noisy Channel Theorem, in Section 4.6. That theorem discloses a remarkable analogy between information flowing through a channel and fluid flowing through a pipe.

3.4.2 Supposing the transition probabilities q_{ij} are known, finding the optimal input frequencies for, and thus the capacity of, a given channel is a straightforward multi-variable optimization problem; we wish to find where $I(A, B)$, as a function of $p_1, \ldots, p_n$, achieves its maximum on

$$K_n = \{(p_1, \ldots, p_n) \in \mathbb{R}^n; \; p_1, \ldots p_n \geq 0 \text{ and } \sum_{i=1}^{n} p_i = 1\}.$$

By convention, the terms in the sum for $I(A, B)$ corresponding to pairs

(i, j) for which $q_{ij} = 0$ do not actually appear in that sum. Note that if $p_t > 0$, $t = 1, \ldots, n$ and $\sum_{t=1}^{n} p_t q_{tj} = 0$, then $q_{tj} = 0$, $t = 1, \ldots, n$. It follows that the formula for $I(A, B)$ defines a differentiable function in the positive part of $\mathbb{R}^n$, $\{(p_1, \ldots, p_n) \in \mathbb{R}^n; \ p_t > 0, \ t = 1, \ldots, n\}$. Consequently, the Lagrange Multiplier Theorem asserts that if $I(A, B)$ achieves a maximum on K_n in $K_n^+ = \{(p_1, \ldots, p_n) \in K_n; \ p_i > 0, \ i = 1, \ldots, n\}$, then the maximum is necessarily achieved at a point where $\frac{\partial}{\partial p_k}(I(A, B) - \lambda \sum_{i=1}^{n} p_i) = 0, k = 1, \ldots, n$, for some λ.

The main content of Theorem 3.4.3, below, is that a sort of converse of this statement holds: if the equations arising from the Lagrange Multiplier Theorem hold at a point $(p_1, \ldots, p_n) \in K_n^+$, then $I(A, B)$ necessarily achieves a maximum, on K_n, at $(p_1, \ldots, p_n)$. The proof of this statement is a bit technical, and is relegated to the next section, which is optional; although it is preferable that even students of applied mathematics understand the theoretical foundations of their subject, in this case it probably won't overly imperil your immortal soul to accept the result without proof.

Let us see where the Lagrange Multiplier method tells us to look for the optimal input frequencies. Setting $F(p_1, \ldots, p_n) = I(A, B) - \lambda \sum_{i=1}^{n} p_i$, considering only points $(p_1, \ldots, p_n)$ where all coordinates are positive, and setting $c = \log(e)$, we have

$$\frac{\partial F}{\partial p_s} = \frac{\partial}{\partial p_s}(I(A, B)) - \lambda$$

$$= \sum_{j=1}^{k} q_{sj} \log \frac{q_{sj}}{\sum_{t=1}^{n} p_t q_{tj}} - c \sum_{i=1}^{n} p_i \sum_{j=1}^{k} \frac{q_{ij} q_{sj}}{\sum_{t=1}^{n} p_t q_{tj}} - \lambda$$

$$= \sum_{j=1}^{k} q_{sj} \log \frac{q_{sj}}{\sum_{t=1}^{n} p_t q_{tj}} - c \sum_{j=1}^{k} \frac{\sum_{i=1}^{n} p_i q_{ij}}{\sum_{t=1}^{n} p_t q_{tj}} q_{sj} - \lambda$$

$$= \sum_{j=1}^{k} q_{sj} \log \frac{q_{sj}}{\sum_{t=1}^{n} p_t q_{tj}} - c \sum_{j=1}^{k} q_{sj} - \lambda$$

$$= \sum_{j=1}^{k} q_{sj} \log \frac{q_{sj}}{\sum_{t=1}^{n} p_t q_{tj}} - (c + \lambda).$$

Replacing $c + \lambda$ by C, and setting the partial derivative equal to 0, we obtain the *capacity equations* for the channel.

3.4.3 Theorem *Suppose a memoryless channel has input alphabet $A = \{a_1, \ldots, a_n\}$, output alphabet $B = \{b_1, \ldots, b_k\}$, and transition probabilities q_{ij}, $i \in \{1, \ldots, n\}$, $j \in \{1, \ldots, k\}$. There are optimal input frequencies for this channel. If $p_1, \ldots, p_n$ are positive real numbers, then $p_1, \ldots, p_n$ are optimal input frequencies for this channel if and only if $p_1, \ldots, p_n$ satisfy the following, for* *some value of C:*

$$\sum_{i=1}^{n} p_i = 1 \quad \text{and} \quad \sum_{j=1}^{k} q_{sj} \log \frac{q_{sj}}{\sum_{t=1}^{n} p_t q_{tj}} = C, \ s = 1, \dots, n.$$

Furthermore, if $p_1, \dots, p_n$ are optimal input frequencies satisfying these equations, for some value of C, then C is the channel capacity.

This theorem may seem, at first glance, to be saying that all you have to do to find the capacity of a channel and the optimal input frequencies is to solve the capacity equations of the channel, the equations arising from the Lagrange Multiplier Theorem, and the condition $\sum_{i=1}^{n} p_i = 1$, for $p_1, \dots, p_n > 0$. There is a loophole, however, a possibility that slips through a crack in the wording of the theorem: it is possible that the capacity equations have no solution. See problems 9 and 14 at the end of this section. Note that in problem 9, it is not just that the equations have no solution $(p_1, \dots, p_n)$ with all the p_i positive; the equations have no solution, period.

From Theorem 3.4.3 you can infer that this unpleasant phenomenon, the capacity equations having no solution, occurs only when the capacity is achieved at points $(p_1, \dots, p_n) \in K_n$ with one or more of the p_i equal to zero. If $p_i = 0$, then a_i is never used; we have thrown away an input character; we are not using all the tricks at our disposal. Problems 9 and 14 show that it can, indeed, happen that there are input characters that we are better off without. Note, however, the result of problem 10, in which the channel quite severely mangles and bullies one of the input letters, a_n, while maintaining seamlessly perfect respect of the others; yet, in the optimal input frequencies, p_n is positive, which shows that we are better off using a_n than leaving it out, in spite of how terribly the channel treats it (provided we accept $I(A, B)$ as a measure of how well off we are). In this respect, note also the results of exercise problems 2, 6 (a special case of problem 10 when $p = 1/2$), and 7. The practical moral to be drawn from these examples seems to be that if the channel respects an input character even a little bit, if you occasionally get some information from the output (upon inputting this character) about the input, then you are better off with the character than without it. The surprising result of Exercise 14 obliterates this tentative conclusion, and shows that we may be in the presence of a mystery.

How will we know when we are in the rare necessity of banishing one or more input characters, and what do we do about determining the optimal input frequencies in such cases? According to Theorem 3.4.3, we are in such a case when and only when the capacity equations of the channel have no solution in K_n^+. In such a situation, the n-tuple $(p_1, \dots, p_n)$ of optimal input frequencies lies on one of the faces of K_n, $F_R = \{(p_1, \dots, p_n) \in K_n; p_i > 0 \text{ for } i \in R \text{ and } p_i = 0 \text{ for } i \notin R\}$, where R is a proper subset of $\{1, \dots, n\}$. For such an R, let $A_R = \{a_i \in A; i \in R\}$, the input alphabet obtained by deleting the a_i indexed by indices not in R. Finding $(p_1, \dots, p_n)$ on F_R amounts to solving the channel capacity problem with A replaced by A_R; if $(p_1, \dots, p_n) \in F_R$ is the n-tuple of optimal input frequencies, then the non-zero p_i, those indexed by $i \in R$, will satisfy the capacity equations associated with this modified problem. (These equations are obtainable from the original capacity equations by omitting those p_i and q_{ij} with $i \notin R$.)

Thus, if the capacity equations for the channel have no solution $(p_1, \ldots, p_n)$ with $p_i > 0$, $i = 1, \ldots, n$, we need merely solve the $2^n - n - 2$ systems of capacity equations associated with the A_R, for R satisfying $2 \leq |R| \leq n - 1$. It is a consequence of Theorem 3.4.3 that we may first consider all A_R with $|R| = n - 1$, and from among the various solutions select one for which the corresponding capacity is maximal. If there are no solutions, move on to A_R with $|R| = n - 2$, and so on. All of this is straightforward, but it is also a great deal of trouble; we hope that in most real situations the optimal input frequencies will be all positive.

3.4.4 As mentioned above, the proof of the main assertion of 3.4.3 is postponed until the next section, the last of this chapter. However, we can give the proof of the last assertion here. If $p_1, \ldots, p_n$ satisfy the equations above, then the value of $I(A, B)$ at $(p_1, \ldots, p_n)$ is

$$I(A, B) = \sum_{i=1}^{n} p_i \sum_{j=1}^{k} q_{ij} \log \frac{q_{ij}}{\sum_{t=1}^{n} p_t q_{tj}} = C \sum_{i=1}^{n} p_i = C.$$

To remember the capacity equations, other than $\sum_{i=1}^{n} p_i = 1$, it is helpful to remember that the left-hand side of

$$\sum_{j=1}^{k} q_{sj} \log \frac{q_{sj}}{\sum_{t=1}^{n} p_t q_{tj}} = C$$

is the thing multiplying p_s in the formula for

$$I(A, B) = \sum_{i=1}^{n} p_i \sum_{j=1}^{k} q_{ij} \log \frac{q_{ij}}{\sum_{t=1}^{n} p_t q_{tj}}.$$

3.4.5 *The capacity of a binary symmetric channel.* Suppose a binary symmetric channel has reliability p. Let p_0, p_1 denote the input frequencies of 0 and 1, respectively. The capacity equations are:

(1) $p_0 + p_1 = 1$,

(2) $p \log \dfrac{p}{p_0 p + p_1(1-p)} + (1-p) \log \dfrac{1-p}{p_0(1-p) + p_1 p} = C$,   and

(3) $(1-p) \log \dfrac{1-p}{p_0 p + p_1(1-p)} + p \log \dfrac{p}{p_0(1-p) + p_1 p} = C$.

Setting the left-hand sides of (2) and (3) equal, and canceling $p \log p$ and $(1-p) \log(1-p)$, we obtain

$$p \log(p_0 p + p_1(1-p)) + (1-p) \log(p_0(1-p) + p_1 p)$$
$$= (1-p) \log(p_0 p + p_1(1-p)) + p \log(p_0(1-p) + p_1 p),$$

whence

$$(2p-1)\log(p_0 p + p_1(1-p)) = (2p-1)\log(p_0(1-p)+p_1 p),$$

so either $p = 1/2$ or

$$p_0 p + p_1(1-p) = p_0(1-p) + p_1 p,$$

i.e.,

$$(2p-1)p_0 = (2p-1)p_1,$$

so $p_0 = p_1 = 1/2$ (in view of (1)) if $p \neq 1/2$, and the channel capacity is $C = p\log 2p + (1-p)\log 2(1-p)$, obtainable by plugging $p_0 = p_1 = 1/2$ into either (2) or (3) above.

If $p = 1/2$, then, since

$$\frac{p}{p_0 p + p_1(1-p)} = \frac{1-p}{p_0(1-p)+p_1 p} = 1$$

for all values of p_0, p_1 satisfying $p_0 + p_1 = 1$, in this case, we have $I(A, B) = 0$ for all values of p_0, p_1. This is as it should be, since when $p = 1/2$, sending a digit through this channel is like flipping a fair coin. We learn nothing about the input by examining the output, the input and output systems are statistically independent, the channel is worthless for communication.

Note that it is not obvious, a priori, that $p\log 2p + (1-p)\log 2(1-p)$ is positive for all values of $p \in [0, 1] \setminus \{1/2\}$, but that this is the case follows from Theorem 2.2.13.

The foregoing shows that when $p \neq 1/2$, p_0, $p_1 = 1/2$ are the unique optimal input frequencies of a binary symmetric channel of reliability p. If we had wished only to verify that $p_0 = p_1 = 1/2$ are optimal—i.e., if the uniqueness is of no interest—then we could have saved ourselves some trouble, and found the capacity, by simply noting that $p_0 = p_1 = 1/2$ satisfy (1) and make the left-hand sides of (2) and (3) equal. The optimality of $p_0 = p_1 = 1/2$, and the expression for C, then follow from Theorem 3.4.3. For a generalization of this observation, see 3.4.7, below.

3.4.6 Here are two questions of possible practical importance that are related, and to which the answers we have are incomplete:

(i) When (under what conditions on Q) are the optimal input frequencies of a channel unique?

(ii) Do the optimal input frequencies of a channel depend continuously on the transition probabilities of the channel?

Regarding (i), the only instances we know of when the optimal input frequencies are not unique are when the capacity of the channel is zero. (Certainly, in this case, any input frequencies will be optimal; but the remarkable thing is that it is *only* in this case that we have encountered non-unique optimal input frequencies.) We hesitantly conjecture that if the channel capacity is non-zero, then the optimal input frequencies are unique. For those interested, perhaps the proof in Section 3.5 will reward study.

Regarding (ii), there is a body of knowledge related to the Implicit Function Theorem in the calculus of functions of several variables that provides an answer of sorts. Regarding the left-hand sides of the capacity equations as functions of both the p_i and q_{ij}, supposing there is a solution of the equations at positive $p_i, i = 1, \ldots, n$, and supposing that a certain large matrix of partial derivatives has maximum rank, then for every small wiggle of the q_{ij} there will be a positive solution of the new capacity equations quite close to the solution of the original system. When will that certain large matrix of partial derivatives fail to have maximum rank? We can't tell you exactly, but the short answer is: almost never. Thus, the answer to (ii) is: yes, except possibly in certain rare pathological circumstances that we haven't worked out yet.

Here is an example illustrating the possible implications and uses of the continuous dependence of the optimal input frequencies on the transition probabilities. Suppose that $A = \{0, 1\}$, $B = \{0, 1, *\}$, and

$$Q = \begin{bmatrix} q_{00} & q_{01} & q_{0*} \\ q_{10} & q_{11} & q_{1*} \end{bmatrix} = \begin{bmatrix} .93 & .02 & .05 \\ .01 & .95 & .04 \end{bmatrix}.$$

Then Q is "close" to $\widetilde{Q} = \begin{bmatrix} 1 & 0 & 0 \\ 0 & 1 & 0 \end{bmatrix}$ which is the matrix of transition probabilities of a BSC. (For the channel associated with $\widetilde{Q}$, $*$ has been removed as an output letter.) Therefore the optimal input frequencies of the original channel are "close" to $p_0 = p_1 = 1/2$ – and the channel capacity is "close" to $\log 2$. Caution: there is a risk involved in rough estimation of this sort. For instance, would you say that the matrix of transition probabilities in Exercise 3.4.14 is

"close" to $\begin{bmatrix} 1/2 & 1/4 & 1/4 \\ 1/4 & 1/2 & 1/4 \\ 1/4 & 1/4 & 1/2 \end{bmatrix}$? If you are in a reckless mood, you might well

do so, yet the optimal input frequencies for the channel with the latter matrix of transition probabilities are $1/3, 1/3, 1/3$ (this will be shown below), while the optimal input frequencies for the channel of problem 14 are $1/2, 0, 1/2$. Disconcerting discrepancies of this sort should chasten our fudging and make us appreciate numerical error analysis of functions of several variables. But we will pursue this matter no further in this text.

3.4.7 n-ary symmetric channels An n-ary symmetric channel of reliability p is a discrete memoryless channel with

$$A = B \quad \text{and} \quad Q = \begin{bmatrix} p & & \frac{1-p}{n-1} \\ & \ddots & \\ \frac{1-p}{n-1} & & p \end{bmatrix};$$

that is, the main diagonal entries of Q are all the same (namely, p), and the off-diagonal entries of Q are all the same. (Their common value will have to be $\frac{1-p}{n-1}$ if the row sums are to be 1.)

It is straightforward to verify that $p_1 = \cdots = p_n = 1/n$ satisfy the capacity equations of such a channel, with

$$C = p \log np + (1-p) \log \frac{n(1-p)}{n-1}$$

$$= \log n + (p \log p + (1-p) \log \frac{1-p}{n-1}),$$

so by Theorem 3.4.3, $(1/n, \ldots, 1/n)$ are optimal input frequencies for the channel and the capacity is C, above. These optimal input frequencies and this capacity are also discoverable by the method explained in the exercise section, after Exercise 3.4.12, and this method has the advantage that by it and the application of a little linear algebra theory, it can easily be seen that $p_i = 1/n$, $i = 1, \ldots, n$ are *unique* optimal input frequencies except in the case $p = 1/n$, which is precisely the case $C = 0$.

Exercises 3.4

1. Verify directly that $f(p) = p \log 2p + (1-p) \log 2(1-p)$ achieves its maximum, $\log 2$, on $[0, 1]$ at the endpoints, 0 and 1, and its minimum, 0, at $1/2$.

2. Verify that the value of $I(A, B)$ at the extreme points $\{(1, 0, \ldots, 0), (0, 1, 0, \ldots, 0), \ldots, (0, \ldots, 0, 1)\}$ of K_n is zero.

3. Suppose $A = B = \{0, 1\}$, but the channel is not symmetric; suppose a transmitted 0 has probability p of being received as 0, and a transmitted 1 has probability q of being received as 1. Let p_0 and p_1 denote the input frequencies. In terms of p, q, p_0, and p_1, write $I(A, B)$, and give the capacity equations for this channel.

4. Give $I(A, B)$ and the capacity equations for the channel described in Exercise 3.3.4.

5. $A = \{0, 1\}$, $B = \{0, 1, *\}$, and the channel treats the input characters symmetrically; for each input, 0 or 1, the probability that it will be received as sent is p, the probability that it will be received as the other digit is q, and the probability that it will be received as $*$ is r. Note that $p + q + r = 1$.

   Find, in terms of p, q, and r, the capacity of this channel and the optimal input frequencies.

6. $A = B = \{a, b\}$; a is always transmitted correctly; when b is transmitted, the probability is p that b will be received (and, thus, $1 - p$ that a will be received). Find, in terms of p, the capacity of this channel and the optimal input frequencies. Verify that even when $p = 1/2$ (a condition of maximum disrespect for the input letter b), the capacity is positive (which is greater than the capacity would be if the letter b were discarded as an input letter— see Exercise 2, above).

7. [Part of this exercise was lifted from [37].] $A = B = \{a, b, c\}$, a is always transmitted correctly, and the channel behaves symmetrically with respect

to b and c. Each has probability p of being transmitted correctly, and probability $1 - p$ of being received as the other character (c or b). (Thus, if a is
received, it is certain that a was sent.)

(a) Find the capacity of this channel, and the optimal input frequencies, as
functions of p.

(b) Suppose that c is omitted from the input alphabet (but not the output
alphabet). Find the capacity of the channel and the optimal input frequencies in this new situation.

(c) Are there any values of p for which the capacity found in (b) is greater
than that in (a)? What about the case $p = 1/2$?

8. Suppose that $A = B = \{a_1, \ldots, a_n\}$, and the channel is perfectly reliable:
when a_i is sent, a_i is certain to be received. Find the capacity of this channel
and the optimal input frequencies.

9. Suppose that $A = \{a_1, \ldots, a_{n+1}\}$, $B = \{a_1, \ldots, a_n\}$, and the channel respects
$a_1, \ldots, a_n$ perfectly; when a_i is sent, a_i is certain to be received, $1 \leq i \leq n$.

(a) Suppose that when a_{n+1} is sent, the output characters $a_1, \ldots, a_n$ are
equally likely to be received. Show that the capacity equations for the
channel have no solution in this case. Find the optimal input frequencies and the capacity of this channel.

(b) Are there any transition probabilities $q_{n+1,j}$, $j = 1, \ldots, n$, for which
there are optimal input frequencies $p_1, \ldots, p_{n+1}$ for this channel with
$p_{n+1} > 0$? If so, find them, and find the corresponding optimal input
frequencies and the channel capacity.

10. Suppose that $n \geq 2$, $A = \{a_1, \ldots, a_n\} = B$, and the channel respects $a_1, \ldots,$
a_{n-1} perfectly. Suppose that, when a_n is sent, the output characters $a_1, \ldots,$
a_n are equally likely to be received. Find the optimal input frequencies and
the capacity of this channel.

11. We have a binary symmetric channel with reliability p, but we take $A =$
$\{000, 111\}$. Let the input frequencies be denoted p_0 and p_1. In terms of
p, p_0, and p_1, write the mutual information between inputs and outputs,
and the capacity equations of this channel. Assuming that $p_0 = p_1 = 1/2$
are the optimal input frequencies, write the capacity of this channel as a
function of p.

12. (a) Show that $I(A, B) \leq H(A)$. (This is a special case of a result in Section 2.4.)

(b) Show that $I(A, B) = H(A)$ if and only if for each letter b_j received,
there is exactly one input letter a_i such that $P(a_i \mid b_j) = 1$ (so $P(a_k \mid$
$b_j) = 0$ for $k \neq i$). [Hint: recall that $H(A \mid B) = H(A) - I(A, B)$; use
Theorem 2.3.5 or its proof.] In other words, $I(A, B) = H(A)$ if and
only if the input is determinable with certainty from the output. In yet

other words, $I(A, B) = H(A)$ if and only if the input system of events is an amalgamation of the output system.

For exercises 13–15, we are indebted to Luc Teirlinck, who observed that

$$I(A, B) = H(B) - H(B \mid A),$$

so that if

$$-H(B \mid A) = \sum_i p_i \sum_j q_{ij} \log q_{ij} \qquad \text{[verify!]}$$

does not depend on $p_1, \ldots, p_n$, as it will not if the sums $S_i = \sum_j q_{ij} \log q_{ij}$ are all the same, $i = 1, \ldots, n$, then $I(A, B)$ is maximized when $H(B)$ is. The obvious way to maximize $H(B)$ is to "make" $P(b_j) = \sum_{t=1}^n p_t q_{tj}$ equal to $1/k$, $j = 1, \ldots, k$. Thus, in these cases, the optimal input frequencies $p_1, \ldots, p_n$ *might* be found by solving the linear system

$$p_1 + \cdots + p_n = 1$$

$$\sum_{t=1}^n p_t q_{tj} = 1/k, \quad j = 1, \ldots, k.$$

[The first equation is redundant: to see this, sum the r equations just above over j.] This method is not certain to succeed because the solutions of this linear system may fail to be non-negative, or may fail to exist.

Notice that the sums S_i will be all the same if each row of Q is a rearrangement of the first row.

13. Find the optimal input frequencies when

$$Q = \begin{bmatrix} 2/3 & 0 & 1/3 \\ 1/3 & 2/3 & 0 \\ 0 & 1/3 & 2/3 \end{bmatrix}.$$

Also, find the capacity of the channel.

14. Find the optimal input frequencies and the channel capacity, when

$$Q = \begin{bmatrix} 1/2 & 1/3 & 1/6 \\ 1/6 & 1/2 & 1/3 \\ 1/6 & 1/3 & 1/2 \end{bmatrix}.$$

15. Suppose that $n \geq 3$, $0 \leq p \leq 1$, and

$$Q = \begin{bmatrix} p & 0 & \cdots & 0 & 1-p \\ 0 & p & \cdots & 0 & 1-p \\ \vdots & \vdots & \ddots & \vdots & \vdots \\ 0 & 0 & \cdots & p & 1-p \\ 1-p & 0 & \cdots & 0 & p \end{bmatrix}.$$

(a) For which values of p does the method of solving a linear system give the optimal input frequencies for this channel?

*(b) What are the optimal input frequencies and the channel capacity, in terms of p, in all cases?

Exercises 14 and 15 are instructive for those interested in the problem of getting conditions on Q under which the optimal input frequencies are unique and positive.

*16. Suppose a channel has input alphabet A, output alphabet B, and capacity C. Suppose we take A^k as the new input alphabet. Show that the new capacity is kC. (This result is a theorem in [81]. You may find the results of 2.4 helpful, as well as the result of exercise 2.3.6.)

3.5* Proof of Theorem 3.4.3, on the capacity equations

By the remarks of the preceding section, what remains to be shown is that (i) $I(A, B)$ does achieve a maximum on K_n and (ii) if the capacity equations are satisfied, for some C, by some $p_1, \ldots, p_n > 0$, then $p_1, \ldots, p_n$ are optimal input frequencies for the channel.

Since K_n is closed and bounded, to prove (i) it suffices to show that $I(A, B)$ is continuous on K_n. This may seem trivial, since $I(A, B)$ appears to be given by a formula involving only linear functions of $p_1, \ldots, p_n$ and log, but please note that this formula is valid at points $(p_1, \ldots, p_n) \in K_n \setminus K_n^+$ only by convention; there is trouble when one or more of the p_i is zero. Still, the verification that $I(A, B)$ is continuous at such points is straightforward, and is left to the reader to sort out. Keep in mind that $x \log x \to 0$ as $x \to 0^+$. See problem 1 at the end of this section.

A real-valued function f defined on a convex subset K of $\mathbb{R}^n$ is said to be *concave* if

$$f(tu + (1-t)v) \geq tf(u) + (1-t)f(v) \text{ for all } u, v \in K, \ t \in [0, 1].$$

If strict inequality holds whenever $u \neq v$ and $t \in (0, 1)$, we will say that f is *strictly concave*.

We shall now list some facts about concave functions to be used to finish the proof of Theorem 3.4.3. Proofs of these facts are omitted. It is recommended that the reader try to supply the proofs. Notice that 3.5.3 and 3.5.4, taken together, constitute the well-known "second derivative test" for concavity and relative maxima of functions of one variable.

3.5.1 Any sum of concave functions is concave, and if one of the summands is strictly concave, then the sum is strictly concave. A positive constant times a (strictly) concave function is (strictly) concave.

3.5.2 Any linear function is concave, and the composition of a linear function with a concave function of one variable is concave.

3.5.3 If $I \subseteq \mathbb{R}$ is an interval, $f : I \to \mathbb{R}$ is continuous, and $f'' \leq 0$ on the interior of I, then f is concave on I. If $f'' < 0$ on the interior of I, then f is strictly concave on I.

3.5.4 If $I \subseteq \mathbb{R}$ is an interval, $f : I \to \mathbb{R}$ is concave on I, and $f'(x_0) = 0$ for some $x_0 \in I$, then f achieves a maximum on I at x_0. If f is strictly concave on I and $f'(x_0) = 0$, then f achieves a maximum on I only at x_0.

Now we are ready to finish the proof of Theorem 3.4.3. Let

$$f(x) = \begin{cases} -x \log x, & x > 0 \\ 0, & x = 0. \end{cases}$$

By 3.5.3, f is strictly concave on $[0, \infty)$. Now,

$$I(A, B) = \sum_{i=1}^{n} p_i \left(\sum_{j=1}^{k} q_{ij} \log q_{ij} \right) - \sum_{j=1}^{k} \left(\sum_{i=1}^{n} p_i q_{ij} \right) \log \left(\sum_{t=1}^{n} p_t q_{tj} \right)$$

$$= \sum_{i=1}^{n} \left(\sum_{j=1}^{k} q_{ij} \log q_{ij} \right) p_i + \sum_{j=1}^{k} f \left(\sum_{i=1}^{n} p_i q_{ij} \right),$$

so by 3.5.1 and 3.5.2, $I(A, B)$ is a concave function on K_n. It is evident that K_n is convex.

If the capacity equations are satisfied, for some C, at a point $(p_1, \ldots, p_n)$ with $p_1, \ldots, p_n > 0$, then $(p_1, \ldots, p_n) \in K_n$ and the gradient of $I(A, B)$ at $(p_1, \ldots, p_n)$ is

$$\nabla I(A, B) \big|_{(p_1, \ldots, p_n)} = (C - \log e, C - \log e, \ldots, C - \log e).$$

That is, the gradient of $I(A, B)$ at $(p_1, \ldots, p_n)$ is a scalar multiple of $(1, \ldots, 1)$, which is normal to the hyperplane with equation $x_1 + \cdots + x_n = 1$, in $\mathbb{R}^n$, of which K_n is a fragment. It follows that the directional derivative of $I(A, B)$, at $(p_1, \ldots, p_n)$, in any direction parallel to this hyperplane, is zero. It follows that the function of one variable obtained by restricting $I(A, B)$ to any line segment in K_n through $(p_1, \ldots, p_n)$ will have derivative zero at the value of the one variable corresponding to the point $(p_1, \ldots, p_n)$. It follows that $I(A, B)$ achieves its maximum on each such line segment at $(p_1, \ldots, p_n)$, by 3.5.4. Therefore, $I(A, B)$ achieves its maximum on K_n at $(p_1, \ldots, p_n)$.

Exercises 3.5

1. Suppose that $(\widetilde{p}_1, \ldots, \widetilde{p}_{n-1}, 0) \in K_n$, and $\widetilde{p}_1, \ldots, \widetilde{p}_{n-1} > 0$. Show that

$$I(A, B) \big|_{(p_1, \ldots, p_n)} \to I(A, B) \big|_{(\widetilde{p}_1, \ldots, \widetilde{p}_{n-1}, 0)}$$

as $(p_1, \ldots, p_n) \to (\widetilde{p}_1, \ldots, \widetilde{p}_{n-1}, 0)$, with $(p_1, \ldots, p_n) \in K_n$. [You may assume that, for each $j \in \{1, \ldots, k\}$, $q_{ij} > 0$ for some $i \in \{1, \ldots, n\}$. (Interpretation?) You may as well inspect the functions $f_{ij}(p_1, \ldots, p_n) = p_i q_{ij} \log(\sum_{t=1}^{n} p_t q_{tj})$. No problem when $q_{ij} = 0$, and no problem when $1 \le i \le n - 1$. When $i = n$, you will need to consider two cases: $q_{ij} = \cdots = q_{n-1,j} = 0$, and otherwise.]

∗2. Under what conditions on the transition probabilities is $I(A, B)$ strictly concave on K_n?

3. Prove the statements in 3.5.1 and 3.5.2.

Chapter 4

Coding Theory

4.1 Encoding and decoding

The situation is this: we have a *source alphabet* $S = \{s_1, \ldots, s_m\}$ and a *code alphabet* $A = \{a_1, \ldots, a_n\}$, which is also the input alphabet of some channel. We would like to transmit text written in the source alphabet, but our channel accepts only code alphabetic characters. Therefore, we aim to associate a code alphabet word to represent each source alphabet word that we might wish to send.

In many real situations, it is not really necessary to represent each member of S^+, the set of all source words, by a code word, a member of A^+. For instance, if the source text is a chunk of ordinary English prose, we can be reasonably certain that we will not have to transmit nonsense words like "zrdfle" or "cccm." However, it does not seem that there is any great advantage to be had by omitting part of S^+ from consideration, and there is some disadvantage—the discussion gets complicated, quarrels break out, anxieties flourish.

Definitions An *encoding function* is a function $\phi : S^+ \to A^+$. We say that such a function *defines*, or *determines*, a *code*. The code determined by ϕ is said to be *unambiguous* if and only if ϕ is one-to-one (injective). Otherwise, the code is *ambiguous*.

A *valid decoder-recognizer* (VDR) for the code determined by ϕ is an algorithm which accepts as input any $w \in A^+$, and produces as output either the message "does not represent a source word" if, indeed, w is not in the range of ϕ, or, if $w \in \mathrm{ran}\,\phi$, some $v \in S^+$ such that $\phi(v) = w$.

The code determined by ϕ is *uniquely decodable* if and only if it is unambiguous and there exists a VDR for it.

Some remarks are in order.

4.1.1 Note that the definitions above do not really say what a code is. It is *something* determined by an encoding function, but what? It might be more satisfying logically to identify the code with the encoding function which determines it, but, unfortunately, that would lead to syntactic constructions that clash

71

with common usage. The definition above stands without apology, but the uses of the word "code" may increase in the future.

4.1.2 A VDR is, as its name indicates, an algorithm that either correctly decodes a code word, or correctly recognizes that the code word cannot be decoded.

We shall be quite informal about describing VDRs, and extremely cavalier about proving that a given algorithm is a VDR for a given code. For instance, suppose that $S = A = \{0, 1\}$, and that ϕ is described by: ϕ doubles each 0 and leaves 1 as is. [Thus, for instance, $\phi(1010) = 100100$.] Then the following describes a VDR for this code: given $v \in \{0, 1\}^+$, scan v, and if any maximal block of consecutive 0's of odd length is found in v, report "does not represent a source word"; otherwise, halve each maximal block of consecutive 0's in v, and output the resulting word w. We leave it to the reader to divine what is meant by "maximal block of consecutive 0's."

The point is that we do not make a fuss about how you scan or how you find a maximal block of consecutive 0's in v and determine its length. Any implementation of the algorithm described would have to handle these and other matters, but the details are not our concern here. Also, it is possible to prove formally that this algorithm is a VDR for the given code, and that the code is uniquely decodable, but a bit of thought will convince anybody that these things are true, so that writing out formal proofs becomes an empty exercise, as well as being no fun. It can be of practical value to attempt proofs of algorithm validity and unique decodability, especially when these matters are in doubt, but we shall not be at all conscientious about such proofs.

4.1.3 In modern naive set theory, it is proven that for any non-empty S and A, there are uncountably many injective functions from S^+ into A^+. The codes determined by two different such functions cannot have the same VDR. It is also proven that there are but countably many algorithms expressible in any natural language. It follows that there are quite a few, in fact, uncountably many codes with no VDR. We certainly want nothing to do with such codes, but don't worry—there is very little danger of encountering such a code.

In most of the codes actually in use in real life, the encoding function is defined in a particularly straightforward way.

Definition An *encoding scheme* for a source alphabet $S = \{s_1, \ldots, s_m\}$ in terms of a code alphabet A is a list of *productions*,

$$s_1 \rightarrow w_1$$
$$\vdots$$
$$s_m \rightarrow w_m,$$

in which $w_1, \ldots, w_m \in A^+$. For short, we will say that such a list is a *scheme for* $S \rightarrow A$.

Each encoding scheme gives rise to an encoding function $\phi : S^+ \to A^+$ by *concatenation*. The concatenation of a sequence of words is just the word obtained by writing them down in order, with no separating spaces, commas, or other marks. Given an encoding scheme, as above, and a word $v \in S^+$, we let $\phi(v)$ be the concatenation of the sequence of the w_i, $1 \le i \le m$, corresponding, according to the scheme, to the source letters occurring in v. For example, if $S = \{a, b, c\}$, $A = \{0, 1\}$, and the scheme is

$$a \to 01$$
$$b \to 10$$
$$c \to 111,$$

then $\phi(acbba) = 01111101001$.

It is sometimes useful to be more formal; given an encoding scheme, we could define the corresponding encoding function by *induction* on the *length* of the source word. For $v \in S^+$, let lgth(v) stand for the length of v, the number of letters appearing in v. If lgth(v) $= 1$, then $v \in S$, so $v = s_i$ for some i, and we set $\phi(v) = w_i$, where w_i is the code word on the right-hand side of the production $s_i \to w_i$ in the scheme. If lgth(v) > 1, then $v = us_i$ for some $s_i \in S$, and some $u \in S^+$ with lgth(u) $=$ lgth(v) $- 1$; $\phi(u)$ has already been defined, so we set $\phi(v) = \phi(u)w_i$.

The formality of this definition of ϕ is unnecessary for most purposes, but it is advisable to keep it in mind. It provides a form for proving by induction statements about the code determined by ϕ. Sometimes the other obvious inductive definition of ϕ, in which source words are formed by adding letters on the left rather than on the right, is more convenient.

4.1.4 When an encoding scheme is given, and thereby an encoding function, the term "the code" can refer to (i) the encoding scheme, (ii) the list $w_1, \ldots, w_m$ of code words appearing in the scheme, or (iii) the set $\{w_1, \ldots, w_m\}$.

4.1.5 Theorem *Every code determined by an encoding scheme has a VDR.*

Proof: Given a scheme $s_i \to w_i$, $i = 1, \ldots, m$, and a word $w \in A^+$, look among all concatenations of the w_i, with length of the concatenation equal to lgth(w). [There are surely systematic ways to go about forming all such concatenations—but it would be tiresome to dwell upon those ways here.] If none of them match w, report "does not represent a source word." If one of them matches w, decode in the obvious way, by replacing each w_i in the concatenation by some corresponding s_i in the encoding scheme. It is left to you to convince yourself that this prescription constitutes a VDR for the given code. $\square$

The algorithm plan described above is a very bad one, extremely slow and inefficient, and should never be used. It is of interest only because it works whatever the encoding scheme.

4.1.6 Given an encoding scheme $s_i \to w_i$, $i = 1, \ldots, m$, *reading-left-to-right* with reference to this scheme is the following algorithm: given $w \in A^+$, scan from left to right along w until you recognize some w_i as an *initial segment* of w. If the w_i you recognize is also w_j for some $i \neq j$ (i.e., if the same code word represents different source letters according to the encoding scheme, heaven forfend), then decide for which i you have recognized w_i by some rule – for instance, you could let i be the smallest of the eligible indices.

If no w_i has been recognized as an initial segment of w, after scanning over $\max_{1 \leq i \leq m} \text{lgth}(w_i)$ letters of w, or if you come to the end of w after scanning fewer letters, without recognizing some w_i, report "does not represent a source word." Otherwise, having recognized w_i, jot down s_i on your decoder pad, to the right of any source letters already recorded, peel (delete) the segment w_i from w, and begin the process anew, with the smaller word replacing w. If there is nothing left after w_i is peeled from w, stop, and declare that the decoding is complete. *Reading-right-to-left* is described similarly.

For example, suppose that $S = \{a, b, c\}$, $A = \{0, 1\}$, and the scheme is

$$a \rightarrow 0$$
$$b \rightarrow 01$$
$$c \rightarrow 001.$$

(This is a particularly stupid scheme, for all practical purposes. Note that 001 represents both ab and c in the code defined by this scheme.) With reference to this scheme, what will be the outcome of applying the reading-left-to-right algorithm to 001? Answer: the output will be "does not represent a source word." (Surprised?) It follows that reading-left-to-right is not a VDR for this code. However, reading-right-to-left *is* a VDR for this code. Verification, or proof, of this assertion is left to you. (Take a look at Exercise 4.2.2.)

4.1.7 If the words w_i appearing in an encoding scheme are all of the same length, the code is said to be a *fixed-length* or *block* code, and the common length ℓ of the w_i is said to be the *length of the code*. Otherwise, the code is said to be a *variable-length* code.

4.1.8 If $A = \{0, 1\}$, or some other two-element set, the code is said to be *binary*.

Exercises 4.1

1. Let S be the set of all English words, let A be the set of letters $a, b, \ldots, z$, and let the encoding scheme be defined by a very complete unabridged dictionary—the O.E.D. will do. Ignore capitalizations. Show that the code defined by this scheme is ambiguous.

2. Suppose that $S = A$ and $\phi : S^+ \rightarrow S^+$ is defined by

$$\phi(w) = \begin{cases} w, & \text{if lgth}(w) \text{ is odd,} \\ ww, & \text{if lgth}(w) \text{ is even.} \end{cases}$$

Show that ϕ is not given by an encoding scheme. Describe a VDR for this code. Is this code uniquely decodable? Justify your answer.

3. Same questions as in 2, except that $\phi(w) = \begin{cases} w, & \text{if lgth}(w) \text{ is even,} \\ ww, & \text{if lgth}(w) \text{ is odd.} \end{cases}$

4. Suppose that $S = \{a, b, c\}$, $A = \{0, 1\}$, and consider the scheme

$$a \to 0, \quad b \to 010, \quad c \to 0110.$$

Show that neither reading-left-to-right nor reading-right-to-left provides a VDR for this code. Describe a VDR for this code—make it a better one than the clunker described in the proof of Theorem 4.1.5. Is this code uniquely decodable? Justify your answer.

5. Give an encoding scheme for a uniquely decodable code for which reading-left-to-right is a VDR, but reading-right-to-left is not.

6. Suppose that $S = \{a, b, c\}$, $A = \{0, 1\}$, and the encoding scheme is

$$a \to 010, \quad b \to 0100, \quad c \to 0010.$$

Is the code defined by this scheme uniquely decodable? Justify your answer.

7. Give an encoding scheme for the code described in 4.1.2.

4.2 Prefix-condition codes and the Kraft-McMillan inequality

An encoding scheme $s_i \to w_i$, $i = 1, \ldots, m$, satisfies the *prefix condition* if there do not exist $i, j \in \{1, \ldots m\}$, $i \neq j$, such that w_i is an initial segment, or *prefix*, (reading left to right) of w_j. The code determined by such a scheme is said to be a *prefix-condition code*. The *suffix condition* is similarly defined.

4.2.1 Theorem *Each prefix-condition code is uniquely decodable, with reading left-to-right providing a VDR.*

Proof: Left to you. □

Remark: a converse of this theorem holds; see Exercise 4.2.2.

4.2.2 Proposition *If $s_i \to w_i$, $i = 1, \ldots, m$, is a fixed-length encoding scheme, then the following are equivalent:*

(a) *the scheme satisfies the prefix condition;*

(b) *the code defined by the scheme is uniquely decodable;*

(c) *$w_1, \ldots, w_m$ are distinct.*

Proof: Left to you. □

4.2.3 Corollary *Given $n = |A|$, $m = |S|$, and a positive integer ℓ, there is a uniquely decodable fixed-length scheme for $S \to A$, of length ℓ, if and only if $m \le n^\ell$.*

Proof: $n^\ell = |A^\ell|$, so $m \le n^\ell$ means that there are m distinct code words of length ℓ available for the desired scheme.   □

4.2.4 The code in Exercise 4.1.4 is neither prefix-condition nor suffix-condition, but is, nonetheless, uniquely decodable.

4.2.5 If the beginning of the code word is on the left—i.e., if the code word is to be fed into the decoder from left to right—then it is clearly a great convenience for reading-left-to-right to be a VDR for the code; you can decode as the code word is being read. By contrast, in order to decode using reading-right-to-left, if the code word starts on the left, you have to wait until the entire code word or message has arrived before you can start decoding. Because of this advantage, prefix-condition codes are also *instantaneous* codes (see [81]). Note Exercise 4.2.2.

4.2.6 Theorem (Kraft's Inequality) *Suppose that $S = \{s_1, \ldots, s_m\}$ is a source alphabet, $A = \{a_1, \ldots, a_n\}$ is a code alphabet, and $\ell_1, \ldots, \ell_m$ are positive integers. Then there is an encoding scheme $s_i \to w_i$, $i = 1, \ldots, m$, for S in terms of A, satisfying the prefix condition, with $\mathrm{lgth}(w_i) = \ell_i$, $i = 1, \ldots, m$, if and only if $\sum_{i=1}^m n^{-\ell_i} \le 1$.*

Proof: For $w \in A^+$ and $\ell \ge \mathrm{lgth}(w)$, let

$$A(w, \ell) = \{v \in A^\ell; w \text{ is a prefix of } v\} \tag{4.1}$$

$$= \{wu; u \in A^{\ell - \mathrm{lgth}(w)}\}. \tag{4.2}$$

Then $|A(w, \ell)| = |A^{\ell - \mathrm{lgth}(w)}| = n^{\ell - \mathrm{lgth}(w)}$. Observe that, if neither of $w_1, w_2 \in A^+$ is a prefix of the other, and $\ell \ge \mathrm{lgth}(w_i)$, $i = 1, 2$, then $A(w_1, \ell)$ and $A(w_2, \ell)$ are disjoint.

Assume, without loss of generality, that $1 \le \ell_1 \le \ell_2 \le \cdots \le \ell_m$. First suppose that $\sum_{i=1}^m n^{-\ell_i} \le 1$. We will choose $w_1, \ldots, w_m \in A^+$ such that no w_i is a prefix of any w_j, $1 \le i < j \le m$, and the choosing will be straightforward. Let w_1 be any member of A^{ℓ_1}. Supposing we have obtained $w_1, \ldots, w_k$ with $w_i \in A^{\ell_i}$, $i = 1, \ldots, k$, and no w_i a prefix of any w_j, $1 \le i < j \le k \le m - 1$, we wonder if there is any $w_{k+1} \in A^{\ell_{k+1}}$ such that none of $w_1, \ldots, w_k$ is a prefix of w_{k+1}. Clearly there is such a w_{k+1} if and only if

$$A^{\ell_{k+1}} \setminus \bigcup_{i=1}^k A(w_i, \ell_{k+1})$$

is non-empty. By the remarks above (and since the $A(w_i, \ell_{k+1})$, $i = 1, \ldots, k$, are pairwise disjoint),

$$\left| \bigcup_{i=1}^{k} A(w_i, \ell_{k+1}) \right| = \sum_{i=1}^{k} |A(w_i, \ell_{k+1})| = \sum_{i=1}^{k} n^{\ell_{k+1} - \text{lgth}(w_i)}$$

$$= n^{\ell_{k+1}} \sum_{i=1}^{k} n^{-\ell_i} < n^{\ell_{k+1}} \sum_{i=1}^{m} n^{-\ell_i} \leq n^{\ell_{k+1}} = |A^{\ell_{k+1}}|.$$

Thus $A^{\ell_{k+1}} \setminus \bigcup_{i=1}^{k} A(w_i, \ell_{k+1})$ is non-empty. Thus we can find $w_1, \ldots, w_m$ as desired, by simple hunting and finding.

On the other hand, if $1 \leq \ell_1 \leq \cdots \leq \ell_m$ and there is a prefix-condition scheme $s_i \rightarrow w_i$, with $\text{lgth}(w_i) = \ell_i, i = 1, \ldots, m$, then

$$w_m \in A^{\ell_m} \setminus \bigcup_{j=1}^{m-1} A(w_j, \ell_m),$$

so

$$1 \leq |A^{\ell_m}| - \sum_{j=1}^{m-1} |A(w_j, \ell_m)| = n^{\ell_m} - n^{\ell_m} \sum_{j=1}^{m-1} n^{-\ell_j},$$

whence $\sum_{j=1}^{m} n^{-\ell_j} \leq 1$. □

4.2.7 The usefulness of being able to prescribe the lengths $\ell_1, \ldots, \ell_m$ of the words $w_1, \ldots, w_m$ in a prefix-condition encoding scheme will become clear in the next two sections. See also Exercise 4.2.5.

Once $\ell_1, \ldots, \ell_m$ satisfying $\sum_{i=1}^{m} n^{-\ell_i} \leq 1$ have been prescribed, there is no obstacle, according to the proof preceding, to choosing the w_i for the scheme, *provided* $\ell_1 \leq \ell_2 \leq \cdots \leq \ell_m$. It may be necessary to reorder $s_1, \ldots, s_m$ to achieve this ordering of the ℓ_i.

In some situations it may be wise to prescribe $\ell_1, \ldots, \ell_m$ satisfying the inequality $\sum_{i=1}^{m} n^{-\ell_i} < 1$, i.e., to avoid $\ell_1, \ldots, \ell_m$ satisfying $\sum_{i=1}^{m} n^{-\ell_i} = 1$, even though the ℓ_i in such a sequence may be more desirable in the short run. The practical reason is that the customer buying the encoding scheme may wish to enlarge the source alphabet at some future time.

4.2.8 When $\ell_1 = \ell_2 = \cdots = \ell_m = \ell$, i.e., when the scheme is to be fixed-length, then the condition given in Kraft's Inequality for the existence of a prefix-condition code simplifies to $n^\ell \geq m$. Since $n^\ell = |A^\ell|$ and $m = |S|$, this condition is also seen to be necessary and sufficient by Proposition 4.2.2.

4.2.9 Theorem (McMillan's Inequality) *If* $|S| = m$, $|A| = n$, *and* $s_i \rightarrow w_i \in A^{\ell_i}, i = 1, \ldots, m$, *is an encoding scheme resulting in a uniquely decodable code, then* $\sum_{i=1}^{m} n^{-\ell_i} \leq 1$.

Proof: Without loss of generality, assume that ℓ_m is the largest of the ℓ_j. For any positive integer k,

$$\left(\sum_{i=1}^{m} n^{-\ell_i}\right)^k = \left(\sum_{i_1=1}^{m} n^{-\ell_{i_1}}\right)\cdots\left(\sum_{i_k=1}^{m} n^{-\ell_{i_k}}\right)$$

$$= \sum_{i_1=1}^{m}\sum_{i_2=1}^{m}\cdots\sum_{i_k=1}^{m} n^{-(\ell_{i_1}+\cdots+\ell_{i_k})} = \sum_{r=1}^{k\ell_m} \frac{h(r)}{n^r},$$

where $h(r)$ is the number of times r occurs as a sum $\ell_{i_1} + \cdots + \ell_{i_k}$, as $i_1, \ldots, i_k$ roam independently over $\{1, \ldots, m\}$.

If $r = \ell_{i_1} + \cdots + \ell_{i_k}$, then $r = \mathrm{lgth}(w_{i_1} \cdots w_{i_k})$. By the assumption of unique decodability, if $(i_1, \ldots, i_k) \neq (i'_1, \ldots, i'_k)$, then $w_{i_1} \cdots w_{i_k} \neq w_{i'_1} \cdots w_{i'_k}$. This means that the function $(i_1, \ldots, i_k) \mapsto w_{i_1} \cdots w_{i_k}$ from $\{(i_1, \ldots, i_k); 1 \leq i_j \leq m, j = 1, \ldots, k$ and $\sum_{j=1}^{k} i_j = r\}$ into A^r is an injection; since the size of the domain is $h(r)$ and the size of A^r is n^r, it follows that $h(r) \leq n^r$, and we have

$$\left(\sum_{i=1}^{m} n^{-\ell_i}\right)^k = \sum_{r=1}^{k\ell_m} \frac{h(r)}{n^r} \leq \sum_{r=1}^{k\ell_m} 1 = k\ell_m,$$

so $\sum_{i=1}^{m} n^{-\ell_i} \leq k^{1/k} \ell_m^{1/k} \to 1$ as $k \to \infty$. □

The elegant proof of McMillan's Inequality given here is due to Karush [38].

4.2.10 The encoding scheme in the hypothesis of Theorem 4.2.9 is not assumed to be a prefix-condition scheme. Thus McMillan's Inequality improves the "only if" assertion of Kraft's Inequality.

4.2.11 Corollary (of 4.2.6 and 4.2.9) *Suppose* $|S| = m$, $|A| = n$, *and* $\ell_1, \ldots,$ ℓ_m *are positive integers. The following are equivalent:*

(a) *there is an encoding scheme* $s_i \to w_i \in A^{\ell_i}$, $i = 1, \ldots, m$, *resulting in a uniquely decodable code;*

(b) *there is a prefix-condition encoding scheme* $s_i \to w_i \in A^{\ell_i}$, $i = 1, \ldots, m$;

(c) $\sum_{i=1}^{m} n^{-\ell_i} \leq 1$.

The moral is that if unique decodability and prescribing the lengths of the w_i in the scheme are the only considerations, then there is no reason to consider anything except prefix-condition (or, in some countries, suffix-condition) codes.

Exercises 4.2

1. Suppose $|S| = m$, $|A| = n$, and we are considering only fixed-length encoding schemes of length ℓ. The resulting code is to be uniquely decodable. Find (a) the smallest value of ℓ possible if $m = 26$ and $n = 2$; (b) the smallest value of ℓ possible if $m = 26$ and $n = 3$; (c) the smallest value of n possible if $m = 80$ and $\ell \leq 4$; (d) the largest value of m possible if $n = 2$ and $\ell \leq 6$.

2. Prove that if a code, given by an encoding scheme, is uniquely decodable, with reading-left-to-right a VDR for the code, then the scheme satisfies the prefix condition. (Hint: prove the contrapositive. That is, start by supposing that the scheme does not satisfy the prefix condition, and prove that either reading-left-to-right is not a VDR for the code, or the code is not uniquely decodable.) Give an example of a scheme that does not satisfy the prefix condition for which reading-left-to-right does provide a VDR.

3. Suppose that $|S| = m \geq 2$ and $|A| = n \geq 2$. For reasons that may become clear later, we will say that a non-decreasing sequence $\ell_1 \leq \cdots \leq \ell_m$ of positive integers is an n-ary Huffman sequence if there is a prefix-condition encoding scheme $s_j \to w_j \in A^{\ell_j}$, $j = 1, \ldots, m$, but if any of the ℓ_j is reduced by one and the new sequence is denoted $\ell'_1, \ldots, \ell'_m$, then there is no prefix-condition scheme $s_j \to w'_j \in A^{\ell'_j}$, $j = 1, \ldots, m$. [Convention: $A^0 = \emptyset$.] Find the n-ary Huffman sequences $\ell_1 \leq \cdots \leq \ell_m$ when

   (a) $n = 2$ and $m = 5$
   (b) $n = 3$ and $m = 5$
   *(c) $n = 2$ and $m = 26$.

4. Suppose $S = \{a, b, c, d, e\}$ and $A = \{0, 1\}$. Find a prefix-condition encoding scheme for S in terms of A, corresponding to each of the sequences you found in 3(a), above.

5. Let $|A| = n$, $|S| = m$, and $L = \max_{1 \leq j \leq m} \text{lgth}(w_j)$. Let us say that a scheme $s_j \to w_j \in A^+$, $j = 1, \ldots, m$, is good if it results in a uniquely decodable code.

   (a) For fixed n and L, what is the largest value of m possible if there is to be a good scheme for S?
   (b) For fixed m and L, what is the smallest value of n possible if there is to be a good scheme for S?
   (c) For fixed m and n, what is the smallest value of L possible if there is to be a good scheme for S?

   [Hint: in every case, the optimum is achieved with a fixed-length encoding scheme.]

4.3 Average code word length and Huffman's algorithm

Suppose that $s_i \to w_i \in A^+$, $i = 1, \ldots, m$, is an encoding scheme for a source alphabet $S = \{s_1, \ldots, s_m\}$. Suppose it is known that the source letters $s_1, \ldots, s_m$ occur with relative frequencies $f_1, \ldots, f_m$, respectively. That is, f_i is to be regarded as the probability that a letter selected at random from the source text will be s_i. It follows that $\sum_{i=1}^{m} f_i = 1$. We will refer to the f_i as the *relative source letter frequencies*, or source frequencies, for short.

Definition In the circumstances described above, the *average code word length* of the code defined by the encoding scheme is

$$\bar{\ell} = \sum_{i=1}^{m} f_i \operatorname{lgth}(w_i).$$

4.3.1 Note that "average code word length" is a bit of a misnomer. The correct term would be "average length of a code word replacing a source letter."

4.3.2 $\bar{\ell}$ is, in fact, the average value of the random variable "length of the code word replacing the source letter" associated with the experiment of randomly selecting a source letter from the source text. On the (dubious?) grounds that reading a section of source text amounts to carrying out the selection of source letters a number of times, it follows from Theorem 1.8.6 that the average, or expected, number of code letters required to encode a source text consisting of N source letters is $\bar{\ell}N$.

Recall that the code letters are also the input letters of a channel. It may be expensive and time consuming to transmit long sequences of code letters; therefore, it may be desirable for $\bar{\ell}$ to be as small as possible. It is within our power to make $\bar{\ell}$ small by cleverly making arrangements when we devise the encoding scheme. What constraints must we observe?

For one thing, we want the resulting code to be uniquely decodable; since $\bar{\ell}$ is a function of the $\ell_i = \operatorname{lgth}(w_i)$, it follows from Corollary 4.2.11 that we may as well confine ourselves to prefix-condition codes.

This is the only constraint we will observe in this section; it is, happily, a simplifying constraint—it makes life easier to be confined to a smaller array of choices. In later sections, however, we will encounter other purposes that might be served in the construction of the encoding scheme. These other purposes are: good approximation of the *optimal input frequencies* of the channel and *error correction*. In no case do these matters require us to abandon prefix-condition codes, but they sometimes do conflict with the minimization of $\bar{\ell}$. When there are more concerns to juggle than just the shortening of the input text, when compromises must be made, the methods to be described in this section may have to be modified or abandoned.

Common sense or intuition suggests that, in order to minimize $\bar{\ell}$, we ought to have the frequently occurring source letters represented by short code words, and to reserve the longer code words of the scheme for the rarely occurring source letters. It is left to the reader to decide whether or not a proof of the validity of this strategy is required. Proofs are available, and, in fact, the validity is enshrined in a famous theorem.

4.3.3 Theorem [Ch. 10, 28] *Suppose that $f_1 \geq f_2 \geq \cdots \geq f_m$ and $\ell_1 \leq \ell_2 \leq \cdots \leq \ell_m$. Then for any rearrangement $\ell_1', \ldots, \ell_m'$ of the list $\ell_1, \ldots, \ell_m$, $\sum_{i=1}^{m} f_i \ell_i \leq \sum_{i=1}^{m} f_i \ell_i'$.*

Recall that, to obtain a prefix-condition encoding scheme $s_j \to w_j$, $j = 1, \ldots, m$, with $\mathrm{lgth}(w_j) = \ell_j$, where $\sum_{j=1}^{m} n^{-\ell_j} \leq 1$, we have no worries provided $\ell_1 \leq \ell_2 \leq \cdots \leq \ell_m$. With ℓ_j in non-decreasing order, we happily choose our w_j avoiding prefixes, without a snag.

It follows that, given the source frequencies, it would be a shrewd first move to reorder the input alphabet, and, correspondingly, the f_i, so that $f_1 \geq f_2 \geq \cdots \geq f_m$. We shall henceforward consider the f_i to be so ordered. To minimize $\bar{\ell}$, we look among sequences $\ell_1, \ldots, \ell_m$ of positive integers satisfying $\ell_1 \leq \cdots \leq \ell_m$ and $\sum_{j=1}^{m} n^{-\ell_j} \leq 1$.

One way to proceed would be to look among the *minimal* such sequences, the n-ary Huffman sequences defined in Exercise 4.2.3, and to select from those the one that makes $\bar{\ell}$ the smallest.

4.3.4 Example Suppose $S = \{a, b, c, d, e\}$, $A = \{0, 1\}$, and the source frequencies are $f_a = 0.25$, $f_b = 0.15$, $f_c = 0.1$, $f_d = 0.2$, and $f_e = 0.3$. We reorder the source alphabet: e, a, d, b, c. We look among minimal sequences (also called n-ary Huffman sequences) $\ell_e \leq \ell_a \leq \ell_d \leq \ell_b \leq \ell_c$. It is hoped that you found three such sequences, in doing problem 3(a) at the end of the preceding section:

(1) $1, 2, 3, 4, 4$;

(2) $1, 3, 3, 3, 3$; and

(3) $2, 2, 2, 3, 3$.

The average code word lengths corresponding to these different sequences are

$$\bar{\ell}_1 = (0.3)1 + (0.25)2 + (0.2)3 + (0.15)4 + (0.1)4 = 2.4,$$
$$\bar{\ell}_2 = (0.3)1 + (0.25 + 0.2 + 0.15 + 0.1)3 = 2.4, \text{ and}$$
$$\bar{\ell}_3 = (0.3 + 0.25 + 0.2)2 + (0.15 + 0.1)3 = 2.25.$$

Thus list (3) is the winner. An optimal encoding scheme:

$$e \to 00$$
$$a \to 11$$
$$d \to 10$$
$$b \to 010$$
$$c \to 011.$$

The process of finding all possible minimal sequences $\ell_1 \leq \cdots \leq \ell_m$ satisfying $\sum_{j=1}^{m} n^{-\ell_j} \leq 1$ can be algorithmized. This approach to minimizing $\bar{\ell}$ is worth keeping in mind, especially since it is adaptable to "mixed" optimization problems, in which we want to keep $\bar{\ell}$ small *and* serve some other purpose—for instance, we might like the input frequency of the code letters to be close to the optimal input frequencies for the channel (see Section 4.4). In such problems we may agree to an encoding scheme that effects a compromise between (or among) the contending requirements; perhaps $\bar{\ell}$ won't be as small as we could get, but it will still be quite small and our other purposes will be served reasonably well.

In making shopper's choices in such mixed problems, it is not at all inefficient or unreasonable to have all the alternative schemes arrayed before us, among which to choose. If the numbers involved are not astronomical, and the time consumed is not prohibitive, especially since we are shopping for a "big ticket" item, it is reasonable to take the trouble to find an encoding scheme which, once chosen, will be installed and used for the foreseeable future.

But mathematicians dislike "brute force" in making choices; the brute-force, shopping-in-the-warehouse approach suggested above may, in fact, be forced upon us in real life for some purposes, but what follows is faster and more elegant in the cases where minimizing $\bar{\ell}$ (with a prefix-condition scheme) is our only objective.

4.3.5 Huffman's algorithm We suppose that $f_1 \geq f_2 \geq \cdots \geq f_m$, and that $m = n + k(n-1)$ for some non-negative integer k. This last requirement can be achieved by adding letters to the source alphabet and assigning source frequency zero to the added letters. Note that when $n = 2 \leq m$, this bothersome preliminary is unnecessary.

Merge. If $m = n$, go to *encode*, below. Otherwise, form a new source alphabet with $n + (k-1)(n-1)$ letters by merging the n source letters with least source frequencies into a single source letter, whose frequency will be the sum of the source frequencies of the merged letters. Thus, the new source alphabet is $S' = \{s_1, \ldots, s_{m-n}, \sigma\}$ and the s_j, $1 \leq j \leq m-n$, have frequencies f_j, while σ has frequency $\sum_{j=m-n+1}^{m} f_j$.

Note which letters were merged into σ, and reorder S' so that source frequencies are in non-increasing order. With S' replacing S and with the new source frequencies replacing $f_1, \ldots, f_m$, go to *merge*.

Encode. We are here initially with a source alphabet $\widetilde{S}$ with n letters. We form a scheme by which these letters are put into one-to-one correspondence with the letters of A, the code alphabet. We will derive from this scheme an optimal encoding scheme for the original source alphabet S, by working our way back through the sequence of source alphabets obtained by *merging*. At each stage of the journey from $\widetilde{S}$ back to S, we obtain, from the current encoding scheme, an encoding scheme for the next alphabet (along the road back to S) by an obvious and straightforward procedure. Suppose that S'' was obtained from S' by merging, and suppose that we have an encoding scheme for S''. Suppose that $\sigma \in S''$ was obtained by merging $s'_{t+1}, \ldots, s'_{t+n} \in S'$. Suppose that, in the encoding scheme for S'', the production involving σ is $\sigma \to w$. Then the scheme for S' is obtained from that for S'' by replacing the single production $\sigma \to w$ by the n productions

$$s'_{t+1} \to wa_1$$
$$\vdots$$
$$s'_{t+n} \to wa_n.$$

4.3.6 Examples Consider the situation in 4.3.4, in which $S = \{a, b, c, d, e\}$, $A = \{0, 1\}$ and $f_e = 0.3 \geq f_a = 0.25 \geq f_d = 0.2 \geq f_b = 0.15 \geq f_c = 0.1$. We run Huffman's algorithm. First merge (b and c merged into σ_1):

$$S_1 = \{e, \quad a, \quad \sigma_1, \quad d\}$$
$$\text{Frequencies:} \quad 0.3, \quad 0.25, \quad 0.25, \quad 0.2$$

Second merge (σ_1 and d merged into σ_2):

$$S_2 = \{\sigma_2, \quad e, \quad a\}$$
$$\text{Frequencies:} \quad 0.45, \quad 0.3, \quad 0.25$$

Last merge (e and a merged into σ_3)

$$S_3 = \{\sigma_3, \sigma_2\}.$$

For the encoding, we obtain

$$S_3 : \begin{cases} \sigma_3 \to 0 \\ \sigma_2 \to 1 \end{cases} \qquad S_2 : \begin{cases} \sigma_2 \to 1 \\ e \to 00 \\ a \to 01 \end{cases}$$

and

$$S_1 : \begin{cases} e \to 00 \\ a \to 01 \\ \sigma_1 \to 10 \\ d \to 11 \end{cases} \qquad S : \begin{cases} e \to 00 \\ a \to 01 \\ d \to 11 \\ b \to 100 \\ c \to 101 \end{cases}$$

Note that the algorithm does, indeed, give a code with minimal $\bar{\ell}$, by the work done in 4.3.4. Note that the encoding scheme is different from that given in 4.3.4, and that, in fact, there is no way to apply Huffman's algorithm to this example to obtain the scheme of 4.3.4. This is because Huffman's algorithm will result in the code words representing e and a having the same first digit.

Let us apply Huffman's algorithm with the same S and source frequencies, but with $A = \{0, 1, *\}$, i.e., with $n = 3$. Note that $5 = 3 + 1 \cdot 2$, so we need not add any source letters with zero frequency (or, equivalently, we need not merge fewer than n letters on the first merge).

The first and only merge: $S_1 = \{e, a, \sigma\}$ [d, b, and c are merged]. The schemes are given by $S_1 : e \to 0, a \to 1, \sigma \to *$ and $S : e \to 0, a \to 1, d \to *0$, $b \to *1, c \to **$.

The proof of the fact that Huffman's algorithm always results in an optimal prefix-condition encoding scheme is outlined in Section 4.3.1 (filling in the details is left to the reader as Exercise 4.3.5).

We conclude this section with the statement of a famous theorem of Shannon which relates the $\bar{\ell}$ achievable by Huffman's algorithm to the source entropy. The proof of this theorem is postponed until Section 5.4 where a sharper statement of the theorem is proven for only the binary case. However, the proof there can be easily modified to give an equally sharp theorem for all $n \geq 2$.

4.3.7 Noiseless Coding Theorem for memoryless sources *Suppose* $|S| = m$, $|A| = n \geq 2$, *and the source frequencies are* $f_1, \ldots, f_m$. *Let* $H = -\sum_{i=1}^{m} f_i \log f_i$. *For every encoding scheme for S, in terms of A, resulting in a uniquely decodable code, the average code word length* $\bar{\ell}$ *satisfies*

$$\bar{\ell} \geq H / \log n.$$

Furthermore, there exists a prefix-condition scheme for which

$$\bar{\ell} < H / \log n + 1.$$

The first inequality above, $\bar{\ell} \geq H / \log n$, has an interpretation that makes the result seem self-evident, if you do not look too closely. Let the base of the logarithm be n, the size of the code alphabet. Then the inequality becomes $\bar{\ell} \geq H$. Also, setting this base for the logarithm defines the *unit* of information: each code letter can carry at most one unit of information. (See Section 2.1.1.)

Now, H is the average number of information units per source letter and $\bar{\ell}$ is the average number of code letters per source letter arising from the encoding scheme. If we have unique decodability, no information is lost; so the average amount of information carried by the code words representing the source letters, which is $\bar{\ell}$ units at most, must be at least as great as H, the average number of units of information per source letter; for if the volume of a vessel is less than that of the fluid that is poured into it, there will be spillage.

The fact that there is a rigorous mathematical proof of this inequality is further evidence that Shannon's definition of information is satisfactory on an intuitive level.

Exercises 4.3

1. Suppose $S = \{a, b, c, d, e, f, g\}$ and the source frequencies are given in the following table:

letter	a	b	c	d	e	f	g
freq	.2	.12	.08	.15	.25	.1	.1

Use the Huffman encoding algorithm to obtain an optimal prefix-condition scheme for S when

(a) $A = \{0, 1\}$
(b) $A = \{0, 1, *\}$.

2. Table 4.1 gives the relative frequencies, in English prose minus punctuation and blanks, ignoring capitalization, of the alphabetic characters a, b, ... , z, estimated by examination of a large block of English prose, believed to be typical. This table is copied, with one small change, from [6, Appendix 1]. Find an optimal (with respect to average code word length) prefix-condition encoding scheme for $S = \{a, b, \ldots, z\}$ if

(a) $A = \{0, 1\}$;

Table 4.1: *Single-letter frequencies in English text.*

Character	% Freq	Character	% Freq
a	8.167	n	6.749
b	1.492	o	7.507
c	2.782	p	1.929
d	4.253	q	0.095
e	12.702	r	5.987
f	2.228	s	6.327
g	2.015	t	9.056
h	6.094	u	2.758
i	6.966	v	0.978
j	0.153	w	2.360
k	0.772	x	0.150
l	4.025	y	1.974
m	2.406	z	0.075

(b) $A = \{0, 1, *\}$.

(c) What are the lengths of the shortest fixed-length encoding schemes, resulting in uniquely decodable codes, for S, in cases (a) and (b), above?

*3. (For those with calculators and some free time.) Verify the conclusion of Theorem 4.3.7 in the circumstances of the preceding exercise.

4. Suppose $S = \{a, b, c, d, e, f\}$ and the source frequencies are given in:

letter	a	b	c	d	e	f
freq	.2	.15	.05	.2	.25	.15

Use Huffman's algorithm to encode $S \to A$ when (a) $A = \{0, 1\}$ and (b) $A = \{0, 1, *\}$. Did you notice that there were choices to be made in running the algorithms in the "merge" part of the process? Run the algorithm in all possible ways, in (a) and (b), if you haven't already. In each case, you should arrive at two essentially different schemes, essentially different in that the sequences of code word lengths are different. However, in each case, $\bar{\ell}$ is minimized by both schemes.

*5. Establish the validity of Huffman's algorithm by filling the gaps in the proof given in Section 4.3.1, below.

*6. Suppose that $\ell_1 \leq \cdots \leq \ell_m$ is an n-ary Huffman sequence. Show that $\bar{\ell} = \sum_{j=1}^{m} f_j \ell_j$ is minimal among the numbers $h(x_1, \ldots, x_m) = \sum_{j=1}^{m} f_j x_j$, where $x_1, \ldots, x_m$ are integers satisfying $\sum_{j=1}^{m} n^{-x_j} \leq 1$, if the f_j are defined by $f_j = n^{-\ell_j} (\sum_{i=1}^{m} n^{-\ell_i})^{-1}$, $j = 1, \ldots, m$. [You may as well assume that $x_1 \geq \cdots \geq x_m$ is an n-ary Huffman sequence. There are only finitely many of these, and thus only finitely many values of $G_0 = \sum_{i=1}^{m} n^{-x_i}$ to consider. Fix one of these, and now suddenly allow the x_i to vary freely, even into non-integer values, but subject to the constraint $G_0 = \sum_{i=1}^{m} n^{-x_i}$.

Use the Lagrange multiplier method to attempt to find where $h(x_1, \ldots, x_m)$ achieves its minimum, subject to this constraint. You will find that when $G_0 \neq G = \sum_{i=1}^{m} n^{-\ell_i}$, the minimum is not achieved at an n-ary Huffman sequence $x_1, \ldots, x_m$, and when $G_0 = G$, the minimum is achieved when $x_i = \ell_i, i = 1, \ldots, m$.]

4.3.1 The validity of Huffman's algorithm

In this section we will try to lead whomever is interested through a proof of the validity of Huffman's algorithm. In fact, we will prove more: not only does Huffman's algorithm always give a "right answer," but, also, every "right answer," in case there is more than one, as in problem 4, above, can be obtained by some instance of Huffman's algorithm. By a "right answer" here we do not mean any actual prefix-condition encoding scheme which minimizes $\bar{\ell}$, but rather the sequence of lengths of the code words in such an encoding scheme. (That Huffman's algorithm always produces a prefix-condition scheme is quite easy to see; we leave it to the reader to work through the proof.)

There is a concise proof of the validity of Huffman's algorithm in the binary case, in Huffman's original paper [36], and this proof can be easily extended to prove the stronger statement given here, when $n = 2$. However, there are some unexpected difficulties that crop up when $n > 2$ that appear to necessitate a much longer proof. We have not seen a proof for $n > 2$ elsewhere. Both Huffman [36] and Welsh [81] give proofs for $n = 2$ and dismiss the cases $n > 2$ as similar. Jones [37] notes that the case $n > 2$ is significantly different from the case $n = 2$ but does not give a proof for $n > 2$.

Thanks are due to Luc Teirlinck for several of the observations on which the proof given here is based. Even more thanks are due to Heather-Jean Matheson, who, while an undergraduate at the University of Prince Edward Island, discovered a serious error in the purported proof in the first edition of this text. (She not only noticed that the logic of a certain inference was wrong, she demonstrated that it could not be made right, by giving a beautiful example. Unfortunately, it would take us too far afield to explain that example here.) Yet further portions of gratitude are due to Maxim Burke for elegantly fixing the error, in a way that improves the entire proof. The statements and proofs of Propositions 4.3.8 and 4.3.9 are entirely due to him.

Recall, from Exercise 4.2.3, that an n-ary Huffman sequence is a sequence $\ell_1 \leq \cdots \leq \ell_m$ of positive integers such that there is a prefix-condition encoding scheme $s_j \rightarrow w_j \in A^{\ell_j}, j = 1, \ldots, n$, for encoding an m-letter source alphabet S with an n-letter code alphabet A, minimal in the sense that if any of the ℓ_j is reduced by one and the new sequence is denoted $\ell'_1, \ldots, \ell'_m$, there is no prefix-condition scheme $s_j \rightarrow w'_j \in A^{\ell'_j}, j = 1, \ldots, m$. [Convention: $A^0 = \emptyset$.]

Notice that, given relative source frequencies $f_1 \geq \cdots \geq f_m > 0$, any sequence $\ell_1 \leq \cdots \leq \ell_m$ of code word lengths for a prefix-condition scheme $s_j \rightarrow w_j \in A^{\ell_j}, j = 1, \ldots, m$, which minimizes $\bar{\ell} = \sum_{j=1}^{m} f_j \ell_j$ is an n-ary Huffman sequence. (Why? In fact, the converse is true, as well: every n-ary Huffman sequence is the sequence of code word lengths in a prefix-condition scheme for $S \to A$ that minimizes $\bar{\ell}$ with respect to *some* sequence $f_1, \ldots, f_m$ of relative source frequencies. See Exercise 4.3.6. But we will not need this fact here.)

By Kraft's Theorem (Theorem 4.2.6), a sequence $\ell_1 \le \cdots \le \ell_m$ of positive integers is an n-ary Huffman sequence if and only if it is minimal with respect to satisfying Kraft's Inequality, $\sum_{j=1}^{m} n^{-\ell_j} \le 1$. Since diminishing the largest of the ℓj increases the sum $\sum_{j=1}^{m} n^{-\ell_j}$ the least, it follows that $\ell_1 \le \cdots \le \ell_m$ is an n-ary Huffman sequence if and only if

$$\sum_{j=1}^{m} n^{-\ell_j} \le 1 < \sum_{j=1}^{m} n^{-\ell_j} + n^{1-\ell_m}.$$

Therefore $\sum_{j=1}^{m} n^{-\ell_j} = 1$ for any positive integers $\ell_1, \ldots, \ell_m$ implies that their non-decreasing rearrangement is n-ary Huffman.

4.3.8 Proposition *If $1 \le \ell_1 \le \cdots \le \ell_m$ and $n \ge 2$ are integers and $\sum_{i=1}^{m} n^{-\ell_i} = 1$, then $m = n + k(n-1)$ for some non-negative integer k, and $\ell_{m-n+1} = \cdots = \ell_m$.*

Proof: We go by induction on $L = \ell_m$. If $L = 1$ then $\sum_{j=1}^{m} n^{-1} = 1$ implies $m = n$ ($k = 0$) and $\ell_1 = \cdots = \ell_n = 1$.

Suppose that $L = \ell_m > 1$ and $\sum_{i=1}^{n} n^{-\ell_i} = 1$. If K is the number of i such that $\ell_i = L$, then $1 = \sum_{\{i; \ell_i \le L-1\}} n^{-\ell_i} + K/n^L$; solving for K shows that K is a multiple of n. Since $K \ge 1$, this establishes the last conclusion of the proposition, that $\ell_i = L$ for the last n values of i. It remains to be shown that $m = n + k(n-1)$.

Set $K = an$ and set $m' = m - an$, the number of indices i such that $\ell_i \le L - 1$. We have

$$1 = \sum_{\{i; \ell_i \le L-1\}} n^{-\ell_i} + \frac{an}{n^L} = \sum_{i=1}^{m'} n^{-\ell_i} + \frac{a}{n^{L-1}}.$$

By the induction hypothesis, $m' + a = n + k'(n-1)$ for some non-negative integer k'. Thus $m = m' + an = n + (k' + a)(n-1)$, which has the desired form. $\square$

4.3.9 Proposition *If $1 \le \ell_1 \le \cdots \le \ell_m = L$ is an n-ary Huffman sequence and $m = n + k(n-1) + t$, where k is a non-negative integer and $1 \le t \le n-1$, then $\sum_{i=1}^{m} n^{-\ell_i} + \frac{n-1-t}{n^L} = 1$, and $\ell_{m-t} = \cdots = \ell_m = L$.*

Proof: The second conclusion follows from the first and Proposition 4.3.8, applied to the longer sequence $\ell_1 \le \cdots \le \ell_{m+n-1-t} = L$.

Since $\ell_1 \le \cdots \le \ell_m$ is an n-ary Huffman sequence, $\sum_{i=1}^{m} n^{-\ell_i} \le 1$ and clearly $\sum_{i=1}^{m} n^{-\ell_i}$ is an integer multiple of n^{-L}. Therefore, for some non-negative integer r, $\sum_{i=1}^{m} n^{-\ell_i} + \frac{r}{n^L} = 1$.

If $r \geq n-1$ then $\sum_{i=1}^{m-1} n^{-\ell_i} + n^{1-L} = \sum_{i=1}^{m} n^{-\ell_i} + \frac{n-1}{n^L} \leq \sum_{i=1}^{m} n^{-\ell_i} + \frac{r}{n^L} = 1$, contradicting that $\ell_1 \leq \cdots \leq \ell_m$ is an n-ary Huffman sequence. Therefore $0 \leq r \leq n-2$.

By Proposition 4.3.8, $m+r = n+k'(n-1)$ for some non-negative integer k'. Thus $m = n+k(n-1)+t = n+k'(n-1)-r = n+(k'-1)(n-1)+(n-1-r)$. Since both t and $n-1-r$ are among $1,\ldots,n-1$, and $t = n-1-r \bmod (n-1)$, it follows that $t = n-1-r$, so $r = n-1-t$, as desired. $\qquad\square$

4.3.10 Corollary If $1 \leq \ell_1 \leq \cdots \leq \ell_m = L$ is an n-ary Huffman sequence, where $m = n+k(n-1)+t$ for integers $k \geq 0$ and $1 \leq t \leq n-1$, then so is the non-decreasing rearrangement of $\ell'_1,\ldots,\ell'_{m-t}$, where $\ell'_j = \ell_j, 1 \leq j < m-t$, and $\ell'_{m-t} = L-1$.

Proof:

$$\sum_{j=1}^{m-t} n^{-\ell'_j} = \sum_{j=1}^{m-t-1} n^{-\ell_j} + n/n^L$$

$$= \sum_{j=1}^{m-t-1} n^{-\ell_j} + \frac{t+1}{n^L} + \frac{n-t-1}{n^L}$$

$$= \sum_{j=1}^{m} n^{-\ell_j} + \frac{n-t-1}{n^L} = 1. \qquad\square$$

Corollary 4.3.10 allows us to provide a relatively easy proof by induction on m that if $f_1 \geq \cdots \geq f_m > 0$, $\sum_{i=1}^{m} f_i = 1$, and integers $\ell_1 \leq \cdots \leq \ell_m$ satisfying $\sum_{i=1}^{m} n^{-\ell_i} \leq 1$ minimize $\sum_{j=1}^{m} f_j \ell_j$, then some instance of Huffman's algorithm applied to $f_1,\ldots,f_m$ with respect to a code alphabet A with n letters will produce an encoding scheme with code word lengths $\ell_1,\ldots,\ell_m$. In these circumstances, if $m \leq n$ we must have $\ell_1 = \cdots = \ell_m = 1$ and Huffman's algorithm trivially gives the desired result. So suppose that $m = n+k(n-1)+t$ for some integers $k \geq 0$ and $t \in \{1,\ldots,n-1\}$; we go by induction on m. Note that although there may well be different instances of Huffman's algorithm applicable to $f_1,\ldots,f_m$, based on different merging choices in the "merge" part of the algorithm, the first merge will invariably merge the $t+1$ source letters $s_{m-t},\ldots,s_m$ into a letter σ, which will be given relative frequency $\sum_{j=m-t}^{m} f_j$.

Let $L = \ell_m$. By the previous observation that $\ell_1 \leq \cdots \leq \ell_m$ is an n-ary Huffman sequence and Proposition 4.3.9, we have that $\ell_m = \cdots = \ell_{m-t} = L$, and by Corollary 4.3.10, the non-decreasing rearrangement of $\ell_1,\ldots,\ell_{m-t-1}$, $L-1$ is an n-ary Huffman sequence. We verify that these code word lengths minimize the average code word length of a possible prefix-condition code for $S' = \{s_1,\ldots,s_{m-t-1},\sigma\} \rightarrow A$ with respect to the relative frequencies $f'_1 = f_1,\ldots,f'_{m-t-1} = f_{m-t-1}, f'_{m-t} = \sum_{j=m-t}^{m} f_j$. Suppose that $\ell'_1,\ldots,\ell'_{m-t}$ are positive integers such that $\sum_{j=1}^{m-t} n^{-\ell'_j} \le 1$ and

$$\sum_{j=1}^{m-t} f'_j \ell'_j = \sum_{j=1}^{m-t-1} f_j \ell'_j + \ell'_{m-t} \sum_{j=m-t}^{m} f_j$$

$$< \sum_{j=1}^{m-t-1} f_j \ell_j + (L-1) \sum_{j=m-t}^{m} f_j. \qquad (*)$$

We have $\sum_{j=1}^{m-t-1} n^{-\ell'_j} + (t+1)n^{-(\ell'_{m-t}+1)} = \sum_{j=1}^{m-t-1} n^{-\ell'_j} + \frac{t+1}{n} n^{-\ell'_{m-t}} \le \sum_{j=1}^{m-t} n^{-\ell'_j} \le 1$, showing that there is a prefix-condition encoding scheme for $S \to A$ with code word lengths $\ell'_1, \ldots, \ell'_{m-t-1}, \ell'_{m-t} + 1, \ldots, \ell'_{m-t} + 1$. But $\sum_{j=1}^{m-t-1} f_j \ell'_j + (\ell'_{m-t}+1) \sum_{j=m-t}^{m} f_j < \sum_{j-1}^{m-t-1} f_j \ell_j + L \sum_{j=m-t}^{m} f_j = \sum_{j=1}^{m} f_j \ell_j$ (by $(*)$), contradicting the assumed minimality of $\sum_{j=1}^{m} f_j \ell_j$.

By the induction hypothesis, there is an instance of Huffman's algorithm resulting in a prefix-condition scheme for $S' \to A$ with code word lengths $\ell_1, \ldots, \ell_{m-t-1}, L-1$ for $s_1, \ldots, s_{m-t-1}, \sigma$, respectively. Let u denote the word of length $L-1$ assigned to σ in this encoding scheme. Then $ua_1, \ldots, ua_{t+1}$ will be the words of length $L = \ell_{m-t} = \cdots = \ell_m$ assigned to $s_{m-t}, \ldots, s_m$ in the scheme obtained by the instance of Huffman's algorithm consisting of preceding that for $S' \to A$ by merging $s_{m-t}, \ldots, s_m$. Thus some instance of Huffman's algorithm results in an encoding scheme for $S \to A$ with code word lengths $\ell_1, \ldots, \ell_m$.

It remains to show that every instance of Huffman's algorithm produces an optimal encoding scheme, with respect to the given source frequencies. In view of what has already been shown, this task amounts to showing that different instances of Huffman's algorithm applied to relative source frequencies $f_1 \ge \cdots \ge f_m$ result in schemes with the same average code word length. We leave the details of this demonstration to the reader. Go by induction on m, and use the observation that if $m = n + k(n-1) + t$, $k \ge 0$, $1 \le t \le n-1$, then every instance of Huffman's algorithm applied to $f_1 \ge \cdots \ge f_m$, up to switching the order of source letters with equal source frequencies, starts with the merging of $s_{m-t}, \ldots, s_m$ into a new letter with relative frequency $\sum_{j=m-t}^{m} f_j$.

Exercises 4.3 (continued)

7. Suppose that $m = n + k(n-1) + t$, $k \ge 0$, $1 \le t \le n-1$, $1 \le \ell_1 \le \cdots \le \ell_m = L$ are integers, and $G = \sum_{i=1}^{m} n^{-\ell_i}$. Show that $\ell_1, \ldots, \ell_m$ is an n-ary Huffman sequence if and only if $G + \frac{n-1-t}{n^L} \le 1 \le G + \frac{n-2}{n^L}$. (Note that Proposition 4.3.9 can be used for part of the proof, and provides the funny corollary that if the two inequalities above hold, then the leftmost is equality.)

4.4 Optimizing the input frequencies

As before, let $S = \{s_1, \ldots, s_m\}$ be the source alphabet, and $A = \{a_1, \ldots, a_n\}$ the code alphabet. A is also the input alphabet of the channel we plan to use. Suppose that the (relative) source frequencies $f_1, \ldots, f_m$ are known, and also the optimal channel input frequencies $\widehat{p}_1, \ldots, \widehat{p}_n$ of the input letters $a_1, \ldots, a_n$. We have the problem of coming up with a "good" encoding scheme, $s_j \to w_j \in A^+$, $j = 1, \ldots, m$. The goodness of the scheme is judged with reference to a number of criteria. We have already seen that for unique decodability, we may as well have a scheme that satisfies the prefix condition. For minimizing $\bar{\ell} = \sum_{j=1}^{m} f_j \, \text{lgth}(w_j)$, we have Huffman's algorithm. Now let us consider the requirement that the input frequencies $p_1, \ldots, p_n$ of the letters $a_1, \ldots, a_n$ should be as close as possible, in some sense, to the optimal input frequencies $\widehat{p}_1, \ldots, \widehat{p}_n$.

In particular circumstances we can wrangle over the metric, the sense of "closeness," to be used, and we can debate the rank of this requirement among the various contending requirements, but it is clear that we will make no progress toward satisfying this requirement if we cannot compute the input frequencies $p_1, \ldots, p_n$ arising from a particular encoding scheme. This computation is the subject of the following theorem.

4.4.1 Theorem *Suppose that* $s_j \to w_j \in A^+$, $j = 1, \ldots, m$, *is an encoding scheme. Suppose that* a_i *occurs exactly* u_{ij} *times in* w_j, $i = 1, \ldots, n$, $j = 1, \ldots, m$. *Then, for* $i = 1, \ldots, m$,

$$p_i = \frac{\sum_{j=1}^{m} u_{ij} f_j}{\sum_{j=1}^{m} f_j \, \text{lgth}(w_j)} = (\bar{\ell})^{-1} \sum_{j=1}^{m} u_{ij} f_j.$$

Proof: We will have a rather informal proof; logicians and philosophers can be hired later to dignify it.

Suppose we have a block of source text with a large number N of source characters, with the marvelous property that, for each $j = 1, \ldots m$, s_j occurs exactly the expected number of times, $N f_j$. After encoding, the total number of characters in the code text is $\sum_{j=1}^{m} (N f_j) \, \text{lgth}(w_j) = N\bar{\ell}$. The number of occurrences of a_i in the code text is $\sum_{j=1}^{m} u_{ij} (N f_j) = N \sum_{j=1}^{m} u_{ij} f_j$. Dividing, we find that the proportion of a_i's in the code text is

$$p_i = \frac{N \sum_j u_{ij} f_j}{N\bar{\ell}} = (\bar{\ell})^{-1} \sum_{j=1}^{m} u_{ij} f_j. \qquad \square$$

4.4.2 Example Suppose that $S = \{a, b, c\}$, $A = \{0, 1\}$ and $f_a = .6$, $f_b = .3$, and $f_c = .1$. Suppose that the encoding scheme is

$$a \rightarrow 00$$
$$b \rightarrow 101$$
$$c \rightarrow 010.$$

This scheme does not minimize average code word length, but it may have compensatory virtues that suit the situation. Letting the alphabet characters serve as indices, we have

$$u_{0a} = 2, u_{0b} = 1, u_{0c} = 2,$$
$$\text{and} \quad u_{1a} = 0, u_{1b} = 2, u_{1c} = 1.$$

Thus the input frequencies will be

$$p_0 = \frac{2(.6) + (.3) + 2(.1)}{2(.6) + 3(.3) + 3(.1)} = \frac{17}{24}$$

and $p_1 = 1 - p_0 = 7/24$. If the channel involved is a binary symmetric channel, then the optimal input frequencies are $\widehat{p}_0 = \widehat{p}_1 = 1/2$ (see 3.4.5), so p_0 and p_1 here are quite far from optimal.

The code designer may have had good reasons for the choice of this scheme. Would the designer agree to changing the second digit in each of the code words of the scheme? This would not change any lengths, nor the relationships among the code words. (You might ponder what "relationships" means here.) The new scheme:

$$a \rightarrow 01$$
$$b \rightarrow 111$$
$$c \rightarrow 000.$$

The new input frequencies: $p_0 = 3/8$, $p_1 = 5/8$. These are not optimal, but they are closer to $1/2$ than were the former input frequencies, $17/24$ and $7/24$. If the new scheme is as good as the original in every other respect, then we may as well use the new scheme.

Optimizing the input frequencies, after minimizing $\bar{\ell}$, with a prefix-condition code

4.4.3 Problem The input consists of S, A, the source frequencies $f_1, \ldots, f_m$, and the optimal input frequencies $\widehat{p}_1, \ldots, \widehat{p}_n$ for the channel of which A is the input alphabet. The output is to be an encoding scheme $s_j \rightarrow w_j \in A^+$, $j = 1, \ldots, m$ such that

(i) the prefix condition is satisfied;

(ii) $\bar{\ell} = \sum_j f_j \operatorname{lgth}(w_j)$ is minimal, among average code word lengths of schemes satisfying (i); and

(iii) the n-tuple $(p_1, \ldots, p_n)$, computed as in Theorem 4.4.1, is as close as possible to $(\widehat{p}_1, \ldots, \widehat{p}_n)$, by some previously agreed upon measure of closeness. If $d(p, \widehat{p})$ denotes the distance from $p = (p_1, \ldots, p_n)$ to $\widehat{p} = (\widehat{p}_1, \ldots, \widehat{p}_n)$, this means that $d(p, \widehat{p})$ is to be minimal among all such numbers computed from schemes satisfying (i) and (ii).

4.4.4 Usually, $d(p, \widehat{p}) = \sum_{j=1}^{n}(p_j - \widehat{p}_j)^2$, but you can take

$$d(p, \widehat{p}) = \sum_{j=1}^{n}|p_j - \widehat{p}_j|^\alpha$$

for some power α other than 2, or $d(p, \widehat{p}) = \max_{1 \le j \le n}|p_j - \widehat{p}_j|$. When $n = 2$, these different measures of distance are equivalent: for any choice of d, above, $d(p, \widehat{p}) \le d(p', \widehat{p})$ if and only if $|p_1 - \widehat{p}_1| \le |p_1' - \widehat{p}_1|$. See Exercise 4.4.4.

It would be nice to have a slick algorithm to solve Problem 4.4.3, especially in the case $n = 2$, when the output will not vary with different reasonable definitions of $d(p, \widehat{p})$. Also, the case $n = 2$ is distinguished by the fact that binary channels are in widespread use in the real world.

We have no such good algorithm! Perhaps someone reading this will supply one some day. However, we do have an algorithm; it's brutish, but it's an algorithm. Here it is: Supposing $f_1 \ge f_2 \ge \cdots \ge f_m$, use Huffman's algorithm to find all n-ary Huffman sequences $\ell_1 \le \cdots \le \ell_m$ that minimize $\bar{\ell} = \sum f_j \ell_j$; for each of these sequences, we find all possible prefix-condition schemes $s_j \to w_j \in A^{\ell_j}$ and compute $p_i = (\bar{\ell})^{-1}\sum_{j=1}^{m}u_{ij}f_j$, $i = 1, \ldots, n$. We choose the scheme for which $(p_1, \ldots, p_n)$ is closest to $(\widehat{p}_1, \ldots, \widehat{p}_n)$.

4.4.5 Example Let's carry out the brute-force program suggested above in the easy circumstances of Example 4.4.2, assuming that the channel is a BSC. We have $S = \{a, b, c\}$, $A = \{0, 1\}$, $f_a = .6$, $f_b = .3$, $f_c = .1$, and $\widehat{p}_0 = \widehat{p}_1 = 1/2$. There is only one sequence of code word lengths to consider: $\ell_a = 1$, $\ell_b = 2 = \ell_c$. We have $\bar{\ell} = 1.4$. There are four different prefix-condition schemes to consider; the two that start with $a \to 0$ are: $a \to 0$, $b \to 10$, $c \to 11$ and $a \to 0$, $b \to 11$, $c \to 10$. For the first of these, $p_0 = (1.4)^{-1}(.6 + .3) = 9/14$, and, for the second, $p_0 = (1.4)^{-1}(.6 + .1) = 1/2$. Clearly the second wins! Alternatively, the scheme $a \to 1$, $b \to 00$, $c \to 01$ gives optimal input frequencies.

With the same S, A, and source frequencies, if the channel had been so oddly constructed that $\widehat{p}_0 = 1/3$, $\widehat{p}_1 = 2/3$, then the optimal scheme of the four candidates would have been $a \to 1$, $b \to 01$, $c \to 00$.

4.4.6 Example $S = \{a, b, c, d, e\}$, $A = \{0, 1\}$, $\widehat{p}_0 = \widehat{p}_1 = 1/2$, $f_e = .35$, $f_a = .3$, $f_d = .2$, $f_b = .1$, and $f_c = .05$. The sequences $(\ell_e, \ell_a, \ell_d, \ell_b, \ell_c)$ satisfying $\sum_{j \in S} 2^{-\ell_j} \le 1$ for which $\bar{\ell} = \sum f_j \ell_j$ is minimal are $(1, 2, 3, 4, 4)$ and $(2, 2, 2, 3, 3)$. [See Exercise 4.2.3 and Example 4.3.4.] The value of $\bar{\ell}$ is 2.15. Both optimal sequences are obtainable from Huffman's algorithm; the difference arises from the choice of the ordering of the alphabet obtained from the second *merge*.

The optimal schemes in this case are associated with $(2, 2, 2, 3, 3)$. (There are quite a number of schemes to look at, but, taking into account that $\widehat{p}_0 = \widehat{p}_1 = 1/2$, the possibilities boil down to only eight or nine essentially different schemes.) Here is one of the optimal schemes:

$$e \to 01$$
$$a \to 10$$
$$d \to 11$$
$$b \to 000$$
$$c \to 001.$$

Verify that $p_0 = 1.05/2.15 = 21/43$, and that this is as close to $1/2$ as you can get in this situation. (Since each f_j is an integer multiple of .05, the numerator of $p_0 = \frac{\sum u_{0j} f_j}{2.15}$ will be an integer multiple of .05. Thus the closest p_0 can be made to $1/2$ is $1.05/2.15$ or $1.10/2.15$.)

Observe that in this case we can have $p_0 = p_1 = 1/2$ exactly, with a prefix-condition scheme, if we sacrifice the minimization of $\bar{\ell}$. For instance, the fixed-length scheme $e \to 0011$, $a \to 1100$, $d \to 0101$, $b \to 1001$, $c \to 1010$ gives unique decodability and $p_0 = p_1 = 1/2$.

4.4.7 In general, whenever the optimal input frequencies $\hat{p}_1, \ldots, \hat{p}_n$ are *rational* numbers, we can achieve exact input frequency optimization, $p_i = \hat{p}_i$, $i = 1, \ldots, n$, with a uniquely decodable block code; just make $\ell = \text{lgth}(w_j)$, $j = 1, \ldots, m$ so large that it is possible to find m distinct words $w_1, \ldots, w_m \in A^\ell$ such that the proportion of the occurrences of a_i *in each* is exactly $\hat{p}_i$. And, if some of the $\hat{p}_i$ are irrational, we can approximate $\hat{p} = (\hat{p}_1, \ldots, \hat{p}_n)$ by a rational vector $(p_1, \ldots, p_n) = p$ (satisfying $\sum_{i=1}^{n} p_i = 1$, $p_i \geq 0$, $i = 1, \ldots, n$) as closely as we wish, and then produce a fixed-length scheme from which the p_i arise as the input frequencies of the a_i. Thus the variables $p_1, \ldots, p_n$ are truly "vary-able," as we promised in Chapter 3, and arrangements can be made in the code, or input, "language," so that the relative input frequencies are as close as desired to optimal.

However, the method suggested in the preceding paragraph for approximating the optimal input frequencies is clearly inpractical; the code words would have to be quite long, so that the *rate* of processing of source text would be quite slow, and increasing that rate is generally reckoned to be of greater consequence than the close approximation of the optimal input frequencies.

In the same vein, one might well question the importance of Problem 4.4.3, although in this problem the approximation of the optimal input frequencies is subordinated to minimizing $\bar{\ell}$ – i.e., to speeding up the processing of source text. As long as the scheme is uniquely decodable and $\bar{\ell}$ is minimized, why fiddle with trying to approximate the optimal input frequencies? Optimizing the average amount of information conveyed by the channel per input letter, with the input stream somewhat artificially regarded as randomly generated, may seem an ivory-tower objective, an academic exercise of doubtful connection to the real world problem of communicating a source stream through a channel.

However, it is an indirect and little-noted consequence of the famed Noisy Channel Theorem, to be explained in Section 4.6, that there is a connection between the practical problems of communication and the problem of encoding the source stream so that the input frequencies are approximately optimal. Not to go into detail, the import of the NCT is that there exist ways of encoding the source stream that simultaneously do about as well as can be done regarding the two most obvious practical problems of communication: keeping pace with the source stream (up to a threshold that depends on the channel capacity), and reducing the error frequency, in the reconstitution of the source stream (decoding) at the receiver of the channel. Although it is not explicitly proven in any of the rigorous treatments of the NCT, the role of the channel capacity in the NCT strongly argues for the information-theoretic folk theorem that the relative input frequencies resulting from those wonderful optimizing coding methods whose existence is asserted by the NCT must be nearly optimal, themselves.

This folk theorem is of particular interest when you realize that all known proofs of the NCT are probabilistic existence proofs; there is no good constructive way known of acquiring those coding methods whose existence is proved. Furthermore, when you understand the nature of those methods, you will understand that they would be totally impractical, even if found. [The situation reminds us of contrived gambling games in which the expected gain per play is infinite, yet the probability of going bankrupt due to accumulated losses is very close to one, even for Bill Gates.] So the problem of effective coding realizing the aspirations expressed in the NCT is still on the agenda, and has been for the 54 years (as this is written) since Shannon's masterpiece [63]. So far as we know, the indirect approach of aiming, among other things, to get close to the optimal relative input frequencies by astute coding has not been a factor in the progress of the past half-century. In part this has to do with the fact that binary symmetric channels are the only channels that have been seriously considered; also, it has been generally assumed that the relative source frequencies are equal (see the discussion, next section, on the equivalence of Maximum Likelihood Decoding and Nearest Code Word Decoding), and the dazzling algebraic methods used to produce great coding and decoding under these assumptions automatically produce a sort of uniformity that makes p_0 and p_1 equal or trivially close to $1/2$. Perhaps the problem of approximating the optimal input frequencies by astute encoding will become important in the future, as communication engineering ventures away from the simplifying assumption of equal source frequencies.

Exercises 4.4

1. We return to 4.3.4: $S = \{a, b, c, d, e\}$, $f_e = 0.3$, $f_a = 0.25$, $f_d = 0.2$, $f_b = 0.15$, and $f_c = 0.1$. Find a scheme which solves the problem in paragraph 4.4.3 when

   (a) $A = \{0, 1\}$, $\widehat{p}_0 = 1/2 = \widehat{p}_1$;
   (b) $A = \{0, 1\}$, $\widehat{p}_0 = 2/3$, $\widehat{p}_1 = 1/3$;
   (c) $A = \{0, 1, *\}$, $\widehat{p}_0 = \widehat{p}_1 = \widehat{p}_* = 1/3$;
   (d) $A = \{0, 1, *\}$, $\widehat{p}_0 = \widehat{p}_1 = 2/5$, $\widehat{p}_* = 1/5$.

2. In each of (a)–(d) in the preceding problem, find a uniquely decodable fixed-length scheme which gives the optimal input frequencies exactly. The shorter the length, the better.

3. When $|S| = m = 26$, find the shortest length of a fixed-length prefix-condition scheme, by which the optimal input frequencies are realized exactly, constructed as suggested in 4.4.7, when the code alphabet and optimal input frequencies are as in 1(a)–(d), above. [Notice that the method suggested in 4.4.7 takes no account of the source frequencies.] Compare with the $\bar{\ell}$ you found in exercise 4.3.2 (a) and (b).

4. Verify the assertion about the case $n = 2$ made in 4.4.4. [Hint: observe that if $p_1 + p_2 = p'_1 + p'_2 = \widehat{p}_1 + \widehat{p}_2$, then if $|p_1 - \widehat{p}_1| \leq |p'_1 - \widehat{p}_1|$, it follows that $|p_2 - \widehat{p}_2| \leq |p'_2 - \widehat{p}_2|$, since $|p_1 - \widehat{p}_1| = |p_2 - \widehat{p}_2|$ and $|p'_1 - \widehat{p}_1| = |p'_2 - \widehat{p}_2|$.]

5. Suppose the source text is encoded by the scheme $s_j \to w_j \in A^+$, $j = 1, \ldots, m$, the source frequencies are $f_1, \ldots, f_m$, and the u_{ij} are as in Theorem 4.4.1. We select a letter at random from the source text and look at it; if it is s_j, we then select a letter at random from w_j. What is the probability that a_i will be selected by this procedure? Is this the same as $(\bar{\ell})^{-1} \sum_{j=1}^m u_{ij} f_j$? If not, why not?

6. This exerise concerns the efficiency of the brute-force algorithm suggested for solving Problem 4.4.3.

   (a) How many prefix-condition binary encoding schemes are there with code word lengths $2, 2, 3, 3, 3, 3$?

   (b) How many prefix-condition binary encoding schemes are there with code word lengths $2, 2, 2, 3, 4, 4$?

   (c) How many prefix-condition ternary ($|A| = 3$) encoding schemes are there with code word lengths $1, 2, 2, 2, 2, 2$?

   *(d) Given $|A| = n \geq 2$ and positive integers $\ell_1 \leq \cdots \leq \ell_m$ satisfying $\sum_{j=1}^m n^{-\ell_j} \leq 1$, give a formula, in terms of n and $\ell_1, \ldots, \ell_m$, for the number of different prefix-condition encoding schemes for $S \to A$, $S = \{s_1, \ldots, s_m\}$, with code word lengths $\ell_1, \ldots, \ell_m$. [See Section 1.6 and the proof of Kraft's Inequality.]

4.5 Error correction, maximum likelihood decoding, nearest code word decoding, and reliability

Let $S = \{s_1, \ldots, s_m\}$ and $A = \{a_1, \ldots, a_n\}$ be as in the preceding sections, and suppose that $B = \{b_1, \ldots, b_k\}$ is the output alphabet of the channel of which A is the input alphabet, the channel that we plan to use. Let $Q = [q_{ij}]$ be the matrix of transition probabilities of this channel. Let $f_1, \ldots, f_m$ be the (relative) source frequencies of $s_1, \ldots, s_m$, respectively. We shall consider only *fixed-length* encoding schemes $s_j \to w_j \in A^\ell$, $j = 1, \ldots, m$, with the w_j distinct, as nature usually demands; so $m \leq n^\ell$.

Suppose we are trying to convey some source letter s_j through the channel. What we really send is the sequence w_j of input characters. The channel does whatever it does to w_j, and what is received is a word $w \in B^+$ of length ℓ. You are at the receiving end. You know the encoding scheme. The input string has been timed and blocked so that you know that the output segment w resulted from an attempt to transmit one of $w_1, \ldots, w_m$. How are you going to guess which w_j (and thus, which s_j) was intended?

4.5.1 We are in a conditional probabilistic situation not unlike that of the man who draws balls from an urn in a dark room, and later wonders which urn he drew from (see 1.3.2 and exercise 1.3.5).

Surely the most reasonable choice of w_j is that for which the conditional probability

$$P(w_j \text{ was sent} \mid w \text{ was received}) = \frac{P(w_j \text{ was sent and } w \text{ was received})}{P(w \text{ was received})}$$

is the greatest; thus it behooves us to inspect the numbers

$$P(w_j, w) = P(w_j \text{ was sent and } w \text{ was received}).$$

These are calculated as follows. Suppose that $w_j = a_{i(1,j)} \cdots a_{i(\ell,j)}$; that is, suppose that $a_{i(s,j)} \in A$ is the sth code letter in w_j, reading left to right. Suppose that $w = b_{t_1} \cdots b_{t_\ell}$. Then

$$P(w_j, w) = P(w_j \text{ was sent}) P(w \text{ was received} \mid w_j \text{ was sent})$$

$$= f_j q_{i(1,j),t_1} \cdots q_{i(\ell,j),t_\ell}$$

$$= f_j \prod_{s=1}^{\ell} q_{i(s,j),t_s}$$

4.5.2 Example Suppose that $S = \{a, b, c\}$, $A = \{0, 1\}$, $B = \{0, 1, *\}$, $f_a = .5$, $f_b = .3$, $f_c = .2$, and the transition probabilities are

$$\begin{bmatrix} q_{00} & q_{01} & q_{0*} \\ q_{10} & q_{11} & q_{1*} \end{bmatrix} = \begin{bmatrix} .9 & .06 & .04 \\ .05 & .92 & .03 \end{bmatrix}.$$

Suppose that the encoding scheme is $a \to 00$, $b \to 11$, and $c \to 01$. Suppose that $w = 0*$ is received. Then

$$P(00, 0*) = f_a q_{00} q_{0*} = (.5)(.9)(.04) = .018,$$
$$P(11, 0*) = f_b q_{10} q_{1*} = (.3)(.05)(.03) = .00045, \text{ and}$$
$$P(01, 0*) = f_c q_{00} q_{1*} = (.2)(.9)(.03) = .0054.$$

Of these, .018 is the greatest; having received $0*$, we would bet that 00, the code word for a, was intended.

The practice of decoding received words $w \in B^\ell$ by choosing the j for which $P(w_j, w)$ is the greatest is called *maximum likelihood decoding*, or MLD, for short. [In case there are two values of j for which $P(w_j, w)$ is maximal, and in case we choose not to decode such w, the practice is sometimes called *incomplete maximum likelihood decoding*, or IMLD. We will stand by MLD, in this text, even though we will not, in fact, decode w in case of ties.]

The MLD Table In a real situation, we would take care of the drudgery of computing and comparing the numbers $P(w_j, w)$ before attempting to decode words received at the receiving end of the channel. The results of this computing and comparing are collected in an *MLD table*, which consists of two columns: on the left we list all $k^\ell = |B^\ell|$ words w that might possibly be received, and on the right the *source* letters s_j corresponding to the w_j for which $P(w_j, w)$ is maximal. That is, opposite each possible received word $w \in B^\ell$, we put the source letter s recommended by MLD for decoding w.[1]

It should be obvious how the MLD table is to be used at the receiving end of the channel. It functions as a dictionary, and for that reason the words $w \in B^\ell$ should be arranged in some reasonable lexicographic order for rapid "looking up." In practice, the process of looking up w and decoding will be electronic.

4.5.3 Example In the situation of Example 4.5.2, the MLD table (with B^ℓ arranged in one of the two obvious lexicographic orders) is

(Receive) w	Decode s
00	a
01	c
0*	a
10	a
11	b
1*	b
*0	a
*1	b
**	a

(Verify that this table is correct.) Thus, if *0 were received, we would quickly decode a.

It may come as a surprise that MLD, as described, is virtually never used in current practice, except when it coincides with Nearest Code Word Decoding, to be described next.

Definition Suppose that B is an alphabet, and that $u, v \in B^\ell$ for some positive integer ℓ. The *Hamming distance* between u and v, denoted $d_H(u, v)$, is the number of places at which u and v differ.

---

[1] In case there are two or more values of j for which $P(w_j, w)$ is maximal, tie-breaking rules can be introduced to choose among the candidates s_j. These rules might arise from considerations peculiar to the particular situation. In this text, we will enter "dnd" for "do not decode" in the decoding column, in case of ties.

For instance, if $B = \{0, 1, *\}$, then $d_H(01101, 111*0) = 3$. Verify that d_H is a *metric* on B^ℓ; that is, for any $u, v, w \in B^\ell$, $d_H(u, v) = d_H(v, u) \geq 0$, with equality if and only if $u = v$, and $d_H(u, v) \leq d_H(u, w) + d_H(w, v)$.

4.5.4 The metric d_H represents one reasonable way of defining the distances between the words of B^ℓ. We could jazz things up considerably, and provide serious mathematicians with hours of fun, by generalizing the Hamming distance as follows. Let δ be any metric on B, and let $\rho : [0, \infty)^\ell \to [0, \infty)$ satisfy:

(i) $0 \leq x_i \leq y_i, i = 1, \ldots, \ell \Rightarrow \rho(x_1, \ldots, x_\ell) \leq \rho(y_1, \ldots, y_\ell)$;

(ii) $\rho(x + y) \leq \rho(x) + \rho(y)$, for $x, y \in [0, \infty)^\ell$; and

(iii) $\rho(x) = 0$ if and only if $x = 0 \in [0, \infty)^\ell$.

Then the *generalized Hamming distance* between words $u = u_1 \ldots u_\ell \in B^\ell$ and $v = v_1 \ldots v_\ell \in B^\ell$, associated with δ and ρ, is $d_{GH}(u, v) = \rho(\delta(u_1, v_1), \ldots, \delta(u_\ell, v_\ell))$. Observe that $d_{GH} = d_H$ when δ is the so-called trivial metric defined by

$$\delta(a, b) = \begin{cases} 1, & \text{if } a \neq b, \\ 0, & \text{if } a = b, \end{cases}$$

and ρ is defined by $\rho(x_1, \ldots, x_\ell) = \sum_{i=1}^{\ell} x_i$.

This sort of thing is amusing, but is it useful? So far as anyone can tell, at this stage of history, no. However, see Exercise 4.5.3.

Nearest Code Word Decoding (NCWD) Suppose that $A \subseteq B$, and we have a fixed length encoding scheme $s_j \to w_j \in A^\ell$. In NCWD, having received $w \in B^\ell$, we decode w as that s_j (if there is exactly one such) for which $d_H(w_j, w)$ is least. If there are two or more j such that $d_H(w_j, w)$ is minimal, we do not decode.

For example, if $S = \{a, b, c\}$, $A = B = \{0, 1\}$, and the encoding scheme is $a \to 000$, $b \to 111$, $c \to 010$, then if 100 is received, we decode a in NCWD. If 011 is received, we do not decode in NCWD, because both 111 and 010 are a (minimal) distance one from 011.

NCWD is a much easier decoding method than MLD. It is not just that it is easier to make up a decoding table or dictionary under the "nearest code word" criterion, than under the "most likely to have been sent" measure; the act of comparing words to determine their Hamming distance apart is so simply algebraic that it lends itself to slick, fast decoding algorithms that run much faster than the "looking up w on a list" method that we use for MLD. Instances of clever choices of $w_1, \ldots, w_m$ leading to clever NCWD algorithms are beyond the scope of this course, but they form the majority of the subject matter of advanced algebraic coding theory (see [30]).

But the simplicity of NCWD comes at a price; NCWD ignores the transition probabilities and the source frequencies, and is therefore possibly unreliable. See Exercise 4.5.2(a), and note the difference between NCWD and MLD in this case. What if the infrequently transmitted source message were "launch the missiles" or some such apocalyptic command? Pretend that you are the manager in charge of communications at a missile silo emplacement in North Dakota, and the source frequencies are not, in fact, known a priori, as is often the case. Pretend that you have a Master's Degree in Applied Mathematics from a large southern state university, and have taken a couple of courses in coding theory that get into sophisticated codes and decoding algorithms that are, in fact, NCWD. Will you "use your education" by resorting to some satisfyingly fancy form of NCWD for transmissions to and from the missile silos, for messages ranging from "Fred, please pick up a bunch of parsley and a pound of Vaseline on the way home" to "Arm and launch immediately!"? Let's hope not. The scene is fanciful, but there is an important point, which we hope you get.

So, when can you depend on NCWD? The following theorem gives a sufficient condition.

4.5.5 Theorem *Suppose that $A = B$, that the relative source frequencies are equal (to $1/m$), that $q_{ii} = q > 1/n$, $i = 1, \ldots, n$, and that the off-diagonal transition probabilities q_{ij}, $i \neq j$, are equal. Then NCWD and MLD are the same, whatever the fixed-length encoding scheme.*

Proof: Under the hypotheses, $q_{ij} = (1-q)/(n-1)$ for $i \neq j$. For any scheme $s_j \to w_j \in A^\ell$, $j = 1, \ldots, m$, and any $w \in B^\ell = A^\ell$,

$$P(w_j, w) = \frac{1}{m} q^{\ell-d} \left(\frac{1-q}{n-1}\right)^d = \frac{q^\ell}{m} \left(\frac{1-q}{q(n-1)}\right)^d,$$

where $d = d_H(w_j, w)$. Since $1/n < q \leq 1$, it follows that $0 \leq \frac{1-q}{q(n-1)} < 1$, and thus $P(w_j, w)$ is a decreasing function of $d = d_H(w_j, w)$. If $q = 1$, the channel is perfect, and NCWD and MLD coincide trivially. If $q < 1$, then $P(w_j, w)$ is a strictly decreasing function of $d = d_H(w_j, w)$; consequently, the unique j, if any, for which $d_H(w_j, w)$ is minimal, is also the unique j for which $P(w_j, w)$ is the greatest. □

4.5.6 Corollary *When the channel is a binary symmetric channel, with reliability greater than $1/2$, and the relative source frequencies are equal, then NCWD and MLD are the same.*

4.5.7 The hypothesis of Theorem 4.5.5, regarding the transition probabilities, says that the channel is an n-ary symmetric channel with reliability $q > 1/n$ (see section 3.4). It is quite common to know, or strongly suspect, a channel to be n-ary symmetric without knowing the reliability.

The hypothesis of Theorem 4.5.5 regarding the source frequencies raises a philosophical question about probability: if the source frequencies are not known a priori, should we take them to be equal? Clearly whatever knowledge we have should affect our estimates of probability—for instance, although we do not know for sure that the sun will rise tomorrow, it would be rash to assign probability $1/2$ to the possibility that it will not. So we are in delicate circumstances when we have a binary symmetric or, more generally, an n-ary symmetric channel and a data communication problem in which we do not know beforehand how frequently the various data are likely to be transmitted. Should we attempt to complicate the decoding process by taking into account our sense of likely bias in the frequencies of the source messages? And, if so, how?

The usual answer is to ignore the problem and to resort to NCWD. The usual theory of binary block codes starts from the implicit assumption that the relative source frequencies are equal; indeed, the source alphabet and the encoding scheme are not elements of the theory, nor even mentioned. We mention this as a caution to future appliers of coding theory.

It can be persuasively argued that the alleged weakness of NCWD—that it leaves the source frequencies and the transition probabilities out of account—is actually a practical strength. The argument rests, not on the *ease* of NCWD, but on the undebatable fact that in many situations the source frequencies and the transition probabilities are fictional quantities, unknown and unknowable.

Reliability and Error Given a source, a channel, a way of encoding the source stream into a string of channel input letters, and a method of decoding the output at the channel receiver, so that for each letter appearing in the original source stream, some source letter will appear in its place in the hopefully resurrected source letter stream emerging from the decoder: the *reliability* R of the given *code-and-channel system* is the probability that a letter randomly selected from the source stream will be decoded correctly at the receiver—i.e., R is the probability that a randomly selected letter from the source stream will be successfully communicated by the code-and-channel system.

The (*average*) *error probability* of the code-and-channel system is $E = 1 - R$. The *maximum error probability* of the system, denoted $\widehat{E}$, is the maximum, over the source letters s, of the probability of an error at a randomly selected spot in the source stream, supposing that s occupied that spot in the original stream.

We will calculate R, E, and $\widehat{E}$ in the circumstances that allow full-fledged MLD. That is, suppose we are given S, $f_1, \ldots, f_m$, A, B, Q, and a fixed-length encoding scheme $s_j \to w_j \in A^\ell$, $j = 1, \ldots, m$, for $S \to A$.

Given an MLD table for the code-and-channel system, the reliability R can be calculated as follows. For each $j = 1, \ldots, m$, let

$$N_j = \{w \in B^\ell; \text{ MLD decodes } w \text{ as } s_j\}.$$

The N_j can be read off from the MLD table; N_j consists of those w in the left-hand column of the table opposite the occurrences of s_j in the right. Then

$$R = \sum_{j=1}^{m} P\left(\begin{array}{c} s_j \text{ is selected for transmission,} \\ \text{and the word } w \text{ received lies in } N_j \end{array}\right)$$

$$= \sum_{j=1}^{m} f_j P(\text{transmission of } w_j \text{ results in a received word } w \in N_j)$$

$$= \sum_{j=1}^{m} f_j \sum_{w \in N_j} P(w \text{ was received} \mid w_j \text{ was sent})$$

$$= \sum_{j=1}^{m} f_j \sum_{w \in N_j} P(w \mid w_j),$$

where, as noted in 4.5.1, $P(w \mid w_j) = \prod_{s=1}^{\ell} q_{i(s,j),t_s}$, when $w_j = a_{i(1,j)} \cdots a_{i(\ell,j)}$ and $w = b_{t_1} \cdots b_{t_\ell}$.

Example In Example 4.5.3, we have $N_a = \{00, 0*, 10, *0, **\}$, $N_b = \{11, 1*, *1\}$, and $N_c = \{01\}$. Thus

$$R = .5[(.9)^2 + (.9)(.04) + (.06)(.9) + (.04)(.9) + (.04)^2]$$
$$+ .3[(.92)^2 + (.92)(.03) + (.03)(.92)] + .2[(.9)(.92)]$$
$$= .90488,$$

and $E = 1 - R = 0.09512$.

The maximum error probability $\widehat{E}$ is calculated as follows:

$$\widehat{E} = \max_{1 \le j \le m} P(\text{incorrect decoding} \mid s_j \text{ was intended})$$

$$= \max_{1 \le j \le m} P(\text{the received } w \in B^\ell \text{ does not lie in } N_j \mid w_j \text{ was transmitted})$$

$$= 1 - \min_{1 \le j \le m} P(w \in N_j \mid w_j \text{ was transmitted}).$$

Notice that, for each j, $P(w \in N_j \mid w_j$ was transmitted) is the quantity multiplied by f_j in the expression for R, above. For instance, again referring to the circumstances of Examples 4.5.2 and 4.5.3, we calculate that c has the least likelihood, 0.828, of being correctly transmitted, and thus, for that code and channel, $\widehat{E} = 1 - .828 = .172$.

Observe that $\widehat{E}$ does not take the source frequencies into account (although they do enter anyway, in the construction of the MLD table). It is a "worst-case" sort of measure of error likelihood.

4.5.8 For many code-and-channel systems we have "do not decode" occurring in the right hand column (the *decode* column) of the MLD table, corresponding to words $w \in B^\ell$ for which there are two or more j for which $P(w_j, w)$ is maximal. With such a table, we have a number of choices to make in assessing the likelihood of error. Should a "do not decode" message, which surely signals some sort of failure of the system, weigh as much in our estimation as an out-and-out error, in which we decode the wrong source message from the received word w? In the definitions of E and $\widehat{E}$, above, the two different sorts of error are treated as the same, but most people would agree that in most situations the "do not decode" message is a less serious sort of error than an incorrect decoding of which we are unaware.

There are an endless number of ways of weighting the significance of the various errors that could occur, using any particular code-and-channel system. You should be aware that the definitions of E and $\widehat{E}$ given here are not graven in stone, and that in the real world you might do well to fashion a measure of error likelihood appropriate to the real situation, a measure which takes your weighting of error significance into account. See the end of this section for exercises on error weighting, and on the computation of reliability and error when NCWD is used.

Reliability of a channel Suppose we have a channel with input alphabet A, output alphabet B, and transition probabilities $q_{ij}, i = 1, \ldots, n, j = 1, \ldots, k$. Let us adjoin a code by taking $S = A$, $f_j = \widehat{p}_j$, $j = 1, \ldots, n$, where $(\widehat{p}_1, \ldots, \widehat{p}_n)$ is some n-tuple of optimal input frequencies, and the encoding scheme $a_j \to a_j$, $j = 1, \ldots, n$. The reliability R (with respect to MLD) of the resulting code-and-channel system will be called the *reliability of the channel*. (Perhaps the definite article is not justified here when $(\widehat{p}_1, \ldots, \widehat{p}_n)$ is not unique; we pass over this difficulty for now.)

In the case of an n-ary symmetric channel, one satisfying the hypothesis of Theorem 4.5.5, we can take $(\widehat{p}_1, \ldots, \widehat{p}_n) = (1/n, \ldots, 1/n)$, and then the equality of the imposed source frequencies $f_1, \ldots, f_n$ implies that MLD and NCWD coincide, by Theorem 4.5.5. Clearly, for each $a_j \in A$, the unique word over A of length 1 closest to a_j is a_j itself. That is, $N_j = \{a_j\}$. Thus

$$R = \frac{1}{n} \sum_{j=1}^{n} P(a_j \text{ is received } | a_j \text{ is sent})$$

$$= \frac{1}{n} \sum_{j=1}^{n} q_{jj} = \frac{1}{n} nq = q,$$

the constant main diagonal entry in the matrix of transition probabilities.

Consequently, the definition of channel reliability given here agrees with the prior definition of the reliability of an n-ary symmetric channel, at least when that reliability is greater than $1/n$.

When the optimal input frequencies are difficult to obtain, and the channel is "close" to being n-ary symmetric (see the discussion in section 3.4), a rough estimate of the reliability of the channel may be obtained by taking the source frequencies to be equal (to $1/n$). For instance, consider the channel described in Example 4.5.2. With respect to the the encoding scheme $0 \to 0$, $1 \to 1$, we have $N_0 = \{0, *\}$ and $N_1 = \{1\}$ (since $P(0, *) = \frac{1}{2}(.04) > P(1, *) = \frac{1}{2}(.03)$), whence $R \approx \frac{1}{2}[.9 + .04] + \frac{1}{2}[.92] = .93$.

The reliability of a discrete memoryless channel, as defined here, appears to be a new index of channel quality. Its relation to channel capacity has not been worked out, and it is not yet clear what role, if any, reliability will play in the theory of communication.

Exercises 4.5

1. In each of the following, you are given a source alphabet S, a code (and input) alphabet A, an output alphabet B, source frequencies $f_1, \ldots, f_m$, an encoding scheme, and the matrix Q of transition probabilities. In each case, produce (i) an MLD table, (ii) the reliability R, and (iii) the maximum error probability of the code-and-channel system.

   (a) $S = \{a, b, c\}$, $A = \{0, 1\}$, $B = \{0, 1, *\}$, $f_a = .4$, $f_b = .35$, $f_c = .25$, $a \to 00, b \to 11, c \to 01$, and

   $$Q = \begin{bmatrix} q_{00} & q_{01} & q_{0*} \\ q_{10} & q_{11} & q_{1*} \end{bmatrix} = \begin{bmatrix} .8 & .15 & .05 \\ .1 & .86 & .04 \end{bmatrix}.$$

   (b) $S = \{a, b, c, d, e\}$, $A = B = \{0, 1, *\}$, $f_a = .25$, $f_b = .15$, $f_c = .05$, $f_d = .15$, $f_e = .4$, $a \to 00, b \to 01, c \to 0*, d \to 10, e \to 11$, and

   $$Q = \begin{bmatrix} q_{00} & q_{01} & q_{0*} \\ q_{10} & q_{11} & q_{1*} \\ q_{*0} & q_{*1} & q_{**} \end{bmatrix} = \begin{bmatrix} .95 & .03 & .02 \\ .04 & .92 & .04 \\ .06 & .04 & .9 \end{bmatrix}.$$

   (c) $S = \{a, b, c, d, e\}$, $A = B = \{0, 1\}$, the source frequencies are as in (b), the encoding scheme is $a \to 000, b \to 001, c \to 010, d \to 011$, $e \to 100$, and the channel is binary symmetric with reliability .9.

2. Suppose the available channel is binary symmetric with reliability .8. Suppose $S = \{a, b\}$, $f_a = .999$, $f_b = .001$, and the encoding scheme is $a \to 000, b \to 111$.

   (a) Verify that MLD will decode every word $w \in \{0, 1\}^3$ as 'a'.
   (b) Calculate the reliability of this code-and-channel system and the maximum error probability.
   (c) Same question as (b), but use NCWD.
   (d) How large must t be so that, if we consider the encoding scheme $a \to 0^t = 0 \cdots 0$ (t zeroes) and $b \to 1^t$, then MLD will decode 1^t as b?
   (e) Find the reliability and the maximum error probability of the code-and-channel system obtained by taking the scheme you found in part (d), when the decoding method is MLD and again when it is NCWD.

3. For each instance of δ and ρ as in 4.5.4, and each code-and-channel system with $A = B$ and a fixed-length encoding scheme $s_j \to w_j$, $j = 1, \ldots, m$

of length ℓ, we can define a nearest-code-word sort of decoding associated with δ and ρ, to be denoted NCWD(δ, ρ), as NCWD was defined, but with the metric d_{GH} arising from δ and ρ playing the role that d_H plays in the definition of NCWD. That is, having received $w \in A^\ell$, we decode w as that s_j for which $d_{GH}(w, w_j)$ is the least, provided there is a unique such j. If there is no unique such j, we report "do not decode."

One reason for considering the metrics d_{GH} is the possibility that, given a code-and-channel system, there may be choice of δ and ρ such that NCWD(δ, ρ) and MLD coincide for that system. The requirements on the system for the existence of such a pair (δ, ρ) await disclosure, but we can see readily that there are cases when there is no such pair.

(a) Show that, with (δ, ρ) and NCWD(δ, ρ) as above, if the code words w_j, $j = 1, \ldots, m$, are distinct, then NCWD(δ, ρ) will decode w_j as s_j.

(b) Conclude that there is no pair (δ, ρ) for which NCWD(δ, ρ) and MLD coincide on the code-and-channel system of Exercise 2(a), above.

4. Estimate the reliability of each of the channels mentioned in exercise 1, above, by imposing equal source frequencies on the input characters. (Of course, in (c) the result will be exact.)

5. The average error probability can be thought of as the *average cost* of an attempted transmission of a source letter (from the choosing of the source letter to the result after decoding), where the cost of an error is one unit, and the cost of no error is zero.

It follows that we can refine the average error probability as a measure of system failure by distinguishing more finely among the outcomes of the "choose s_j – transmit w_j – decode" experiment, assigning different appropriate costs to these outcomes, and then calculating average cost. This type of refinement was alluded to in 4.5.8.

For example, in the circumstances of Exercise 2 above, it may be that source message b is extremely grave and that it would be very costly to mistake a for b. Let us suppose that, whatever the decoding method, decoding b when a was intended costs 1000 units, decoding a when b was intended costs 100 units, getting "do not decode" when a was intended costs one unit, and "do not decode" when b was intended costs 50 units. A correct transmission costs nothing. Find the average cost of an attempted transmission when

(a) the decoding method is MLD and the scheme is as in Exercise 2(a), above;

(b) the decoding method is NCWD and the scheme is as in 2(a);

(c) the decoding method is MLD and the scheme is the one you found in 2(d);

(d) the decoding method is NCWD and the scheme is that of 2(d).

6. In each part of Exercise 1, compute the reliability, the average error probability, and the maximum error probability of the system in which decoding
is by NCWD and in which

   (i) "do not decode" counts as an error;
   (ii) "do not decode" does not count as an error;
   (iii) "do not decode" counts as one-half of an error.

   In each case, assume that the relative source frequencies are known, and
   are as given.

7. The channel is binary and symmetric, with reliability p, $1/2 < p < 1$.
There are two source messages, a and b, with equal frequency. For a positive integer t, consider the encoding scheme $a \rightarrow 0^t$, $b \rightarrow 1^t$. Let $R(t)$
and $E(t) = 1 - R(t)$ denote the reliability and average error probability,
respectively, of this code-and-channel system, using MLD (= NCWD, in
this case); note that $E(t) = \widehat{E}(t)$ by the symmetry of the situation. Let
$R_0(t)$ and $E_0(t)$ denote the corresponding probabilities if "do not decode"
is considered a success, not a failure.

   (a) Express $R(t)$ and $R_0(t)$ explicitly as functions of p and t.
   (b) Show that $R(t+1) = R_0(t) - \binom{t}{\lfloor t/2 \rfloor} p^{\lceil t/2 \rceil} (1-p)^{\lfloor t/2 \rfloor + 1}$ for each positive integer t.
   (c) Show that $R(t) \leq R(t+2)$ and that $R_0(t) \leq R_0(t+2)$ for each positive
   integer t.
   (d) Show that $R(t) \rightarrow 1$ as $t \rightarrow \infty$.

   [Hints: consider the cases where t is odd or even separately, for (a), (b),
   and (c). For (d), use the fact that $p > 1/2$, and the Law of Large Numbers;
   see section 1.9.]

8. Suppose that $|S| = m = 35$, and we have a shortest possible fixed-length
encoding scheme for a uniquely decodable code. How long will the MLD
table be when

   (a) $|A| = |B| = 2$;
   (b) $|A| = 2, |B| = 3$;
   (c) $|A| = |B| = 3$?

9. In Exercises 1 (b) and (c), above, note that the encoding schemes are as
short as possible, but not thoughtfully conceived. For instance, in 1(b),
the code word for e, the most commonly encountered source letter, is a
Hamming distance 2 from the word for c, the least common code word, but
a distance 1 from each of the words for b and d. Surely it would increase
the reliability R if we interchange the code words representing c and d, or
c and b.

   Verify that this is so. Also, find a fixed-length scheme, of length 3, to
   replace the scheme in 1(c), which increases the reliability.

4.6 Shannon's Noisy Channel Theorem

The theorem referred to describes a beautiful relationship between the competing goals of (a) transmitting information as rapidly as possible, and (b) making the average error probability as small as possible, given a source and a channel.

For us, a "source" consists of a source alphabet and a probability distribution over the source alphabet, the relative source frequencies. In the "model" that we have been using, the source letters are emitted, randomly and independently, with the given relative frequencies (which may be arrived at by observing the source for a long time). Shannon's Noisy Channel theorem applies to a more general sort of source, one which emits source letters, but not necessarily randomly and independently. We will have a more thorough discussion of these sources at the end of Chapter 7.

It is curious that the Noisy Channel Theorem is widely regarded as the centerpiece of information theory, yet both the statement and the proof of the theorem are largely useless for practical purposes. Nor can it be said that the theorem has worked indirectly as an inspiration in the actual devising of efficient "error-correcting" codes, nor in the theory of such things, which could be, and usually is, laid out without a single occurrence of the word *entropy*. Yes, the greats of coding theory were aware of Shannon's theory and the Noisy Channel Theorem, but so are professors of accounting or finance aware of the Unique Factorization Theorem for the positive integers.

The rightful acclaim that the Noisy Channel Theorem enjoys arises, we think, from its beauty. Shannon's definitions of information and entropy were audacious and not immediately convincing. Of course, a definition is not usually required to be convincing, but when you attempt a definition of a word that carries a prior connotation, as do *information* and *entropy*, the definition should have implications and overtones that agree with the prior connotation. Shannon was very aware of this informal requirement; he went so far as to show ([63] and [65]) that if the entropy of a system is to be the average information contained in the system (of events), and if entropy is to satisfy certain plausible axioms, then information *must* be defined as it is. (As we saw in Section 2.1, a simpler and more convincing demonstration of the inevitability of Shannon's quantification of information was later discovered by Aczél and Daroczy [1].)

Still, the newcomer to the theory might be forgiven a bit of queasiness as conditional entropy and mutual information are added to the list of fundamentals, with channel capacity coming along as a corollary. As touched on in Section 3.4, Shannon's interpretation of channel capacity as measuring the maximum possible rate of information flow through a channel, which is what channel *capacity* sounds like it ought to measure, was supported by examples and extremal considerations—but is that enough?

The beauty of the Noisy Channel Theorem lies at least partly in the validation it provides of the interpretation of maximum mutual information between inputs and outputs as measuring maximum possible information flow. In brief, the main statement of the theorem says that if the rate of information flow from the source is less than the channel capacity, then you have enough room to afford the luxury of error correction—you can make the maximum error probability as small as you please, if you are willing to take the trouble, while accommodating the information flow from the source with no accumulated delays or backlog. Does that not make the channel capacity sound like a maximum possible rate of information flow? Another similarly telling assertion in Shannon's original formulation of the theorem is that if the rate of information flow from the source is greater than the channel capacity, then the average error probability cannot be reduced below a certain positive amount, a function of the source frequencies and time rate and of the channel's transition probabilities.

Think of a flash flood bearing down on a culvert. The culvert pipe can convey a certain maximum volume of water per unit time. If the rate at which the flood is arriving at the pipe entrance is below that maximum rate, then, in ideal principle, the water can be directed through the pipe without a drop sloshing over the roadway above the culvert, and without a pond of unconveyed water building up on the flood side of the culvert. In practice, the directing of the flood waters into the pipe, a civil engineering problem, will not be perfect – some water will be lost by sloshing. But as long as the flood flow is below the theoretical maximum that the pipe can handle (the pipe's *capacity*), steps can be taken to reduce the sloshing loss (error) below any required positive threshold, while maintaining flow and avoiding backup. If the flood rate exceeds the pipe capacity, then no engineering genius will be able to avoid some combination of water loss and backup; the flood volume per unit time in excess of the pipe capacity has to wind up somewhere other than the pipe.

These common sense observations regarding floods and culverts serve as a good analogy to the conclusions of the NCT, with the source stream playing the role of the flood waters, the channel playing the role of the culvert pipe, and the coding/decoding method playing the role of the hypothetical engineering measures taken to direct the flood water into the pipe, so as to keep slosh tolerable while maintaining flow. The beauty and inevitability of the NCT reside in the closeness of this analogy, in our opinion.

The geometry of the ancient Greeks could have been used by ancient Greek craftsmen to make measurements and designs, but the historical evidence apparently indicates that it was not so used. Geometry was propagated in the intellectual world over two millenia purely because of its beauty; utility was not a factor. The Noisy Channel Theorem inserts a promise of inevitability, of immortality, in information theory. It is analogous to the theorem in geometry that the sum of the interior angles of a planar triangle is a straight angle. There is more to geometry than that, and we hope that there will be more to information theory than the Noisy Channel Theorem; nonetheless, the theorem alone, and its immediate consequences, are of considerable weight.

Preliminaries Suppose a channel, (A, B, Q), is given, and also a source alphabet $S = \{s_1, \ldots, s_m\}$ with relative source frequencies $f_1, \ldots, f_m$, all positive. As usual, $H(S) = \sum_j f_j \log 1/f_j$.

Suppose that the source emits r characters per unit time. Since the average information content per character is $H(S)$, it follows that the source is emitting information at the rate $rH(S)$ information units per unit time, on average. Let C denote the channel capacity. Suppose that the channel transmitter can send off ρ input letters per unit time. (So, by Shannon's interpretation of $I(A, B)$, the channel can convey a maximum of ρC units of information per unit time, on average.)

4.6.1 The Noisy Channel Theorem *If $rH(S) < \rho C$, then for any $\varepsilon > 0$ it is possible to encode the source stream and arrange a decoding method so that on average each source character is represented by no more than ρ/r input characters, and so that the maximum, over the source characters, of the probability of an error at an occurrence of the character in the source stream, is less than ε.*

If $rH(S) > \rho C$, then there is a positive number ε_0 such that no matter how the source stream is encoded and decoded, if, on average, each source character is represented by no more than ρ/r input characters, the average error probability of the code-and-channel system will be greater than or equal to ε_0.

We will give a synopsis of the proof shortly, but first a few comments are in order. Since input letters are transmitted at the rate of ρ per unit time and source letters appear at the rate of r per unit time, clearly the requirement that the source stream be encoded so that the average number of input letters per source letter is not greater than ρ/r is meant to insinuate that the flow of information from the source, through the channel to the receiver, and then to the decoder, proceeds smoothly, with no backlog of unprocessed source letters. It can be, and often is, objected that this view of things leaves out of account the time spent encoding and decoding. However, this objection is not entirely fair.

In the proof of the NCT, the encoding is to be by a fixed length scheme applied, not to S, but to S^N, for some large integer N. The length of the fixed-length scheme is to be approximately $\rho N/r$. Another objection to this procedure is that the encoder has to wait N/r time units for a sequence of N source letters to accumulate. But the two objections, about time spent encoding and decoding on the one hand, and time spent waiting for N source letters to accumulate on the other, cancel each other out, if we can be quick enough in encoding and decoding—and by "quick enough" we do not mean instantaneous, we mean taking no longer than N/r time units to encode a source word of length N and to decode an output word, at the receiver, of length $\rho N/r$. If we can encode and decode that quickly, then we can spend the time waiting for the next source word of length N to accumulate by encoding the most recently emerged source word, transmitting the encoded version of the one before that, and decoding the one before that. It is true that there will be a hiatus of up to N/r time units between reports from the decoder, but the reward for the wait will be, not a lone source letter, but a great hulking source word of length N. So the source stream lurches rather than flows, but the average rate at which source letters emerge from the decoder will be the same as the rate at which they entered the encoder: r per unit time.

If you followed this discussion, you might then look back at the statement of the NCT and wonder what that business is about the maximum error probability over the source characters. If we are going to encode S^N, shouldn't we think about the maximum error probability over S^N? Well, no; and, in fact, that maximum error probability will not be made small. What is meant by the probability of an error at an occurrence of a source letter s is the probability that, when s occurs in the source stream, the place in the stream emerging from the decoder that was occupied by s originally, is occupied by something other than s. If we are encoding source words of length N, such an error can occur only if s occurred in some source block of length N that got misconstrued by the code-channel-decode system; that is, every such error is part of a larger catastrophe.

In asserting that the maximum error probability over the single source letters can be made as small as desired, with a code that keeps up the rate of information flow when $r H(S) < \rho C$, we are departing from the usual statement of this part of the NCT. The usual statement these days (see [4] and [81]) is mathematically stronger, or no weaker, than our statement, but suffers from opacity. We think it is wise to sacrifice strength for friendliness in a theorem that is not really *used* for anything. We will mention the usual conclusion in the proof synopsis, below.

In the last assertion of the NCT, we also depart from Shannon's version, again for esthetic reasons. We will indicate how Shannon put it below.

Synopsis of the proof of the NCT As mentioned above, the idea is to encode S^N, with N a large integer, by a fixed-length scheme of length $\ell = \lfloor \rho N / r \rfloor$.

By the Law of Large Numbers, if N is large, the source words of length N in which the proportions of the source letters within the word differ markedly from the f_j have very small probability, collectively. [For instance, if $S = \{a, b, c\}$, $f_c = .2$, and $n = 10,000$, the probability that a source word of length 10,000 will contain fewer than 1,000 c's, or more than 3,000 c's, is minuscule.] We take N large and divide S^N into two sets of words, L (for *likely*), in the words of which the source letters occur in proportions quite close to the f_j, and U (for *unlikely*), $U = S^N \setminus L$. How large N is, and how close those proportions are to the f_j, depend on $\varepsilon, r, H(S), \rho$, and C. In any case, $P(U) = \sum_{w \in U} P(w)$ is very small.

Now the idea will be to assign to each word in L a code word $w \in A^\ell$, $\ell = \lfloor \rho N / r \rfloor$. As for the words in U—ignore them! If you must encode them, assign them any which way to the code words for L. This means that when a word in U actually occurs, after encoding and transmission the word received is almost certain to be misdecoded—but the likelihood of a word in U emerging from the source is, by arrangement, so small that this certainty of error in these cases will have very little effect on the probabilities of errors at occurrences of the source characters.

Now, how do we find an encoding scheme for $L \to A^\ell$? This is where we will be *very* synoptic; we don't actually "find" an effective encoding scheme. There is a *probabilistic proof of the existence* of $w_1, \ldots, w_{|L|} \in A^\ell$ such that if the w_j are transmitted and decoded by MLD, assuming they have approximately equal likelihood $L^{-1} \approx \left(\prod_{j=1}^{m} f_j^{f_j} \right)^N$, then $\max_{1 \leq j \leq L} P$(the received word is misdecoded $| \ w_j$ was transmitted) $< \varepsilon/2$, say. (This is the stronger conclusion in this part of the NCT alluded to in earlier discussion. Provided $P(U) < \varepsilon/2$, it certainly implies that the maximum probability of an error at a single source character is less than ε.)

It is in this probabilistic existence proof that the inequality $r H(S) < \rho C$ enters strongly—of course, it has already subtly influenced the foresightful choice of N to be sufficiently large to make everything work. We omit all details of this proof—we hope to stimulate the reader's curiosity. Be warned, however, in looking at the proofs in [4], [37], and [81], that the tendency is to first convert the source S to a *binary* source, which murks everything up.

As for the other conclusion of the NCT, when $r H(S) > \rho C$, it can be shown by general monkeying around as in Shannon's original proof in [63] that the conditional entropy or *equivocation*, $H(S \mid B)$, the measure of uncertainty about the source, given the output, no matter what the coding method, can be no less that $(r/p)H(S) - C$. Shannon would have left it at that. To get the conclusion we desire, we need a connection between the equivocation and the average error probability. There is one. It is called Fano's Inequality. See, e.g., [81], p. 43.

This concludes our synopsis. You can see that the horrible thing about the proof is that, when $r H < C$, it does not tell you how to encode S^N so that maximum single-letter error probability is made less than ε. Sharper formulations do give estimates of N, but the estimates are discouragingly large. Small wonder that NCT is solemnly saluted by coding theorists far and wide, and then put in a drawer; in the next section we will look at the fields where practical coders really play.

Exercises 4.6

1. Suppose that $S = \{a, b, c\}$, $f_a = .5$, $f_b = .3$, $f_c = .2$, and the channel is binary and symmetric with reliability .95. Suppose that the channel can transmit 100 bits (binary digits) per second. According to the NCT, what is the upper limit on the number of source characters per second that this channel can theoretically handle without backlog and with maximum single letter error probability as small as desired (but not zero)?

2. Suppose that the source alphabet S, the relative source frequencies, and the channel are as in problem 1. Suppose that S^2 is encoded as follows:

$$
\begin{array}{lll}
aa \rightarrow 0000 & ba \rightarrow 0101 & ca \rightarrow 1001 \\
ab \rightarrow 1111 & bb \rightarrow 1100 & cb \rightarrow 0110 \\
ac \rightarrow 1010 & bc \rightarrow 0011 & cc \rightarrow 1110
\end{array}
$$

Find the maximum single letter error probability and the average single letter error probability. [This will involve more than making an MLD table, but you may as well make one as an aid. Assume that the source letters are emitted randomly and independently, so that the relative frequency of the two-letter sequence ac, for instance, would be $(.5)(.2) = .1$.]

3. Again, S and the relative source frequencies are as in problem 1, and it is assumed that the source letters are emitted randomly and independently. Find a binary encoding scheme for S^2 using Huffman's algorithm, and compute the average number of code letters per source letter if this scheme is used to encode the source stream.

4.7 Error correction with binary symmetric channels and equal source frequencies

The case of equal source frequencies and a binary symmetric channel is a very important special case because it is the case we think we are in, in a great number of real, practical situations in the world today. Or, perhaps we just hope and assume that we are in this situation; see 4.5.7 and Section 3.1.

By Corollary 4.5.6, when the channel is binary and symmetric with reliability $p > 1/2$ and the source frequencies are equal, MLD and NCWD coincide. Thus, for each $w \in \{0, 1\}^\ell$ received, we decode by examining the words $w_1, \ldots, w_m \in \{0, 1\}^\ell$ in the encoding scheme and picking the one, if any, closer to w in the Hamming distance sense than are any of the other w_i. In this section we will see a way that this procedure might be simplified, at the cost of some reliability.

As remarked in 4.5.7, the situation described in the title of this section is the setting of most of coding theory, which is mainly about binary block codes. We shall not go far into that theory, but during our excursion we shall observe its customs. For one thing, we shall refer to the set $C = \{w_1, \ldots, w_m\} \subseteq \{0, 1\}^\ell$ of code words appearing in the encoding scheme as *the code*, and all mention of the source alphabet and of the encoding scheme will be suppressed. This is not unreasonable in the circumstances, since our decoding method is NCWD; the only thing we need to know about the source alphabet is its size, $m \leq 2^\ell$.

Definitions The operation $+$ is defined on $\{0, 1\}$ by $0+0 = 0, 0+1 = 1+0 = 1$, and $1+1 = 0$. The operation $+$ is then defined on $\{0, 1\}^\ell$ coordinatewise, given the definition above. [For example, with $\ell = 5$, $01101 + 11110 = 10011$.]

The *Hamming weight* of a word $w \in \{0, 1\}^\ell$ is $\text{wt}(w) = $ number of ones appearing in w. [For example, $\text{wt}(10110) = 3$.]

If $w_1, \ldots, w_m \in \{0, 1\}^{\ell}$ are distinct words, the *distance* of the code $C = \{w_1, \ldots, w_m\}$ is

$$d(C) = \min_{1 \le i < j \le m} d_H(w_i, w_j) = \min_{\substack{w, v \in C \\ w \ne v}} d_H(w, v)$$

4.7.1 For $u, v \in \{0, 1\}^{\ell}$, note that $u + v$ has ones precisely where u and v differ. Thus $d_H(u, v) = \text{wt}(u + v)$.

4.7.2 Verify that $\text{wt}(u + v) \le \text{wt}(u) + \text{wt}(v)$, for all $u, v \in \{0, 1\}^{\ell}$.

Definition We will say that a code $C \subseteq \{0, 1\}^{\ell}$ *corrects the error pattern* $u \in \{0, 1\}^{\ell}$, if and only if, for each $w \in C$, NCWD will decode $w + u$ as w (or, as whatever source letter w represents).

The u appearing in the last definition above could be any binary word of length ℓ. When we call u an *error pattern* we are thinking that, during the transmission of a binary word of length ℓ through the channel, errors occurred at precisely those places in the word marked by 1's in u. Thus, by the definition of $+$, if w was transmitted and the *error pattern* u *occurred*, the word received at the receiving end of the channel would be $w + u$. Thus the definition above says that C corrects u if and only if, whenever the error pattern u occurs and the code C is in use, NCWD (= MLD) will correctly decode the received word, whichever $w \in C$ was sent.

4.7.3 Example Let $C = \{00000, 11100, 01111\}$. Verify that C corrects: 00000, 10000, 01000, 00100, 00010, 00001, and no other error patterns. Note that if 11100 is transmitted and the error pattern 01010 occurs, then 10110 will be received, which is closer to 11100 than to either of the other two code words; but if 00000 or 01111 is transmitted and that error pattern occurs, NCWD will decide not to decode. Thus that error pattern is not corrected by the code.

4.7.4 Let $C = \{0^6, 0^3 1^3, 1^3 0^3, 1^6\}$. Verify that the set of error patterns corrected by C is $\{0^6\} \cup \{\text{all 6 binary words of length 6, of Hamming weight 1}\} \cup \{100100, 100010, 100001, 010100, 010010, 010001, 001100, 001010, 001001\}$.

4.7.5 Theorem *Suppose* $C \subseteq \{0, 1\}^{\ell}$ *and* $|C| \ge 2$. *Then* C *corrects all error patterns of length* ℓ, *of Hamming weight* $\le (d(C) - 1)/2$.

Proof: Suppose $u \in \{0, 1\}^{\ell}$ and $\text{wt}(u) \le (d(C) - 1)/2$. Suppose that $w, v \in C$ and $w \ne v$. Then

$$d(C) \le d_H(w, v) \le d_H(w, w + u) + d_H(w + u, v)$$
$$= \text{wt}(w + (w + u)) + d_H(w + u, v)$$
$$= \text{wt}(u) + d_H(w + u, v)$$

which implies

$$d_H(w+u,v) \geq d(C) - \mathrm{wt}(u) \geq d(C) - \frac{d(C)-1}{2}$$

$$= \frac{d(C)+1}{2} > \frac{d(C)-1}{2} \geq \mathrm{wt}(u) = d_H(w, w+u).$$

Thus, for each $w \in C$, w is the unique word in C closest to $w+u$, so NCWD will decode $w+u$ as w. □

4.7.6 Corollary *Let* $t = \lfloor \frac{d(C)-1}{2} \rfloor$. *Then, with* $\widehat{E}$ *denoting the maximum error probability with C in use, with a binary symmetric channel with reliability $p >$ 1/2,*

$$\widehat{E} \leq 1 - \sum_{j=0}^{t} \binom{\ell}{j} p^{\ell-j} (1-p)^j.$$

Proof: Let $U = \{u \in \{0,1\}^{\ell}; \mathrm{wt}(u) \leq t\}$, and, for each $v \in C$, $N_v = \{w \in \{0,1\}^{\ell};$ NCWD decodes w as $v\}$. By the theorem,

$$v + U = \{v+u; u \in U\} \subseteq N_v, \text{ for each } v \in C.$$

Therefore, for each $v \in C$, with w denoting "the received word,"

$$P(w \in N_v \mid v \text{ is sent}) \geq P(w \in v+U \mid v \text{ is sent})$$

$$= P \text{ (the error pattern } u \text{ lies in } U \mid v \text{ is sent)}$$

$$= P(u \in U)$$

$$= P\binom{t \text{ or fewer errors occurred, in } \ell \text{ trials,}}{\text{with probability } 1-p \text{ of error on each trial}}$$

$$= \sum_{j=0}^{t} \binom{\ell}{j} p^{\ell-j} (1-p)^j, \text{ by Theorem 1.5.7.}$$

Since $v \in C$ is arbitrary, the desired conclusion follows. □

Definition Suppose $C \subseteq \{0,1\}^{\ell}$ and $|C| \geq 2$. Let $d = d(C)$. In *simplified nearest code word decoding* (SNCWD), a received word $w \in \{0,1\}^{\ell}$ is decoded as $v \in C$ if and only if $d_H(v, w) \leq \frac{d-1}{2}$. If there is no such $v \in C$, do not decode w.

By the proof of Theorem 4.7.5, for each $w \in \{0,1\}^{\ell}$ there is at most one $v \in C$ such that $d_H(v, w) = \mathrm{wt}(v+w) \leq (d(C)-1)/2$. Observe that if $u = v+w$ then $v = w+u$, because of the peculiar definition of $+$. Consequently, the carrying out of SNCWD can proceed as follows: given w, start calculating the words $v+w, v \in C$, until you run across one of weight $\leq (d(C)-1)/2$. If you have saved the v, report that as the intended code word. Alternatively, v can be recovered from $v+w$ and w by addition. If there is no $v \in C$ for which $\mathrm{wt}(v+w) \leq (d(C)-1)/2$, report "do not decode."

Is this procedure any easier than plain old NCWD? From the naive point of view, no. In both procedures you have to calculate $d_H(v, w), v \in C$ until either a v is found for which $d_H(v, w) \leq (d(C) - 1)/2$, or until all $v \in C$ have been tried, at which point, with NCWD, the numbers $d_H(v, w)$ must be compared. You save a little trouble with SNCWD by omitting this last comparison; but surely, you might think, this saving would not compensate us sufficiently for the loss of reliability incurred by forsaking NCWD for SNCWD.

But the fact is that SNCWD is very commonly used. The details are beyond the scope of this course. Suffice it to say that knowing exactly which error patterns will be corrected sometimes leads, in the presence of certain algebraic properties of C, to very efficient decoding procedures.

If we define "C corrects the error pattern u" for SNCWD as it was defined for NCWD, it is easy to see that, with SNCWD, the error patterns corrected by C are precisely the words of weight $\leq (d(C) - 1)/2$; furthermore, the error patterns corrected correctly are the same for different code words—see Example 4.7.3 to see that this is not necessarily the case with NCWD. Note also Example 4.7.4 and compare the error patterns corrected there by NCWD with the error patterns corrected by SNCWD.

Reliability For $C \subseteq \{0, 1\}^\ell$, let $R(C, p)$ denote the reliability of the code-and-channel system obtained by using C, a binary symmetric channel with reliability p, and NCWD. (As elsewhere in this section, the source frequencies are assumed to be equal.) Let $R_S(C, P)$ denote the reliability when SNCWD is used. Corollary 4.7.6 implies that

$$R(C, p) \geq \sum_{j=0}^{\lfloor \frac{d(C)-1}{2} \rfloor} \binom{\ell}{j} p^{\ell-j}(1-p)^j.$$

4.7.7 Proposition $R_S(C, P) = \sum_{j=0}^{\lfloor \frac{d(C)-1}{2} \rfloor} \binom{\ell}{j} p^{\ell-j}(1-p)^j.$

The proof, after that of 4.7.5 and the remarks above, is straightforward.

$R_0(C, p)$ and $(R_S)_0(C, p)$ will denote, as in Exercise 4.5.7, the relaxed reliabilities obtained by not considering a "do not decode" message to be an error.

Exercises 4.7

1. Express explicitly, as formulas in p, the reliabilities $R(C, p)$, $R_S(C, p)$, $R_0(C, p)$, and $(R_S)_0(C, p)$, when C is the code of Example 4.7.3.

2. Same question for Example 4.7.4.

3. There are two famous binary block codes, of lengths 23 and 24, called the *Golay code* and the *extended Golay code*, respectively. Let us denote the Golay code by C_{23}, and the extended Golay code by C_{24}. Their distances are $d(C_{23}) = 7$ and $d(C_{24}) = 8$.

Both codes have the remarkable property that NCWD and SNCWD coincide when these codes are in use. C_{23} has the further property that there is never a "do not decode" result. This is not the case with C_{24}, however, and, in fact, every error pattern of weight 4 occurring when C_{24} is in use will result in a "do not decode" message.

(a) Express $R(C_{23}, p)$ and $R(C_{24}, p)$ explicitly as functions of p, assuming $p > 1/2$.
(b) Show that $R(C_{23}, p) > R(C_{24}, p)$ for $1/2 < p < 1$, but that $R_0(C_{23}, p) = R(C_{23}, p) < R_0(C_{24}, p)$ for $1/2 < p < 1$.
(c) $|C_{23}| = |C_{24}|$, both codes come equipped with very fast NCWD algorithms, and the slightly greater length of C_{24} is a negligible drawback; so, what do the results of part (b), above, suggest to you about which of the two codes you would choose to use? In some circumstances you would take C_{23} over C_{24}, and in other circumstances it would be the other way around. What sort of consideration decides the choice? Be brief.

4. $C \subseteq \{0, 1\}^{\ell}$ and the channel is binary and symmetric with reliability p. Show that $t = \lfloor \frac{d(C)-1}{2} \rfloor$ is the largest integer among the integers i with the property that C will correct (using NCWD) all error patterns of weight $\leq i$.

4.8 The information rate of a code

For a binary code $C \subset \{0, 1\}^{\ell}$, the *information rate* of C is generally defined to be $(\log_2 |C|)/\ell$. To see what this really means, and how to generalize it to variable-length encoding schemes over possibly non-binary code alphabets, suppose that C is used to encode a source alphabet S with $m = |C|$ letters, of equal relative frequencies. Then $H(S) = \log_2 m = \log_2 |C|$; this is the number of bits of information carried by each code word. Since each code word is ℓ bits long, the number of bits per bit, so to speak, carried by the code words is $(\log_2 |C|)/\ell$. To put it another way, $(\log_2 |C|)/\ell$ is the rate at which the code words are carrying information, in bits per input (code) letter.

The account preceding rests on the assumption that the relative source frequencies are equal. In the more general situation when we have a source alphabet S, $|S| = m$, with possibly unequal relative frequencies, and a uniquely decodable scheme for $S \to A$, where A is a code alphabet with $|A| = n \geq 2$, the discussion above is adaptable to give the result that the information rate of the code, interpreted as the average amount of (source) information carried by the code words, per code letter, is $H(S)/\bar{\ell}$, where $\bar{\ell}$ is the average code word length, computed from the scheme and the relative source frequencies. The units of information are determined by the choice of the base of the log appearing in the computation of $H(S)$. To accord with the binary case, we may as well adopt the convention that that base is to be $n = |A|$.

If A is the input alphabet of a channel, at what rate, then, is information originating from the source appearing at the receiver of the channel, per output letter, given a uniquely decodable scheme for $S \to A$? Since there is one output letter per input letter, and $I(A, B)$ gives the rate of flow of information through the channel, i.e., the average number of units of information arriving at the receiver for each unit of information transmitted, it follows that the average amount of information from the source to the receiver, per output letter, is $(H(S)/\bar{\ell})(I(A, B)|_{p_1,\dots,p_n})$, with $\log = \log_n$ and $p_1, \dots, p_n$ computed from the encoding scheme and the relative source frequencies, as in section 4.4.

Both $H(S)/\bar{\ell}$ and this quantity multiplied by $I(A, B)|_{p_1,\dots,p_n}$ are indicators of an encoding scheme's efficacy, but let us be under no illusions as to the sensitivity of these indices. Notice that $H(S)/\bar{\ell}$ is increased only by decreasing $\bar{\ell}$, within the requirement of unique decodability; clearly different uniquely decodable encoding schemes with the same $\bar{\ell}$ can have very different qualities. In particular, in the case of fixed-length encoding schemes, this index leaves error-correction facility and encoding/decoding speed and efficiency completely out of account. The index $(H(S)/\bar{\ell})(I(A, B)|_{p_1,\dots,p_n})$ is somewhat more interesting—it goes up as $\bar{\ell}$ decreases and/or as the $p_1, \dots, p_n$ resulting from the scheme better approximate the optimal input frequencies of the channel. But it still leaves out of account code qualities of practical interest. (See Exercise 4.8.3.)

This does not mean that these indices are useless! Consider that knowing the area of a planar figure tells you nothing about the shape or other geometric and topological properties of the figure. Does that mean that we should give up on the parameter we call "area?" Just so, the two parameters we are discussing here will have their uses in the discussion and comparison of code-and-channel systems. We need to be aware of the limitations of these discussions and comparisons, but if we are aware, then let's proceed! From here on, we will refer to $H(S)/\bar{\ell}$ as the (pretransmission) information rate of the code involved, and $(H(S)/\bar{\ell})(I(A, B)|_{p_1,\dots,p_n})$ as the (post-transmission) information rate of the code-and-channel system.

Given $S, f_1, \dots, f_m, A, B, Q$, and a fixed-length uniquely decodable scheme for $S \to A$, $s_j \to w_j \in A^\ell$, $j = 1, \dots, m$, as in Section 4.5, there is another parameter associable to the code-and-channel system that might be preferable to $(H(S)/\ell)(I(A, B))$ as a measure of information flow from the source to the channel receiver, per output letter: $I(S, B^\ell)/\ell$. Here S and B^ℓ stand for the obvious systems of events associable to the multistage probabilistic experiment described in section 4.5. The mutual information $I(S, B^\ell)$ is divided by ℓ, above, to make the result comparable to $(H(S)/\ell)(I(A, B)|_{p_1,\dots,p_m})$ as a measure of information conveyed per output letter.

Shannon's interpretation of $I(A, B)$ as measuring the average amount of information conveyed by the channel (given certain relative input frequencies) per input letter (see Section 3.4) transfers to an interpretation of $I(S, B^\ell)$, in this more complicated situation, as the average amount of information conveyed by the code-and-channel system, per source letter, given $f_1, \ldots, f_m$ and a fixed-length encoding scheme for $S \to A$. Thus $I(S, B^\ell)/\ell$ would seem to be a measure of rate of information flow from the source to the channel receiver, per output letter, that is more reflective of error correction concerns and a generally more sensitive index of the efficacy of the code-and-channel system than is $(H(S)/\ell)(I(A, B))$.

However, $I(S, B^\ell)/\ell$ as an index of goodness suffers from a grave defect: it is frightfully difficult to calculate, even in the simplified circumstances of binary block codes with equal source frequencies and a binary symmetric channel. Suppose we are in these circumstances and, in addition, the relative input frequencies generated by the use of the code are $p_0 = p_1 = 1/2$. (This is a common circumstance in practice. For instance, 0 and 1 occur equally often when the Golay codes, mentioned in Exercise 4.7.3, are used with equal source frequencies. The same holds for any linear block code – see [30] for definitions – containing the word with all ones, and almost all of the commonly used binary block codes are of this sort.) Then $(H(S)/\ell)I(A, B) = \frac{\log_2 |C|}{\ell}(p \log_2 2p + (1 - p) \log_2 2(1 - p))$, where $C \subseteq \{0, 1\}^\ell$ is the code ($|C| = m = |S|$) and p is the reliability of the channel. That is, the post-transmission information rate is just the conventional information rate times the channel capacity. Meanwhile, $I(S, B^\ell)$ is a daunting sum of $m \cdot 2^\ell$ terms – see Exercise 4.8.2. For particular binary block codes this expression can be greatly simplified – but not enough to put it in the category of $(\log_2 |C|)/\ell$, $d(C)$, or ℓ itself as easily consulted indices of the quality of a binary block code, in standard circumstances. One could argue that the difficulty of calculating $I(S, B^\ell)$ is the price you pay for the subtle power of this index. But an indicator that is harder to calculate than the items of interest that it might be an indicator of, like reliability and error probability, is not a useful indicator.

Still, $I(S, B^\ell)$ is an important and interesting number associated with a fixed-length code-and-channel system, and an academic study of its behavior and its relation to other indicators may bring some rewards. Here is a question for anyone who might be interested: by Corollary 2.4.6, $I(S, B^\ell) \leq H(S)$; is it necessarily the case that $I(S, B^\ell) \leq H(S)I(A, B)$? $I(A, B)$ here is, of course, calculated with the relative input frequencies produced by the encoding scheme and the relative source frequencies.

Exercises 4.8

1. Calculate $H(S)/\ell$ and $(H(S)/\ell)(I(A, B))$ for each of the fixed-length code-and-channel systems in Exercise 4.5.1.

2. Given S, $f_1, \ldots, f_m$, A, B, Q, and a fixed-length encoding scheme $s_j \to w_j = a_{i_{(1,j)}}, \ldots, a_{i_{(\ell,j)}} \in A^\ell$; verify that

$$I(S, B^\ell) = \sum_{j=1}^{m} f_j \sum_{1 \leq t_1, \ldots, t_\ell \leq k} \prod_{z=1}^{\ell} q_{i(z,j),t_z} \log \frac{\prod_{z=1}^{\ell} q_{i(z,j),t_z}}{\sum_{u=1}^{m} f_u \prod_{u=1}^{\ell} q_{i(z,u),t_z}}.$$

Write this out as a function of p, using $\log = \log_2$, in case $m = 4$, $f_j = 1/4$, $j = 1, \ldots, m$, $\{w_1, w_2, w_3, w_4\} = \{0000, 0011, 1100, 1111\}$, and the channel is binary symmetric with reliability p. Compare $I(S, B^\ell)$ to $H(S)$ and to $H(S)I(A, B)$ in this case.

3. Describe a binary block code C of length 23 with the same information rate and post-transmission information rate as the Golay code, C_{23}, mentioned in Exercise 4.7.3, with $d(C) = 1$. [You need to know that $|C_{23}| = 2^{12}$ and that 0 and 1 occur equally often, overall, in the code words of C_{23}. Assume equal source frequencies.]

Chapter 5

Lossless Data Compression by Replacement Schemes

Most (but not all) modern data compression problems are of the following form: you have a long binary word (or "file") W which you wish to transform into a shorter binary word U in such a way that W is recoverable from U, or, in ways to be defined case by case, *almost* or *substantially* recoverable from U. In case W is completely recoverable from U, we say we have *lossless* compression. Otherwise, we have *lossy* compression. The *compression ratio* is $\mathrm{lgth}(W)/\mathrm{lgth}(U)$. The "compression ratio achieved by a method" is the average compression ratio obtained, using that method, with the average taken over all instances of W in the cases where the method is used. (This taking of the average is usually hypothetical, not actual.)

Sometimes the file W is sitting there, available for leisurely perusal and sampling. Sometimes the file W is coming at you at thousands of bits per second, with immediate compression required and with no way of foretelling with certainty what the bit stream will be like 5 seconds from now. Therefore, our compression methods will be distinguished not only by how great a compression ratio they achieve, together with how much information they preserve, but also by how fast they work, and how they deal with fundamental changes in the stream W (such as changing from a stream in which the digits $0, 1$ occur approximately randomly to one which is mostly 0's).

There is another item to keep account of in assessing and distinguishing between compression methods: *hidden costs*. These often occur as instructions for recovering W from U. Clearly it is not helpful to achieve great compression, if the instructions for recovering W from U take almost as much storage as W would. We will see another sort of hidden cost when we come to arithmetic coding: the cost of doing floating-point arithmetic with great precision. Clearly hidden costs are related to speed, adaptability, compression ratio, and information recovery in a generally inverse way; when you think you have made a great improvement in a method, or you think you have a new method with improved performance over what went before, do not celebrate until you have looked for the hidden costs!

In the compression method to be described in this chapter, the hidden costs are usually negligible—an encoding scheme very much smaller than the file has to be stored—and the method is lossless with very fast decompression (recovering W from U). On the down side, the method is applicable mainly in the cases where the file W is lying still and you have all the time in the world, and the compression ratios achieved are not spectacular. Roughly, "spectacular" begins at 10-to-1, and the compression ratios we will see by this method are nowhere near that.

From now on, the code alphabet will be $\{0, 1\}$, unless otherwise specified. We will survey extensions of our methods and results to the non-binary cases from time to time, but the binary alphabet is king at this point in history, so it seems more practical to do everything binary-wise and occasionally mention generalizations, rather than to struggle with the general case in everything.

5.1 Replacement via encoding scheme

In a nutshell, the game is to choose binary words $s_1, \ldots, s_m$ into which the original file can be *parsed* (divided up), and then to replace each occurrence of each s_i in the parsed file with another binary word w_i; the w_i are to be chosen so that the new file is shorter than the original, but the original is recoverable from the new. This kind of game is sometimes called *zeroth-order replacement*. You will see how "zero" gets into it later, when we consider higher order replacement. The assignment of the w_i to the s_i is, as in Chapter 4, called an *encoding scheme*.

For example, suppose we take

$$\begin{aligned} s_1 &= 0 \\ s_2 &= 10 \\ s_3 &= 110 \\ s_4 &= 1110 \\ s_5 &= 1111, \end{aligned} \qquad (*)$$

and the original "file" is 111110111111011101110110. (Of course this is an unrealistic example!) The file can be parsed into the string $s_5 s_2 s_5 s_4 s_4 s_5 s_1 s_4 s_3$. (Notice that there was no choice in the matter of parsing; the original binary word is *uniquely parsable* into a string of the s_i. Notice also that we are avoiding a certain difficulty in this example. If we had one, two, or three more 1's at the end of the original file on the right, there is no way that we could have incorporated them into the *source word*, the string of s_i's.)

Now, suppose we encode the s_i according to the *encoding scheme*

$$\begin{aligned} s_1 &\to 1111 \\ s_2 &\to 1110 \\ s_3 &\to 110 \\ s_4 &\to 10 \\ s_5 &\to 0. \end{aligned} \qquad (**)$$

The resulting new file is 01110010100111110110. Notice that this file is 20 bits long, while the original is 30 bits long, so we have a compression ratio of $3/2$.

(Not bad for an unrealistically small example! But, of course, for any positive number R it is possible to make up an example like the preceding in which the compression ratio is greater than R. See Exercise 5.1.3.)

But is the old file recoverable from the new file? Ask a friend to translate the new file into a string of the symbols s_i, according to (**). Be sure to hint that scanning left to right is a good idea. Your friend should have no trouble in translating the new file into $s_5 s_2 s_5 s_4 s_4 s_5 s_1 s_4 s_3$, from which you (or your friend!) can recover the old file by replacing the s_j with binary words according to (*).

Those who have perused chapter 4, or at least the first two sections, will not be surprised about the first stage of the recovery process, because the encoding scheme (**) satisfies the prefix condition. You might notice, as well, that the definition of the s_i in (*), regarded as an encoding scheme, satisfies the prefix condition, as well. This is no accident!

We shall review what we need to know about the prefix condition in the next section. For now, we single out a property of $s_1, \ldots, s_5$ in the preceding example that may not be so obvious, but which played an important role in making things "work" in the example.

Definition A list $s_1, \ldots, s_m$ of binary words has the *strong parsing property* (SPP) if and only if every binary word W is *uniquely* a concatenation,

$$W = s_{i_1} \cdots s_{i_t} v,$$

of some of the s_i and a (possibly empty) word v, with no s_i as a prefix (see Section 5.2), satisfying $\mathrm{lgth}(v) < \max_{1 \leq i \leq m} \mathrm{lgth}(s_i)$.

The word v is called the *leave* of the parsing of W into a sequence of the s_i. The uniqueness requirement says that if $W = s_{i_1} \cdots s_{i_t} v = s_{j_1} \cdots s_{j_r} u$, with neither u nor v having any of the s_i as a prefix and $\mathrm{lgth}(v), \mathrm{lgth}(u) < \max_i \mathrm{lgth}(s_i)$, then $t = r$, and $i_1 = j_1, \ldots, i_t = j_r$, and $v = u$.

Notice that in any list with the SPP the s_i must be distinct (why?), and any rearrangement of the list will have the SPP as well. Therefore, we will allow ourselves the convenience of sometimes attributing the SPP to finite *sets* of binary words; such a set has the SPP if and only if any list of its distinct elements has the SPP.

To see that $s_1, \ldots, s_5$ in (*) in the preceding example have the SPP, think about trying to parse a binary word W into a string of the s_i, scanning left to right. Because of the form of $s_1, \ldots, s_5$, the parsing procedure can be described thus: scan until you come to the first zero, or until you have scanned four ones. Jot down which s_j you have identified and resume scanning. Pretty clearly this procedure will parse any W into a string of the s_i with leave $v = \lambda, 1, 11,$ or 111 (with λ denoting the empty string). It becomes clear that this parsing is always possible, and most would agree that the parsing is unique, on the grounds that there is never any choice or uncertainty about what happens next during the parsing. We will indicate a logically rigorous proof of uniqueness, in a general setting, in the exercises at the end of the next section.

Which sets of binary words have the SPP? We shall answer this question fully in the next section. But there is a large class of sets with the SPP that ought to be kept in mind, not least because these are the *source alphabets* that are commonly used in current data compression programs, not only with the zeroth-order replacement strategy under discussion here, but also with all combinations of higher order, adaptive, and/or arithmetic methods, to be looked at later. All of these methods start by parsing the original file W into a *source string*, a long word over the source alphabet S. The methods differ in what is then done with the source string.

The most common sort of choice for source alphabet is: $S = \{0, 1\}^L$, the set of binary words of some fixed length L. Since computer files are commonly organized into bytes, binary words of length 8, the choice $L = 8$ is very common. Also, $L = 12$, a byte-and-a-half, seems popular.

If $S = \{0, 1\}^L$, the process of parsing a binary word W into a source string amounts to chopping W into segments of length L. If $L = 8$ and the original file is already organized into bytes, that's that; the parsing is immediate. There is another good reason for the choice $L = 8$. When information is stored byte by byte, it very often happens that you rarely need all 8 bits in the byte to record the datum, whatever it is. For instance, suppose there are only 55 different basic data messages to store, presumably in some significant order. You need only 6 bits (for a total of $2^6 = 64$ possibilities) to accommodate the storage task, yet it is customary to store 1 datum per byte. Thus one can expect a compression ratio of at least $8/6 = 4/3$ in this situation, just by deleting the 2 unused bits per byte. Thus the historical accident that files are, sometimes inefficiently, organized into bytes, makes the choice $L = 8$ rather shrewd. The best zeroth-order replacement method, Huffman encoding, takes advantage of this inefficiency, and more. In the hypothetical situation mentioned above, we might expect something more like $8/\log_2 55$ as a compression ratio, using $S = \{0, 1\}^8$ and zeroth-order simple Huffman encoding. Details to follow!

Even though $S = \{0, 1\}^L$ is the most common sort of choice of source alphabet, we do not want to limit our options, so we will continue to allow all S with the SPP; these S will be completely characterized in the next section.

A problem that may have occurred to the attentive anxious reader is: what effect should the *leave* have in the calculation of the compression ratio? In real life, the original binary word W to be parsed and compressed is quite long; saving and pointing out the leave may require many more bits than the leave itself, but the added length to the compressed file will generally be negligible, compared to the total length of the file. For this reason, we shall ignore the leave in the calculation of the compression ratio. Therefore, if $S = \{s_1, \ldots, s_m\}$ and a file W parses into $W = s_{i_1} \cdots s_{i_t} v$, $\mathrm{lgth}(v) < \max_i \mathrm{lgth}(s_i)$, and if the s_i are replaced according to the encoding scheme $s_i \to w_i$, $i = 1, \ldots, m$, the compression ratio will be $\sum_{j=1}^t \mathrm{lgth}(s_{i_j}) / \sum_{j=1}^t \mathrm{lgth}(w_{i_j})$, regardless of v, by convention.

Exercises 5.1

1. Suppose $S = \{0, 1\}^L$ is the source alphabet. What are the possible leaves in the parsing of files by S?

2. Show that each of the following does *not* possess the SPP.

   (a) $S = \{0, 01, 011, 0111, 1111\}$. [Hint: let $W = 01$.]
   (b) $S = \{00, 11, 01\}$
   (c) $S = \{0, 1\}^L \setminus \{w\}$, for any binary word w of length L.

3. Invent a situation, with a source alphabet $S = \{s_1, \ldots, s_m\}$ satisfying the SPP, an encoding scheme $s_i \to w_j$, $j = 1, \ldots, m$, satisfying the prefix condition (see the next section), and an original file W, such that the compression ratio achieved by parsing and then encoding W is at least 5 to 1. [Hint: you could take the example of this section as a model, with $s_1 = 0$, $s_2 = 10$, ..., with an encoding scheme in the mode of the example, and with a silly W; what if W has all 1's?]

4. Find the compression ratio if the original file in the example in this section is parsed using $S = \{0, 1\}^3$ and encoded using the scheme

$$
\begin{array}{ll}
000 \to 1111111 & 100 \to 1110 \\
001 \to 1111110 & 101 \to 110 \\
010 \to 111110 & 110 \to 10 \\
011 \to 11110 & 111 \to 0
\end{array}
$$

5.2 Review of the prefix condition

We collect here some of the definitions and results from Chapter 4 and apply them to our current purpose, the characterization of lists of binary words with the strong parsing property.

A binary word u is a *prefix* of a binary word w if and only if $w = uv$ for some (possibly empty) word v. A list $w_1, \ldots, w_m$ of binary words *satisfies the prefix condition* if and only if whenever $1 \le i, j \le m$ and $i \ne j$, then w_i is not a prefix of w_j. An encoding scheme $s_i \to w_i$, $i = 1, \ldots, m$, satisfies the prefix condition if and only if the list $w_1, \ldots, w_m$ does. It is common usage to say that such a scheme defines a prefix-condition code, or simply a prefix code. Take note that this terminology is somewhat misleading, because "prefix code" is characterized by an *absence* of prefix relations in its encoding scheme.

The practical importance of prefix codes is encapsulated in Theorem 4.2.1, which says:

5.2.1 *Every prefix code is uniquely decodable, with reading-left-to-right serving as a valid decoder-recognizer.*

In case you have not read Section 4.1, here is a translation:

5.2.2 *If* $s_j \rightarrow w_j$, $j = 1, \ldots, m$, *is a prefix-condition encoding scheme, and if for some subscripts* $1 \leq i_1, \ldots, i_t, j_1, \ldots, j_r \leq m$, $w_{i_1} \cdots w_{i_t} = w_{j_1} \cdots w_{j_r}$, *then* $t = r$ *and* $i_k = j_k$, $k = 1, \ldots, t$. *Furthermore, the source word* $s_{i_1} \cdots s_{i_t}$ *can infallibly be recovered from the code word* $w_{i_1} \cdots w_{i_t}$ *by scanning left to right and noting the source letter* s_{i_k} *each time a word* w_{i_k} *from the encoding scheme is recognized.*

(Actually, the assertion that "reading-left-to-right is a valid decoder-recognizer" for a code given by a particular scheme is a stronger statement than is given in the last part of 5.2.2, because of the word "recognizer"—the last assertion in 5.2.2 says that reading-left-to-right is a valid *decoder* for a prefix code—but we will not tarry further over this point.)

In the example in the preceding section, the encoding scheme (∗∗) satisfies the prefix condition, and if you followed the example you experienced directly the pleasures of reading-left-to-right with respect to the scheme (∗∗). This sort of decoding is also called *instantaneous* decoding, for reasons that should be obvious.

Pretty clearly, 5.2.2 is telling us that lists of words with the prefix condition satisfy something resembling the uniqueness provision of the strong parsing property. We will give the full relation between the prefix condition and the SPP after restating the Kraft and McMillan Theorems for binary codes. More general statements and proofs are given in Section 4.2.

5.2.3 Kraft's Theorem for binary codes *Suppose* m *and* $\ell_1, \ldots, \ell_m$ *are positive integers. There is a list* $w_1, \ldots, w_m$ *of binary words, satisfying the prefix condition and* $\text{lgth}(w_i) = \ell_i$, $i = 1, \ldots, m$, *if and only if* $\sum_{i=1}^{m} 2^{-\ell_i} \leq 1$.

5.2.4 McMillan's Theorem for binary codes *Suppose* m *and* $\ell_1, \ldots, \ell_m$ *are positive integers, and* $w_1, \ldots, w_m$ *are binary words satisfying* $\text{lgth}(w_i) = \ell_i$, $i = 1, \ldots, m$. *If* $s_i \rightarrow w_i$, $i = 1, \ldots, m$ *is a uniquely decodable encoding scheme (see the first part of 5.2.2 for the meaning of this) then* $\sum_{i=1}^{m} 2^{-\ell_i} \leq 1$.

McMillan's Theorem has virtually no practical significance at this point. We repeat it here just because it is a beautiful theorem. Together with Kraft's Theorem, what it says of pseudo-practical importance is that if your primary criteria for goodness of an encoding scheme are unique decodability, first, and prescribed code word lengths, second, then there is no need to leave the friendly family of prefix codes.

Kraft's Theorem brings us to the main result of this section. The proof is outlined in Exercise 5.2.2.

5.2.5 Theorem *A list* $w_1, \ldots, w_m$ *of binary words has the strong parsing property if and only if the list satisfies the prefix condition and* $\sum_{i=1}^{m} 2^{-\text{lgth}(w_i)} = 1$.

Exercises 5.2

1. Complete the following to lists with the SPP, adding as few new words to the lists as possible.

   (a) 00, 01, 10.

   (b) 00, 10, 110, 011.

   (c) 0, 10, 110, 1110.

2. Prove Theorem 5.2.5 by completing the following.

   (a) Show that if binary words $w_1, \ldots, w_m$ do not satisfy the prefix condition, then the list cannot possess the SPP. [If $w_i = w_j v$, $i \neq j$, then $w_i = w_j v$ can be parsed in at least two different ways by $w_1, \ldots, w_m$.]

   (b) Prove this part of the assertion in 5.2.2: If $w_1, \ldots, w_m$ satisfy the prefix condition, and if $w_{i_1} \cdots w_{i_t} = w_{j_1} \cdots w_{j_r}$, then $r = t$ and $i_k = j_k$, $k = 1, \ldots, t$. [Hint: if not, let k be the smallest index such that $i_k \neq j_k$. Then $w_{i_k} \cdots w_{i_t} = w_{j_k} \cdots w_{j_r}$; since the two strings agree at every position, they must agree in the first $\min(\text{lgth}(w_{i_k}), \text{lgth}(w_{j_k}))$ places. But then $w_1, \ldots, w_m$ do *not* satisfy the prefix condition, contrary to supposition. Why don't they?]

   (c) Suppose $w_1, \ldots, w_m$ are binary words satisfying the prefix condition, and $\sum_{i=1}^{m} 2^{-\ell_i} < 1$, where $\ell_i = \text{lgth}(w_i)$, $i = 1, \ldots, m$. Let $\ell \geq \max_i \ell_i$ be sufficiently large that $2^{-\ell} + \sum_{i=1}^{m} 2^{-\ell_i} \leq 1$. By the proof of Kraft's Theorem (see Section 4.2), there is a binary word W of length ℓ that has none of $w_1, \ldots, w_m$ as a prefix. Conclude that W cannot be parsed into a string $w_{i_1} \cdots w_{i_t} v$ for some indices $i_1, \ldots, i_t$ and leave v satisfying $\text{lgth}(v) < \max_i \text{lgth}(w_i)$ (why not?) and that, therefore, $w_1, \ldots, w_m$ does not have the SPP.

   (d) Suppose $w_1, \ldots, w_m$ are binary words satisfying the prefix condition and $\sum_{i=1}^{m} 2^{-\ell_i} = 1$, where $\ell_i = \text{lgth}(w_i)$, $i = 1, \ldots, m$. If W is a binary word of length $\ell \geq \max_i \ell_i$, then W must have one of the w_i as a prefix, for, if not, then the list $w_1, \ldots, w_m, W$ satisfies the prefix condition, yet $2^{-\text{lgth}(W)} + \sum_{i=1}^{m} 2^{-\text{lgth}(w_i)} = 2^{-\ell} + 1 > 1$, which is impossible, by Kraft's Theorem. Conclude that every such W can be written $W = w_{i_1} \cdots w_{i_t} v$ for some indices $i_1, \ldots, i_t$ and v satisfying $\text{lgth}(v) < \max_i \ell_i$. [How do you arrive at this conclusion?]

   Put (a)–(d) together for a proof of Theorem 5.2.5.

3. What are the analogues of the SPP and Theorem 5.2.5 for non-binary code alphabets?

4. Show that any list $w_1, \ldots, w_k$ of binary words satisfying the prefix condition can be completed to a list $w_1, \ldots, w_m$ with the SPP. [You will probably need to use a result, mentioned in part (c) of Exercise 2, above, which is embedded in the proof of Kraft's Theorem. Incidentally, the result of this exercise holds for non-binary alphabets, as well.]

5.3 Choosing an encoding scheme

Suppose that binary words $s_1, \ldots, s_m$ with the SPP have been settled upon, and the original file has been parsed (at least hypothetically) into what we will continue to call a *source string*, $Z = s_{i_1} \cdots s_{i_N}$, together with, possibly, a small leave at the end. [How does one select the best source alphabet, $S = \{s_1, \ldots, s_m\}$, for the job? Not much thought has been given to this question! Recall from previous discussion that the blue-collar solution to the problem of choosing S is to take $S = \{0, 1\}^L$, usually with $L = 8$.]

Now consider the problem of deciding on an encoding scheme, $s_i \to w_i$, $i = 1, \ldots, m$, for the replacement of the s_i by other binary words, the w_i, so that the resulting file U is as much shorter than the original as possible, and so that the original file is recoverable from U. It suffices to recover Z from U. (The leave, if any, is handled in separate arrangements.)

Pretty clearly the length of U will be completely determined by the lengths of the w_i. McMillan's Theorem now enters to assure us that we lose nothing by requiring that $w_1, \ldots, w_m$ satisfy the prefix condition—which takes care of the recoverability of Z from U. Plus, the recovery will be as rapid as can reasonably be expected.

So we are looking for $w_1, \ldots, w_m$, satisfying the prefix condition, such that $w_{i_1} \cdots w_{i_N}$ is of minimal length. Common sense says, find a prefix code that assigns to the most frequently occurring s_i the shortest w_i, and to the least frequently occurring s_i the longest w_i (which will be no longer than necessary). Common sense is not mistaken, in this instance. (See Theorem 4.3.3, after reading what follows.) But common sense still leaves us wondering what to do.

Let us focus on the phrase "most frequently occurring"; let, as in Chapter 4, f_i stand for the relative frequency of the source letter s_i in the source text. To put it another way, f_i is the proportion of s_i's in the source string Z. For instance, if $Z = s_1 s_4 s_4 s_2 s_3 s_1 s_4$, then $f_1 = 2/7$, $f_2 = 1/7 = f_3$, and $f_4 = 3/7$. (And what about f_5, in case $m > 4$? $f_5 = 0$ in this case.) Note again that $\sum_{i=1}^{m} f_i = 1$. Recall that $f_1, \ldots, f_m$ can be thought of as the probabilities assigned to the distinct outcomes of a probabilistic experiment; the experiment is choosing a source letter "at random" from the source string Z, and f_i is the probability that the letter chosen will be s_i.

Recall that if $s_i \to w_i \in \{0, 1\}^{\ell_i}$ is an encoding scheme, the average code word length of the scheme is $\bar{\ell} = \sum_{i=1}^{m} f_i \ell_i$. The length of the file U obtained by replacing the s_i by the w_i, in Z, will be $\bar{\ell} N$, where N is the length of Z as a word over $S = \{s_1, \ldots, s_m\}$. (Verify! This assertion arises from the definitions of the f_i and of $\bar{\ell}$.) If the binary word s_i has length L_i, and $\bar{L} = \sum_{i=1}^{m} f_i L_i$, then the length of the original file W, parsed by the s_i into Z (we neglect the leave), has length $\bar{L} N$. Thus the compression ratio would be $\bar{L} N / \bar{\ell} N = \bar{L} / \bar{\ell}$.

The problem we face here is the same as that enunciated in Section 4.3: given $s_1, \ldots, s_m$ and $f_1, \ldots, f_m$, find a prefix-condition scheme $s_i \to w_i \in \{0, 1\}^{\ell_i}$

which minimizes $\bar{\ell} = \sum_{i=1}^m f_i \ell_i$. We know how to solve this problem; the solution is Huffman's algorithm, discovered in 1951 (published in 1952) by David Huffman, then a graduate student at M.I.T. He discovered the algorithm in response to a problem—the very problem to which his algorithm is the solution—posed in a course on communication theory by the great mathematician, R. M. Fano. (Professor Fano did not tell the class that the problem was unsolved!)

In Section 4.3 Huffman's algorithm is described approximately as Huffman described it in his paper of 1952 [36]. Later in this section we will take another look at it and give an equivalent description, involving tree diagrams, which turns out to be much easier to implement and lends itself to more clever improvements than the merge-and-rearrange, merge-and-rearrange procedure described in 4.3.

Before that we will look at two other algorithmic methods for approximately solving the problem of finding an optimal prefix-condition encoding scheme, given $s_1, \ldots, s_n$ and $f_1, \ldots, f_n$. The first is due to Claude Shannon, who is to the theory of communication as Euclid is to geometry; the second is due to the aforementioned R. M. Fano. Both are given in Shannon's opus [63], published in 1948. They do not always give optimal encoding schemes, but they are of historical and academic interest, they leave unanswered questions pursuit of which might be fruitful, and Shannon's method provides a proof of an important theorem that tells you approximately how good a compression ratio can be achieved by zeroth order replacement, before you do the hard work of achieving it.

5.3.1 Shannon's method

Given $s_1, \ldots, s_m$ and $f_1, \ldots, f_m$, rename, if necessary, so that $f_1 \geq \cdots \geq f_m > 0$. (Any source letters that do not appear are deleted from the source alphabet.) Define $F_1 = 0$ and $F_k = \sum_{i=1}^{k-1} f_i$, $2 \leq k \leq m$. Let ℓ_k be the positive integer satisfying $2^{-\ell_k} \leq f_k < 2^{-\ell_k+1}$. In other words, $\ell_k = \lceil \log_2 f_k^{-1} \rceil$. (Verify! Incidentally, $\lceil \cdot \rceil$ stands for "round up.")

In Shannon's method, the encoding scheme is $s_j \rightarrow w_j$, $j = 1, \ldots, m$, with w_j consisting of the first ℓ_j bits in the binary expansion of F_j. Two remarks are necessary before we look at an example.

Dyadic fractions are rational numbers of the form $p/2^m$, with m a non-negative integer and p an integer. If $1 \leq p < 2^m$, then $p = \sum_{j=0}^{m-1} a_j 2^j$ for some (binary) digits $a_0, \ldots, a_{m-1} \in \{0, 1\}$; it follows that $p/2^m = \sum_{j=0}^{m-1} a_j 2^{j-m} = \sum_{k=1}^m a_{m-k} 2^{-k} = (.a_{m-1} \cdots a_0)_2$. The point is, every dyadic fraction between 0 and 1 has a "finite" binary representation, meaning a binary expansion with an infinite string of zeros at the end. And conversely, every number between 0 and

1 with a finite binary representation is a dyadic fraction:

$$(.b_1 \cdots b_m)_2 = \sum_{j=1}^{m} b_j 2^{-j} = \frac{1}{2^m} \sum_{j=1}^{m} b_j 2^{m-j} = p/2^m,$$

with p an integer between 0 and $2^m - 1$, inclusive, if $b_1, \ldots, b_m \in \{0, 1\}$.

Because $1 = \sum_{k=1}^{\infty} 2^{-k}$, the non-zero dyadic fractions have the unpleasant property that they each have two different binary representations, one ending in an infinite string of ones. For instance,

$$\frac{3}{4} = \frac{1}{2} + \frac{1}{4} = (.11)_2 = (.1011\cdots)_2 = (.10\bar{1})_2$$

$$= \frac{1}{2} + \sum_{k=3}^{\infty} 2^{-k}.$$

(We will sometimes use the common bar notation, as in $(.10\bar{1})_2$, to indicate infinite repetition. Thus $(.11\overline{101})_2 = (.11101101\cdots)_2$.) This is a nuisance because, in the description of Shannon's method, we have the reference to "*the* binary expansion of F_j." What if F_j is a dyadic fraction? The answer is that, here and elsewhere, we take "the" binary expansion of a dyadic fraction to be the "finite" one, the one that ends in an infinite string of zeros.

Incidentally, if you have never checked such a thing before, it might be a salutary exercise to prove that the non-zero dyadic fractions are the only numbers with two distinct binary expansions. Or, you can take our word for it.

The second remark concerns the following question: given a number $r \in (0, 1)$, how do you determine its binary expansion if r is given as a fraction or as a decimal? Here is a trick we lifted from Neal Koblitz [41]. Start doubling. Every time the "answer" is less than one, the corresponding binary digit is 0. Every time the "answer" is ≥ 1, the next binary digit is 1; then subtract 1 from the "answer" before doubling again.

For instance, if $r = 2/7$, the process looks like:

double after adjustment	4/7	8/7	2/7	4/7	
binary digit	0	1	0	0	...

So, $2/7 = (.\overline{010})_2$.

Another instance: suppose $r = .314$, in decimal form. The process looks like

double after adjustment	.628	1.256	.512	1.024	.048
binary digit	0	1	0	1	0

.096	.192	.384	.768	1.536	1.072	.144
0	0	0	0	1	1	0

and we mercifully stop here, since the "period" of the binary expansion could be 499 bits long, and we are not really interested in finding out just how long it is. Anyway, we have the first 12 bits of the binary expansion:

$$.314 = (.010100000110\cdots)_2$$

We leave it to those interested to ponder why Koblitz's method works. Think about the effect of doubling a number between 0 and 1 already in binary form.

Examples of Shannon's method Let's first recall the example in Section 5.1. After parsing, the source string was $s_5 s_2 s_5 s_4 s_4 s_5 s_1 s_4 s_3 = Z$. So $f_1 = f_2 = f_3 = 1/9$, $f_4 = f_5 = 3/9$. But wait! Shannon's method applies to source frequencies in non-increasing order, the reverse of what we have here. So put $s_i' = s_{5-i+1}$, $f_i' = f_{5-i+1}$ and let's follow Shannon's method with the F_k in the recipe renamed F_k' and the ℓ_k renamed ℓ_k'.

It is easy to see that $\ell_1' = \ell_2' = 2$ and $\ell_3' = \ell_4' = \ell_5' = 4$. [It is easy to calculate the ℓ_k in Shannon's recipe if you bear in mind that ℓ_k is the "first" positive integer exponent such that that power of $1/2$ is $\leq f_k$. So, for instance, if $f_k = 1/9$, the first power of $1/2$ less than or equal to $1/9$ is $1/16 = (1/2)^4$, so $\ell_k = 4$.] Using our method for calculating binary expansions, we have

$$F_1' = 0 = (.00\ldots)_2$$
$$F_2' = 3/9 = (.01\ldots)_2$$
$$F_3' = 6/9 = (.1010\ldots)_2$$
$$F_4' = 7/9 = (.1100\ldots)_2$$
$$F_5' = 8/9 = (.1110\ldots)_2$$

Thus the encoding scheme given by Shannon's method is

$$s_5 = s_1' \to 00$$
$$s_4 = s_2' \to 01$$
$$s_3 = s_3' \to 1010$$
$$s_2 = s_4' \to 1100$$
$$s_1 = s_5' \to 1110$$

This gives $\bar{\ell} = 2(3/9) + 2(3/9) + 4(1/9) + 4(1/9) + 4(1/9) = 8/3$; we also have $\bar{L} = 1/9 + 2/9 + 3/9 + 4(3/9) + 4(3/9) = 10/3$, for a compression ratio of $\bar{L}/\bar{\ell} = 5/4$ (which you can verify directly by plugging the code words for the s_j into Z to obtain a compressed string U with 24 bits, compared to the 30 bits we started with). Since we achieved a compression ratio of $3/2$ with the prefix code (**) in Section 5.1, this example shows that Shannon's method may fail to give an optimal encoding scheme. The shrewd will notice that we can modify our encoding scheme by lopping off the final zero of the code words for s_3, s_2, and s_1, and still have a prefix code. Even this modification does not optimize compression; we would have $\bar{\ell} = 7/3$ and $\bar{L}/\bar{\ell} = 10/7 < 3/2$.

Another example: $m = 6$ and $f_1 = .25 > f_2 = f_3 = .2 > f_4 = f_5 = .15 > f_6 = .05$. (Here we have already renamed, or rearranged, the source letters $s_1, \ldots, s_6$, if necessary, so that their relative frequencies are in non-increasing order.) We compute $\ell_1 = 2$, $\ell_2 = \ell_3 = \ell_4 = \ell_5 = 3$, $\ell_6 = 5$,

$$F_1 = 0 = (.00...)_2 \qquad F_4 = .65 = (.101...)_2$$
$$F_2 = .25 = (.010...)_2 \qquad F_5 = .80 = (.110...)_2$$
$$F_3 = .45 = (.011...)_2 \qquad F_6 = .95 = (.11110...)_2$$

so the encoding scheme is

$$s_1 \to 00 \qquad s_4 \to 101$$
$$s_2 \to 010 \qquad s_5 \to 110$$
$$s_3 \to 011 \qquad s_6 \to 11110$$

with $\bar{\ell} = 2.85$ (check!). We cannot compute the compression ratio without knowing $\bar{L}$; still, we will see later that $\bar{\ell} = 2.85$ is far from optimal in this situation. Note that we could obtain a shorter prefix code by lopping off the last two bits of $w_6 = 11110$, but the new value of $\bar{\ell}$, 2.75, is still far from optimal.

So, the experimental evidence is that Shannon's method is not so great. We will see that it has certain charms, and is worth knowing about for "academic" reasons, in the next section.

5.3.2 Fano's method

Once again, assume that $s_1, \ldots, s_m$ and $f_1 \geq \cdots \geq f_m$ are given. In Fano's method, you start by dividing $s_1, \ldots, s_m$ into two "blocks" with consecutive indices, $s_1, \ldots, s_k$ and $s_{k+1}, \ldots, s_m$, with k chosen so that $\sum_{j=1}^{k} f_j$ and $\sum_{j=k+1}^{m} f_j$ are as close as possible to being equal. The code representatives $w_1, \ldots, w_k$ of $s_1, \ldots, s_k$ start (on the left) with 0, and $w_{k+1}, \ldots, w_m$ start with 1. At the next stage, the two blocks are each divided into two smaller blocks, according to the same rule, and so on. At each new division, a zero is appended to the code representatives of the letters in one of the new blocks, and one is appended to the code representatives of the letters in the other. Blocks consisting of a single letter cannot be further subdivided.

Examples of Fano's method We will apply Fano's method in the two cases to which we applied Shannon's method. In the first case, we will not fuss about the renaming, as we did in 5.3.1.

For the first example, consider

Frequency	Letter		Code word	
3/9	s_1	$\to$	00	
3/9	s_2	$\to$	01	2
1/9	s_3	$\to$	10	1
1/9	s_4	$\to$	110	2
1/9	s_5	$\to$	111	3

The number on the right indicates the order in which the divisions were drawn. Observe that we do not, for instance, group s_1 and s_3 together in the first block, with s_2, s_4, s_5 in the other, even though 4/9 and 5/9 are more nearly equal than 2/3 and 1/3. Once the s_i are arranged in order of non-increasing frequency, the blocks must consist of consecutive s_i's.

Could we have started by putting s_1 in a block by itself, and $s_2, \ldots, s_5$ in the other block? Yes, and, interestingly, the encoding scheme obtained would have been like (∗∗) in 5.1. The average code word length for each scheme is $\bar{\ell} = 20/9$.

For the second example, consider

Frequency	Letter		Code word	
.25	s_1	→	00	2
.20	s_2	→	01	1
.20	s_3	→	10	2
.15	s_4	→	110	3
.15	s_5	→	1110	4
.05	s_6	→	1111	

There is a choice at the second division of blocks in Fano's method applied to this source alphabet with these relative frequencies. The encoding scheme resulting from the other choice has the same average code word length, $\bar{\ell} = 2.55$, considerably better than that achieved with Shannon's method. (In fact, this is the best $\bar{\ell}$ achievable, with a prefix code.)

In some cases when there is a choice in the execution of Fano's method, schemes with different values of $\bar{\ell}$ are obtained. For instance, try Fano's method in two different ways on a source alphabet with relative frequencies $4/9, 1/9, 1/9, 1/9, 1/9, 1/9$.

Therefore, it is not the case that Fano's method always solves the problem of minimizing $\bar{\ell}$ with a prefix code. In Exercise 3 at the end of this section we offer some examples (for two of which we are indebted to Doug Leonard) which show that sometimes no instance of Fano's method solves that problem, and sometimes the one that does, when there is a choice, is not the one arrived at by the obvious make-the-upper-block-as-small-as-possible convention.

Since Huffman's algorithm, coming up next, solves the problem of minimizing $\bar{\ell}$ with a prefix code, Fano's method is apparently only of historical and academic interest. But sometimes pursuit of academic questions leads to practical advances. Here are some academic questions to ponder, for those so inclined:

(i) For which relative source frequencies $f_1 \geq \cdots \geq f_m$ does Fano's method minimize $\bar{\ell}$?

(ii) Same question, for Shannon's method.

(iii) Does Fano's method always do at least as well as Shannon's?

We leave to the reader the verification that Fano's method always results in a prefix code.

5.3.3 Huffman's algorithm

We repeat, in abbreviated form, the account of the binary case of Huffman's algorithm given in 4.2, which is essentially that given by Huffman [36]. Given

$s_1, \ldots, s_m$ and $f_1 \geq \cdots \geq f_m$, start by merging s_{m-1} and s_m into a new letter, σ, with relative frequency $f_{m-1} + f_m$. Now rearrange the letters $s_1, \ldots, s_{m-2}, \sigma$ into $s'_1, \ldots, s'_{m-1}$ with relative frequencies $f'_1 \geq \cdots \geq f'_{m-1}$ and repeat the process. Continue until you have merged down to two letters, τ_0 and τ_1. Begin encoding by $\tau_0 \to 0$, $\tau_1 \to 1$. Follow through the merging in reverse, coding as you go; if, for instance, σ is encoded as w, then $s_{m-1} \to w0$, $s_m \to w1$.

Here is another account of the algorithm which seems to lead to implementations which take care of all that rearranging in the merging part of the algorithm. In the language of graph theory, we start constructing a *tree*, called a *Huffman tree*, the *leaf nodes* of which are the letters $s_1, \ldots, s_m$, sporting weights $f_1, \ldots, f_m$, respectively. The nodes s_m, s_{m-1} now become *siblings* and are attached to a *parent node* σ, with weight $f_m + f_{m-1}$. At the next stage, two nodes with least weight that have not yet been used as siblings are paired as siblings and attached to a parent node which gets the sum of their weights as its weight. Continue this process until the last two siblings are paired. The final parent node, with weight 1, is called the *root* of the tree.

From each parent there are two edges going to siblings. Label one of these 0, the other 1. (Which gets which label does not really matter, but there might be practical reasons for establishing conventions governing this label assignment, as we shall see later when we consider "adaptive" or "dynamic" coding.)

The code word representing each s_j is obtained by following the path from s_j to the root node, writing down each edge label, right to left.

Examples of Huffman's algorithm We will do the two examples already treated by the methods of Shannon and Fano. In the first example, a Huffman tree can be created as follows:

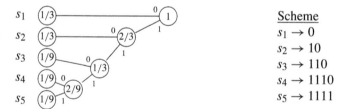

Scheme

$s_1 \to 0$
$s_2 \to 10$
$s_3 \to 110$
$s_4 \to 1110$
$s_5 \to 1111$

and $\bar{\ell} = 20/9$. The careful reader will note that there was a certain choice involved at the third "merge" in the construction of the tree above. If we make the other essentially different choice, we get

Scheme

$s_1 \to 00$
$s_2 \to 01$
$s_3 \to 10$
$s_4 \to 110$
$s_5 \to 111$

and, again, $\bar{\ell} = 20/9$.

For the second example, consider

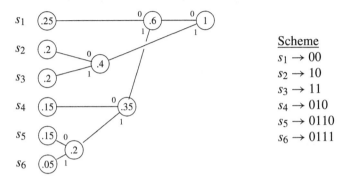

Scheme
$s_1 \to 00$
$s_2 \to 10$
$s_3 \to 11$
$s_4 \to 010$
$s_5 \to 0110$
$s_6 \to 0111$

with $\bar{\ell} = 2.55$.

There is one essentially different tree obtainable by making a different choice at the second merge, which results in a scheme with code word lengths $2, 2, 3, 3, 3, 3$; again, we have $\bar{\ell} = 2.55$.

In fact, by the proof outlined in Section 4.3.1, Huffman's algorithm always results in a prefix condition scheme which minimizes $\bar{\ell}$, and for every p.c. scheme minimizing $\bar{\ell}$ (given $f_1, \ldots, f_m$), there is an instance of Huffman's algorithm that will produce a scheme with the same code word lengths as the given scheme.

Exercises 5.3

1. (a) $s_1 = 00, s_2 = 01, s_3 = 10, s_4 = 11; f_1 = .4, f_2 = .25 = f_3, f_4 = .1$. Find the compression ratio if the s_j are encoded by Huffman's algorithm.

   (b) Same question, with

   $$
   \begin{array}{llll}
   s_1 = 00 & f_1 = .1 & s_4 = 101 & f_4 = .2 \\
   s_2 = 01 & f_2 = .2 & s_5 = 110 & f_5 = .15 \\
   s_3 = 100 & f_3 = .1 & s_6 = 111 & f_6 = .25
   \end{array}
   $$

2. Find the encoding schemes and the compression ratios when the methods of Shannon and Fano are applied in 1(a) and 1(b), above.

3. Process the following by the method of Fano, taking every possibility into account. When are the results optimal? (You could use Huffman's algorithm on the same data to determine optimality.) The relative source frequencies are:

   (a) .38, .24, .095, .095, .095, .095

   (b) 5/13, 2/13, 2/13, 2/13, 2/13

   (c) .4, .15, .1, .09, .09, .09, .08

5.4 The Noiseless Coding Theorem and Shannon's bound

Here are two good things about Shannon's method.

5.4.1 *Shannon's method always results in a prefix-condition encoding scheme.*

Proof: Suppose $s_1, \ldots, s_m$ and $f_1 \geq \cdots \geq f_m > 0$ are given. Since $\ell_k = \lceil \log_2 f_k^{-1} \rceil$, we have $\ell_1 \leq \cdots \leq \ell_m$. Therefore, since ℓ_k is the length of the code word w_k to which s_k is assigned in the encoding scheme arrived at by Shannon's method, it will suffice to show that for $1 \leq k < r \leq m$, w_k is not a prefix of w_r.

For such k and r,

$$F_r \geq F_{k+1} = F_k + f_k \geq F_k + 2^{-\ell_k}.$$

If the binary expansions of F_r and F_k were to agree in the first ℓ_k positions, then the most they could differ by would be $\sum_{j=k+1}^{\infty} 2^{-j} = 2^{-k}$, and they could only differ by this much if the binary expansion of one of them were all zeroes from the $(k+1)$st place on, and the binary expansion of the other were all ones from that point on. None of our binary expansions end with an infinite string of ones, by convention. Therefore, if the binary representations of F_r and F_k agree in the first ℓ_k places, they are less than $2^{-\ell_k}$ apart. Since they are, in fact, at least $2^{-\ell_k}$ apart, it follows that their binary representations do not agree in the first ℓ_k places, so w_k is not a prefix of w_r.   □

5.4.2 *Given $S = \{s_1, \ldots, s_m\}$ and $f_1, \ldots, f_m$, the encoding scheme resulting from Shannon's method has average code word length $\bar{\ell}$ satisfying $\bar{\ell} < H + 1$, where $H = H(S) = \sum_{k=1}^{m} f_k \log_2 f_k^{-1}$.*

Proof:

$$\bar{\ell} = \sum_{k=1}^{m} f_k \ell_k = \sum_{k=1}^{m} f_k \lceil \log_2 f_k^{-1} \rceil$$

$$< \sum_{k=1}^{m} f_k (\log_2 f_k^{-1} + 1)$$

$$= H + \sum_{k=1}^{m} f_k = H + 1. \qquad \square$$

Those who have read certain parts of the first four chapters will recognize $H(S)$ as the *source entropy*. More exactly, $H(S)$ is the source entropy when the "source" is a random emitter of source letters, with s_j being emitted with relative frequency (probability) f_j. We will consider slightly more sophisticated models of the "source," that mysterious and, legend has it, imaginary entity that produces source text, in Chapter 7.

Assertion 5.4.2 is half of the binary version of the Noiseless Coding Theorem for Memoryless Sources, stated in full, for code alphabets of any size, as Theorem 4.3.7. We will give the proof for the binary case here and relegate the proof of the more general theorem to the problem section. We owe our proof of the second half of the theorem to Dominic Welsh [81] or to Robert Ash [4].

5.4.3 Noiseless Binary Coding Theorem for Memoryless Sources *Given* S $= \{s_1, \ldots, s_m\}$ *and* $f_1, \ldots, f_m > 0$, *with* f_j *being the relative frequency of* s_j *in the source text; then the smallest average code word length* $\bar{\ell}$ *achievable by a uniquely decodable scheme* $s_j \rightarrow w_j \in \{0, 1\}^{\ell_j}$ *satisfies* $H(S) \leq \bar{\ell} < H(S)+1$. *Furthermore,* $\bar{\ell} = H(S)$ *is achievable if and only if each* f_j *is an integral power of* $1/2$.

Proof: $\bar{\ell} < H(S)+1$ follows from 5.4.1 and 5.4.2. Now suppose $s_j \rightarrow w_j \in \{0,1\}^{\ell_j}$ is any uniquely decodable scheme, and $\bar{\ell} = \sum_{j=1}^{m} f_j \ell_j$. By McMillan's Theorem, $G = \sum_{j=1}^{m} 2^{-\ell_j} \leq 1$. Set $q_k = 2^{-\ell_k}/G$, $k = 1, \ldots, m$. Note that $\sum_{k=1}^{m} q_k = 1$.

We will use Lemma 2.1.1, which says that $\ln x \leq x - 1$ for all $x > 0$, with equality only if $x = 1$. It follows that

$$\sum_{j=1}^{m} f_j \log_2 f_j^{-1} - \sum_{j=1}^{m} f_j \log_2 q_j^{-1}$$

$$= \log_2(e) \sum_{j=1}^{m} f_j \ln \frac{q_j}{f_j} \leq \log_2(e) \sum_{j=1}^{m} f_j (\frac{q_j}{f_j} - 1)$$

$$= \log_2(e)(\sum_{j=1}^{m} q_j - \sum_{j=1}^{m} f_j) = \log_2(e)(1 - 1) = 0.$$

Thus $H(S) = \sum_{j=1}^{m} f_j \log_2 f_j^{-1} \leq \sum_{j=1}^{m} f_j \log_2 q_j^{-1}$ with equality if and only if $q_j = f_j$, $j = 1, \ldots, m$.

We also have

$$\sum_{j=1}^{m} f_j \log_2 q_j^{-1} = \sum_{j=1}^{m} f_j \log_2(2^{\ell_j} G)$$

$$= \sum_{j=1}^{m} f_j \ell_j + (\log_2 G) \sum_{j=1}^{m} f_j \leq \sum_{j=1}^{m} f_j \ell_j = \bar{\ell},$$

since $G \leq 1$ implies $\log_2 G \leq 0$. We have equality in this last inequality if and only if $G = 1$. Thus $H(S) \leq \bar{\ell}$, and equality implies $f_j = q_j = 2^{-\ell_j}/G = 2^{-\ell_j}/1 = 2^{-\ell_j}$, $j = 1, \ldots, m$.

This proves everything except that $\bar{\ell} = H$ can be achieved if $f_j = 2^{-\ell_j}$, $j = 1, \ldots, m$, for positive integers $\ell_1, \ldots, \ell_m$. However, in this case the code word lengths in the scheme resulting from Shannon's method are precisely $\ell_1, \ldots, \ell_m$, so the $\bar{\ell}$ for that scheme is

$$\bar{\ell} = \sum_{j=1}^{m} f_j \ell_j = \sum_{j=1}^{m} f_j \log_2 f_j^{-1} = H(S). \qquad \square$$

For instance, returning to the example in 5.1, with $f_1 = f_2 = f_3 = 1/9$, $f_4 = f_5 = 3/9$, and $\bar{L} = 10/3$, the Noiseless Coding Theorem says that the best $\bar{\ell}$ we can get with a prefix-condition scheme satisfies

$$2.113 \approx H \le \bar{\ell} < H + 1 \approx 3.113,$$

so the best compression ratio we can hope for by the method of this chapter with this choice of source alphabet satisfies

$$1.07 \approx \frac{10/3}{H+1} < \frac{\bar{L}}{\bar{\ell}} \le \frac{10/3}{H} \approx 1.577.$$

(In fact, the compression ratio of 1.5 achieved in 5.1 is the best possible, because the encoding scheme there was generated by Huffman's algorithm, which always gives the smallest $\bar{\ell}$ among those arising from prefix-condition schemes.)

The Noiseless Coding Theorem has practical value. If you have chosen s_1, ..., s_m binary words with the SPP, and determined, or estimated, $f_1, \ldots, f_m$, their relative frequencies in the source text obtained by parsing the original file, then you know $\bar{L} = \sum_j f_j \operatorname{lgth}(s_j)$ and $H = \sum_j f_j \log_2 f_j^{-1}$; the Noiseless Coding Theorem says that you cannot achieve a greater compression ratio, by replacement of the s_j according to some prefix-condition encoding scheme, than $\bar{L}/H$. If $\bar{L}/H$ is not big enough for your purposes, then you can stop wasting your time, back up, and either try again with a different source alphabet, or try some entirely different method of data compression.

In the context of this chapter, with $s_1, \ldots, s_m$ and $f_1, \ldots, f_m$ given so that $\bar{L}$ and H are calculable, $\bar{L}/H$ is called the *Shannon bound* on the compression ratio. Shannon noted that, under a certain assumption about the source, there is a trick that enables you to approach the Shannon bound as closely as desired, for long, long source strings. This trick is contained in the proof of the next result.

5.4.4 Theorem *Suppose that $S = \{s_1, \ldots, s_m\}$ is a set of binary words with the SPP, and suppose we confine our efforts to files which are parsed by S into source strings in which the s_j occur randomly and independently with relative frequencies f_j. Then for any $\epsilon > 0$, it is possible to achieve a compression ratio greater than $(\bar{L}/H) - \epsilon$, with lossless compression, on sufficiently long files, where*

$$\bar{L} = \sum_{j=1}^{m} f_j \operatorname{lgth}(s_j) \quad and \quad H = \sum_{j=1}^{m} f_j \log_2 f_j^{-1}.$$

Proof: The trick is, instead of encoding $s_1, \ldots, s_m$, we take as source alphabet S^N, the set of all words of length N over S, where N is a positive integer that we will soon take to be large enough for our purposes, depending on ϵ.

The assumption about the source (which is essentially unverifiable in real life) implies that the relative frequency of the word $s_{i_1} \cdots s_{i_N} \in S^N$, among all source words of length N in long source texts, will be the product $f_{i_1} \cdots f_{i_N}$. Therefore,

$$H(S^N) = \sum_{1 \le i_1,\dots,i_N \le m} f_{i_1} \cdots f_{i_N} \log_2 (f_{i_1} \cdots f_{i_N})^{-1}$$

$$= \sum_{1 \le i_1,\dots,i_N \le m} f_{i_1} \cdots f_{i_N} \sum_{j=1}^{N} \log_2 f_{i_j}^{-1}$$

$$= \sum_{i_1=1}^{m} f_{i_1} \log_2 f_{i_1}^{-1} \Big(\sum_{1 \le i_2,\dots,i_N \le m} f_{i_2} \cdots f_{i_N} \Big) + \cdots +$$

$$+ \sum_{i_N=1}^{m} f_{i_N} \log_2 f_{i_N}^{-1} \Big(\sum_{1 \le i_1,\dots,i_{N-1} \le m} f_{i_1} \cdots f_{i_{N-1}} \Big).$$

Now, for instance, $\sum_{1 \le i_2,\dots,i_N \le m} f_{i_2} \cdots f_{i_N} = (\sum_{i_2} f_{i_2}) \cdots (\sum_{i_N} f_{i_N}) = 1$, so

$$H(S^N) = \sum_{i_1=1}^{m} f_{i_1} \log_2 f_{i_1}^{-1} + \cdots + \sum_{i_N=1}^{m} f_{i_N} \log_2 f_{i_N}^{-1}$$

$$= N \sum_{i=1}^{m} f_i \log_2 f_i^{-1} = NH(S) = NH.$$

Also, by the statistical principle that the average of the sum is the sum of the averages (see Section 1.8), the binary words represented by the source words $s_{i_1} \cdots s_{i_N} \in S^N$ will have average length $N\bar{L}$.

By, say, Shannon's method (see 5.3.1), S^N can be encoded with a prefix-condition scheme with average code word length $\bar{\ell}(N) < H(S^N)+1 = NH+1$. Thus the compression ratio achieved over original files so long that lengths of the source strings obtained by parsing them are considerably greater than N will be

$$\frac{N\bar{L}}{\bar{\ell}(N)} > \frac{N\bar{L}}{NH+1} = \frac{\bar{L}}{H + \frac{1}{N}} > \frac{\bar{L}}{H} - \epsilon$$

for N sufficiently large. $\square$

The trick contained in the proof, of jazzing up the source alphabet by replacing it by the set of all words of length N over it, with introduced probabilities defined by

$$(\text{freq. of } s_{i_1} \cdots s_{i_N}) = f_{i_1} \cdots f_{i_N},$$

is worth remembering. The assumption about the source in Theorem 5.4.4 is essentially the same as the assumption that these defined probabilities are valid. When would they not be? Well, for instance, if $S = \{s_1, s_2, s_3, s_4\}$ and the source string consists of $s_1s_1s_3s_2s_4$ repeated over and over, then $f_1 = 2/5$, $f_2 = f_3 = f_4 = 1/5$, but the probability of, say, with $N = 4$, $s_2s_1s_2s_3$, is *zero*, not $2/5^4$.

However, in practice, even if there is some orderliness in the source which violates the assumption in Theorem 5.4.4, it often happens that imposing probabilities on S^N by multiplication is not a bad approximation to reality, and the compression ratios obtained by applying, say, Huffman's algorithm to S^N with those assumed probabilities results in improved compression.

A last note on the Shannon bound: does it have anything to tell us about how to choose a source alphabet? The choice of a source alphabet affects both $\bar{L}$ and H. In principle, we want $\bar{L}$ to be large and H to be small. Roughly speaking, $\bar{L} = \sum_j f_j \operatorname{lgth}(s_j)$ will be large when the longer s_j have larger relative frequency f_j; and from the basics about entropy, in Chapter 2, H will be small when f_j is negligibly small except for a very few f_j. [H is zero when one f_j is 1 and the rest are zero.] So Theorems 5.4.4 and 5.4.3 verify common sense: we can achieve large, handsome compression ratios if we can choose $s_1, \ldots, s_m$ such that the longer binary words s_j occur with great frequency, relatively. This does not tell us how to find $s_1, \ldots, s_m$, it just gives some targets to shoot for.

Exercises 5.4

1. Compute the Shannon bound $\bar{L}/H$ on the compression ratio in both parts of Exercise 5.3.1.

2. Let $S = \{s_1, s_2, s_3, s_4\}$ with

$$
\begin{array}{ll}
s_1 = 111 & f_1 = .4 \\
s_2 = 110 & f_2 = .3 \\
s_3 = 10 & f_3 = .2 \\
s_4 = 0 & f_4 = .1
\end{array}
$$

   (a) Find the compression ratio if the s_j are encoded using Huffman's algorithm.
   (b) Find the compression ratio if $S^2 = \{s_is_j; 1 \le i, j \le 4\}$ is encoded using Huffman's algorithm, assuming that the relative frequency of s_is_j is f_if_j.
   (c) Find the Shannon bounds on the compression ratios in (a) and (b). [Hint: if they are not the same, then something is wrong!]

3. Suppose that $S = \{0, 1\}^L$ for some positive integer L and all source characters are equally likely. Compute the Shannon bound on the compression ratio in this case, and the compression ratios actually achieved by the methods of Shannon, Fano, and Huffman.

4. Suppose $S = \{s_1, \ldots, s_m\}$ is a source alphabet with relative source frequencies $f_j = (1/2)^{\ell_j}$, where $\ell_1, \ldots, \ell_m$ are positive integers. Show that Shannon's method results in an encoding scheme with average code word length $\bar{\ell} = H(S)$. [Hint: this demonstration appears somewhere in Section 5.4.] Do Fano's and Huffman's methods do as well in such a case?

5. Prove the Noiseless Coding Theorem for code alphabets $A = \{a_1, \ldots, a_n\}$, $n \geq 2$. The statement is just the same as for the binary case, except that $\log_2$ is replaced by $\log_n$. In the version of Shannon's method used in one part of the proof, binary expansions are replaced by n-ary expansions. The version of McMillan's theorem to be used in the other part of the proof is the general one, to be found in 4.2.

Chapter 6

Arithmetic Coding

As in the preceding chapter, we have a source alphabet $S = \{s_1, \ldots, s_m\}$ and relative source frequencies $f_1, \ldots, f_m$, presumably estimated by a statistical study of the source text. However, in arithmetic coding it is not the case that individual source letters, or even blocks of source letters, are replaced by binary code words (although replacing blocks of source letters by binary words derived arithmetically is an option; see Section 6.3). Rather, the entire source text, $s_{i_1} \cdots s_{i_N}$, is assigned a codeword arrived at by a rather complicated process, to be described below.

Methods of arithmetic coding vary, but they all have certain things in common. Each source word $s_{i_1} \cdots s_{i_N}$ is assigned a subinterval $A(i_1, \ldots, i_N)$ of the unit interval $[0, 1)$. This assignment takes place in such a way that $A(1), \ldots, A(m)$ are disjoint subintervals of $[0, 1)$, and for $N > 1$, the m intervals $A(i_1, \ldots, i_{N-1}, 1), \ldots, A(i_1, \ldots, i_{N-1}, m)$ are disjoint subintervals of $A(i_1, \ldots, i_{N-1})$; the lengths of these subintervals are to be proportional, or roughly proportional, to $f_1, \ldots, f_m$.

Having determined (in principle) the interval $A(i_1, \ldots, i_N) = A$, the arithmetic encoder chooses a number $r = r(i_1, \ldots, i_N) \in A$ and represents the source word $s_{i_1} \cdots s_{i_N}$ (which is usually the entire source text) by some finite segment of the binary expansion of r. Arithmetic coding methods differ in how r is arrived at, and in how much of the binary expansion of r is taken to encode the source word. Enough of the binary expansion of r will have to be taken so that the decoder will be able to figure out (in principle) in which of the intervals $A(i_1, \ldots, i_N)$ the number r lies; from this the decoder can recover the source word $s_{i_1} \cdots s_{i_N}$. Usually, the smaller the interval $A(i_1, \ldots, i_N)$ is, the farther out you will have to go in the binary expansion of any number in it to let the decoder know which interval, and thus which source word, is signified by the code. For this reason, the larger the intervals $A(i_1, \ldots, i_N)$ are, the better the compression, because the code representative of the source text will be shorter, on average.

Therefore, in "pure" arithmetic coding, the intervals $A(i_1, \ldots, i_N)$, $1 \le i_1, \ldots, i_N \le m$, are not only disjoint, they partition $[0, 1)$. Also, you can see the justification for making the lengths of $A(i_1, \ldots, i_k, j)$, $1 \le j \le m$, proportional to or, at least, increasing functions of, the relative source frequencies f_j, for each k. This policy will result in the more likely source texts $s_{i_1} \cdots s_{i_N}$, in which letters of higher relative source frequency predominate, being assigned longer intervals $A(i_1, \ldots, i_N)$, and will therefore achieve better compression, on the average, than any perverse policy which inverts the order of the lengths of the $A(i_1, \ldots, i_k, j)$ relative to the f_j. More about this later.

The first variety of arithmetic coding that we will consider is probably the best zeroth-order non-adaptive lossless compression method that will ever exist, if such methods are to be judged by the compression ratios achieved over a wide range of source texts. The particular version that we will present in Section 6.1 is not, in fact, in use in the real world. We think that it conveys the main idea of arithmetic coding better than the real world implementations, in which the main idea is somewhat obscured by practical tinkering. Further, since this is a textbook and not a how-to manual, it is appropriate that general paradigms be presented whenever possible. The method of Section 6.1 can be modified in a number of ways to be more practicable, but it might be difficult to go from one of these offspring to another, without understanding their parent.

We will look at the compression ratio achievable by the method of Section 6.1 in Section 6.2, and then consider some of the drawbacks of the method, and possible modifications to overcome those drawbacks, in 6.3. In Section 6.4 we will present a full-fledged practical implementation of arithmetic coding which overcomes every problem with the "pure" method of Section 6.1, at the cost of a certain amount of fudging and approximation that may diminish the compressive power of arithmetic coding, but which seems to compress as well as or better than Huffman encoding, in practice.

6.1 Pure zeroth-order arithmetic coding: dfwld

The 'dfwld' in the title of this section is the acronym for *dyadic fraction with least denominator*. The plan will be to select the dyadic fraction $r = \frac{p}{2^L}$, p an odd integer (or $p = 0$ when $r = 0$), with L as small as possible, as the representative of the interval $A = A(i_1, \ldots, i_N)$. To see why we do this, observe that if the decoder is supplied the source word length N and a number in A, then the decoder can recover the sequence $i_1, \ldots, i_N$, and thus the source word $s_{i_1} \cdots s_{i_N}$. (The decoder knows how the intervals $A(i_1', \ldots, i_N')$, $1 \leq i_1', \ldots, i_N' \leq m$, are calculated, so the decoder could, in principle, calculate them all and then pick the one containing the given number. We shall see a more efficient method of calculating $i_1, \ldots, i_N$ from r and N later in this section.) If $b_1 \cdots b_t$ is a binary word and $\rho = (.b_1 \cdots b_t)_2 \in A$, then ρ is a dyadic fraction in A, with denominator 2^q, where $L \leq q \leq t$; $L \leq q$ because r is the dfwld in A. Therefore, the binary expansion of r, $r = (.a_1 \cdots a_{L-1}1)_2$, supplies the shortest possible code word, namely $a_1 \cdots a_{L-1}1$, from which the decoder can recover the source word.

Thus, choosing the dfwld to represent the interval $A = A(i_1, \ldots, i_N)$ is always a good idea in arithmetic coding, no matter how the intervals are generated.

About notation: when $w = s_{i_1} \cdots s_{i_k}$, a source word, it will sometimes be convenient to denote $A(i_1, \ldots, i_k)$ alternatively as $A(s_{i_1} \cdots s_{i_k})$ or $A(w)$.

Subdividing to find $A(w)$

Suppose that $S = \{s_1, \ldots, s_m\}$ and the relative source frequencies are $f_1 \geq \cdots \geq f_m > 0$. Our intervals will always be closed on the left, open on the right, starting with $[0, 1)$. If $[\alpha, \beta) = [\alpha, \alpha + \ell)$, $\ell = \beta - \alpha$, is our "current interval," either $[0, 1)$ or $A(s_{i_1} \cdots s_{i_{k-1}})$ for some $k \geq 2$ and $i_1, \ldots, i_{k-1} \in \{1, \ldots, m\}$, we want to subdivide $[\alpha, \beta)$ into half-open intervals with lengths proportional to $f_1, \ldots, f_m$. Here are the endpoints of the sought-for subintervals of $[\alpha, \beta)$: $\alpha, \alpha + f_1\ell, \alpha + (f_1 + f_2)\ell, \ldots, \alpha + (\sum_{i=1}^{m-1} f_i)\ell, \beta$. That is, the jth subinterval, $1 \leq j \leq m$, is $[\alpha + (\sum_{i<j} f_i)\ell, \alpha + (\sum_{i\leq j} f_i)\ell)$, where, when $j = 1$, the sum over the empty set of indices is interpreted as zero. It is recommended that the reader verify that the length of the jth subinterval is $f_j\ell$ and that the right-hand endpoint of the mth subinterval is β. [Recall that $\ell = \beta - \alpha$ and that $\sum_{i=1}^m f_i = 1$.]

The process of subdividing these intervals is illustrated below, with $S = \{a, b, c, d\}$, $f_a = .4$, $f_b = .3$, $f_c = .2$, and $f_d = .1$.

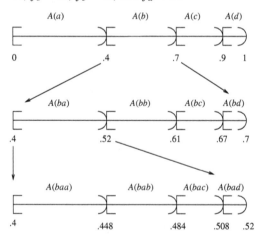

Meanwhile, $A(a)$ and $A(d)$ are subdivided as follows:

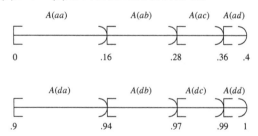

$A(i_1, \ldots, i_N)$ is specified by its left-hand endpoint α and its length ℓ, and

these are straightforward to compute iteratively. If the computation were organized into a table with columns for "next (source) letter", "left-hand endpoint
α", and "length ℓ", the table would look like this:

next letter	α	ℓ
	0	1
s_{i_1}	$\sum_{j<i_1} f_j$	f_{i_1}
$\vdots$	$\vdots$	$\vdots$
$s_{i_{k-1}}$	α	ℓ
s_{i_k}	$\alpha + (\sum_{j<i_k} f_j)\ell$	$f_{i_k}\ell$
$\vdots$	$\vdots$	$\vdots$

Examples

Suppose that $S = \{a, b, c, d\}$, $f_a = .4$, $f_b = .3$, $f_c = .2$, and $f_d = .1$. (As usual,
in the absence of subscripts on the source letters, we use the letters themselves
to subscript the relative source frequencies.) We will calculate the intervals
assigned to the source words $bacb$ and $ccda$.

For $bacb$, the table is

next letter	left-hand endpoint α	length ℓ
	0	1
b	.4	.3
a	.4	.12
c	$.4 + (.12)(.7) = .484$	.024
b	$.484 + (.024)(.4) = .4936$	.0072

Thus $A(bacb) = [.4936, .4936 + .0072) = [.4936, .5008)$. Notice that it is very
easy to find the dfwld in $A(bacb)$, since clearly $.5 = 1/2$ is in this interval,
and $1/2$ is the dfwld in all of $(0, 1)$; in $[0, 1)$ only $0 = 0/1$ beats $1/2$ for least
denominator, among the dyadic fractions. Thus the source word $bacb$ would be
encoded by 1, a single bit, by the method of arithmetic coding of this section.

For the source word $ccda$, the table is

next letter	α	ℓ
	0	1
c	.7	.2
c	$.7 + (.2)(.7) = .84$	.04
d	$.84 + (.04)(.9) = .876$	.004
a	.876	.0016

Thus $A(ccda) = [.876, .8776)$. This time we are unlucky, and the dfwld in this
interval is not immediately apparent.

Finding the dfwld in a subinterval

Suppose that the interval is $[\alpha, \beta)$. We consider two methods to find the best
representative for the interval.

Method 1 Find the smallest integer t such that $\frac{1}{2^t} \leq \ell = \beta - \alpha$; i.e., find the integer t satisfying $2^{-t} \leq \ell < 2^{-t+1}$. Then solve the inequalities $\alpha \leq \frac{x}{2^t} < \beta$ for integers x. There will be at least one, and at most two, integers x satisfying this inequality. In case there are two, they will be consecutive. Take the even one. In any case, $r = \frac{x}{2^t}$ (reduce to lowest terms!) is the dfwld in the interval.

In the first example above, suppose we did not notice that the dfwld is $1/2$. Applying this method, we would find $t = 8$ and set about solving

$$.4936 \leq \frac{x}{256} < .5008 \quad \text{or} \quad 126.3616 \leq x < 128.2048.$$

There are two whole numbers x satisfying these inequalities, 127 and 128. We take $x = 128$ and get $r = \frac{128}{256} = 1/2$.

In the second example, by this method, we find $t = 10$ and set about solving $.876 \leq \frac{x}{1024} < .8776$; $897.024 \leq x < 898.6624$. This time there is only one x, namely $x = 898$. We find

$$r = \frac{898}{1024} = \frac{449}{512} = \frac{256 + 128 + 64 + 1}{512} = (.111000001)_2.$$

Thus the code for $ccda$ is 111000001.

Method 2 Carry out the binary expansions of α and β until they differ. At the first place they differ, there will be a 0 in the expansion of α, and a 1 in the expansion of β; i.e., $\alpha = (.a_1 \cdots a_{t-1} 0 a_{t+1} \cdots)_2$ and $\beta = (.a_1 \cdots a_{t-1} 1 b_{t+1} \cdots)_2$. In most cases, the dfwld in $[\alpha, \beta)$ will then be $r = (.a_1 \cdots a_{t-1} 1)_2$. There are two annoying exceptions to this rule.[1]

Exception 1. If $\alpha = (.a_1 \cdots a_{t-1})_2$, then α itself is the dfwld in $[\alpha, \beta)$.

Exception 2. If $\alpha > (.a_1 \cdots a_{t-1})_2$ (i.e., $a_i = 1$ for some $i > t$) and $\beta = (.a_1 \cdots a_{t-1} 1)_2$ (i.e., $b_i = 0$ for all $i > t$), then $(.a_1 \cdots a_{t-1} 1)_2 = \beta$ is not actually in $[\alpha, \beta)$, and so cannot be the dfwld in that interval. In this case the dfwld r in $[\alpha, \beta)$ is found by continuing the binary expansion of α until a 0 is found among $a_{t+1}, a_{t+2}, \cdots$. If all a_i beyond that point are zero, then, again, α itself is the dfwld in $[\alpha, \beta)$. Otherwise, change that zero to a one and truncate the binary expression at that point to obtain the dfwld. Examples: if $\alpha = (.101010111)_2$ and $\beta = (.101011)_2$, then $r = \alpha$; if $\alpha = (.10101011001 \cdots)_2$ and $\beta = (.101011)_2$ then $r = (101010111)_2$.

In spite of these exceptions, Method 2 is the more "machinable" of the two methods. Note that, with α and β as above, regardless of everything else, the

---

[1]Doug Leonard points out that if $\alpha = (.a_1 \cdots a_{t-1} 0 \cdots)_2$ and $\beta = (.a_1 \cdots a_{t-1} 1 \cdots)_2$ then $r = (.a_1 \cdots a_{t-1} 1)_2$ is the dfwld in $(\alpha, \beta]$, in all circumstances. Therefore, we could avoid those pesky exceptions in Method 2 by changing our way of subdividing intervals so that the intervals wind up closed on the right, open on the left, for the most part. In fact, there is a reasonable way to do this so that for all N, $A(w)$ is of the form $(\alpha, \beta]$ for every $w \in S^N$, with two exceptions: $A(s_1^N) = [0, f_1^N]$ and $A(s_m^N) = (1 - f_m^N, 1)$. However, in real live arithmetic coding, to be described in Section 6.4, it is conventional to take intervals closed on the left, and finding the exact dfwld at the end of the process is not insisted upon. We decided that it would overly complicate the transition from the finicky academic version of arithmetic coding of this section to the implementation version in 6.4 to have our intervals here closed on the right, and our intervals there closed on the left.

binary word $a_1 \cdots a_{t-1}$ will be part of the code stream. The next subsection is about how to take advantage of this fact.

Notice that the annoying exceptional cases occur only when at least one of α, β is a dyadic fraction. Therefore, when neither is, we can forget about those exceptions and take $r = (.a_1 \cdots a_{t-1}1)_2$ as the dfwld. For instance, in the first example on page 144, we had $\alpha = .4936 = (.0\cdots)_2$ and $\beta = .5008 = (.1\cdots)_2$ and neither α nor β is a dyadic fraction, clearly, so $r = (.1)_2 = 1/2$. Similarly, in the second example, leaving out the details of finding the binary expansions, $\alpha = .876 = (.111000000\cdots)_2$, $\beta = .8776 = (.111000001\cdots)_2$, neither dyadic fractions, so $r = (.111000001)_2$.

6.1.1 Rescaling while encoding

The encoding method described in the preceding subsections is, in abbreviation: find the interval $A(i_1, \ldots, i_N) = A$, and then the dfwld r in A. The procedure thus described is wasteful in that initial segments of the final code, the binary expansion of r, are stored twice in the endpoints α_k and $\alpha_k + \ell_k$ of the intermediate intervals $A(i_1, \ldots, i_k)$. Furthermore, there is a waste of time involved: you have to wait until the entire source string is scanned and processed before you have the code for it. Recall that in compression by replacement schemes, you can encode as you go along, without waiting to see what is up ahead in the source string. There are some situations in which this is a great advantage—in digital communications, for instance, when speed is required and typically decoding of the code string starts before encoding of the source string is finished.

But now Method 2 of calculating dfwlds suggests a way of beginning the arithmetic encoding of a source string without waiting for the end of the string, while lessening the burden of computation of the endpoints of the intervals $A(i_1, \ldots, i_k)$, as those endpoints get closer and closer together.

If $\alpha = (.a_1 \cdots a_{t-1}0\cdots)_2$ and $\alpha + \ell = (.a_1 \cdots a_{t-1}1\cdots)_2$, then the first $t-1$ digits in the binary expansion of any number in an interval with endpoints α and $\alpha + \ell$ will be $a_1 \cdots a_{t-1}$. Consequently, if $r = r(i_1, \ldots, i_N)$ is in that interval somewhere, then we know the first $t-1$ bits of the code for $s_{i_1} \cdots s_{i_N}$. We extract those $t-1$ bits and multiply α and $\alpha + \ell$ by 2^{t-1} mod 1 to obtain the endpoints of the new current interval. Multiplying α and $\alpha + \ell$ by 2^{t-1} mod 1 means that we subtract the integer $(a_1 \cdots a_{t-1})_2$ from $2^{t-1}\alpha$ and $2^{t-1}(\alpha + \ell)$. Note that this amounts to shifting $a_1 \cdots a_{t-1}$ in each of α and $\alpha + \ell$ to the left of the binary point and out of the picture. By storing $a_1 \cdots a_{t-1}$ in the code word being constructed (tack it on to the right of whatever part of the code word had been found previously) we are really keeping track of α and $\alpha + \ell$, but we do not need to leave $a_1 \cdots a_{t-1}$ in the binary expressions of these numbers—these extra bits complicate the calculations needlessly.

This process of finding $a_1 \cdots a_{t-1}$ and then replacing the interval by a new interval obtained by multiplying the old endpoints by 2^{t-1} mod 1 is called *rescaling*. Rescaling does not affect the final code word because we go from

$A(i_1, \ldots, i_{k-1})$ to the intervals $A(i_1, \ldots, i_k)$, $1 \le i_k \le m$ by dividing the large interval into subintervals with lengths *proportional* to $f_1, \ldots, f_m$, and because we rescale by multiplying by a power of 2, which preserves binary expressions beyond the part that gets shifted away. Let's try it with $S = \{a, b, c, d\}$, $f_1 = .4$, $f_b = .3$, $f_c = .2$, $f_d = .1$, and $w = ccda$.

Next letter or rescale	α	ℓ	code so far
	0	1	
c	.7	.2	
Rescale (find the binary expansions of $.7 = (.10\ldots)_2$ and $.9 = (.11\ldots)_2$ and shift out the part where they agree)	$.4 = 2(.7) - 1$	$.4 = 2(.2)$	1
c	$.68 = .4 + (.4)(.7)$	$.08 = (.2)(.4)$	1
Rescale $[.68 = (.10\ldots)_2$ and $.76 = (.11\ldots)_2]$	$.36 = 2(.68) - 1$	$.16 = 2(.08)$	11
d	$.504 = .36 + (.9)(.16)$	$.016 = .16(.1)$	11
Rescale $[.504 = (.100000\ldots)_2$ and $.520 = (.100001\ldots)_2]$	$.128 = 2^5(.504) - 16$	$.512 = 32(.016)$	1110000
a	.128	$.2048 = (.512)(.4)$	1110000
Rescale $[.128 = (.00\ldots)_2$ and $.3328 = (.01\ldots)_2]$	$.256 = 2(.128)$	.4096	11100000

Now find the dfwld in $[.256, .6656)$; it is $1/2 = (.1)_2$. Tack this on to the last "code so far" to obtain 111000001 as the code word for $ccda$.

This process superficially may seem more complicated than what we went through before to encode $ccda$ because of the added rescaling steps, but the lines on the first table for $ccda$ that we generated were filled in at the cost of increasingly onerous arithmetic. Rescaling lifts the burden of that arithmetic somewhat and, as an important bonus, gives us initial segments of the final code word early on. (There is one extreme case in which this is not true; i.e., in one case the partial code words supplied by rescaling are not prefixes of the final code word. See Exercise 6.1.5.)

However, as the preceding example shows, rescaling does not supply those initial segments at a regular pace, as the encoder reads through the source word. Furthermore, the example of $bacb$ shows that you may not get any initial segments of the code word at all; in that example, the binary expansions of the endpoints α and $\alpha + \ell$ always differ in the very first position, so there is no rescaling and no partial construction of the code word until the very end of the calculation. A little thought shows that this sort of unpleasantness—no rescaling for a long time—occurs when the dfwld in $A(s_{i_1} \cdots s_{i_k})$ is the same for many values of k. It is slightly ironic that those occasions when the compression with dfwld arithmetic coding is great—when the dfwld in $A(s_{i_1} \cdots s_{i_N})$ has a small denominator—are occasions when there is certain to be a long run of no rescaling in the computation of the intervals and the code word, because $r = r(i_1, \ldots, i_N)$, the dfwld in $A(s_{i_1} \cdots s_{i_N})$, will also be the dfwld in $A(s_{i_1} \cdots s_{i_k})$ for many values of $k \le N$.

There are two inconveniences to be dealt with when there is a long run of no rescaling. The first is that we have to carry on doing exact computations of the endpoints α and $\alpha + \ell$ with smaller and smaller values of ℓ. This is a major drawback, the Achilles heel of arithmetic coding, and we will consider some ways to overcome this difficulty in later sections of this chapter. The other inconvenience is that initial segments of the code representative of the source text are not available beyond a certain point.

We will deal with the first of these problems by a device called the *underflow expansion*, which we found in the work of Cleary, Neal, and Witten [84]. It is essential for their arithmetic coding and decoding algorithm, to be presented in Section 6.4. Before describing the underflow expansion, however, we will make the rescaling operation more practical.

Rescaling one bit at a time

In the account of rescaling given above, the binary expansions of the endpoints α and β of the current interval arc worked out until they disagree. This computation is wasteful and unnecessary.

If the binary expansions of α and β agree at all in the first few bits, then they agree in the first bit. This bit will be 1 if and only if $1/2 \leq \alpha < \beta$, and will be 0 if and only if $\alpha < \beta < 1/2$. We may as well enlarge this second case to $\alpha < \beta \leq 1/2$, since if the current interval is $[\alpha, 1/2)$ then the eventually-to-be-discovered dfwld r in the eventually-to-be-discovered final interval is $< 1/2$, so the first bit in its binary expansion—in other words, the next bit in the code stream—will be 0.

If $1/2 \leq \alpha < \beta \leq 1$, shifting out the initial bit, 1, in the binary expansions of α and β and adding it to the code stream results in $[2\alpha - 1, 2\beta - 1)$ as the new current interval (and the new dfwld that we are seeking is $2r - 1$, if r was the old dfwld, somewhere in $[\alpha, \beta)$. If $0 \leq \alpha < \beta \leq 1/2$, shifting out the bit 0 into the code stream results in $[2\alpha, 2\beta)$ as the new current interval. Thus the rules for rescaling one bit at a time are: if $0 \leq \alpha < \beta \leq 1/2$, replace $[\alpha, \beta)$ by $[2\alpha, 2\beta)$ and add 0 to the code stream; if $1/2 \leq \alpha < \beta \leq 1$, replace $[\alpha, \beta)$ by $[2\alpha - 1, 2\beta - 1)$ and add 1 to the code stream.

Notice that the length $\ell = \beta - \alpha$ is multiplied by 2, in each case. Notice also that rescaling will *not* be possible when and only when $\alpha < 1/2 < \beta$, i.e., when $1/2$ is in the interior of the current interval.

Although rescaling one bit at a time superficially seems to increase the number of operations in dfwld encoding, it in fact provides a "machinable" and efficient way of carrying out the computation of the new code stored in the endpoints of the current interval. Here is the encoding of $ccda$, with source letters and relative frequencies as before, with rescaling one bit at a time. We use "$x \to 2x$" and "$x \to 2x - 1$" to indicate which of the rescaling transformations is being applied.

Next letter or rescale	α	ℓ	New code
	0	1	
c	.7	.2	
$x \to 2x - 1$	.4	.4	1
c	$.68 = .4 + (.4)(.7)$	$.08 = (.2)(.4)$	
$x \to 2x - 1$	.36	.16	1
d	$.504 = .36 + (.9)(.16)$	$.016 = (.16)(.1)$	
$x \to 2x - 1$	.008	.032	1
$x \to 2x$	.016	.064	0
$x \to 2x$	.032	.128	0
$x \to 2x$	.064	.256	0
$x \to 2x$	.128	.512	0
a	.128	$.2048 = (.512)(.4)$	
$x \to 2x$	.256	.4096	0

As before, $1/2$ is the dfwld in the last interval, so the code obtained is again 111000001; of course!

The underflow expansion

Suppose that $1/2$ is the interior of the current interval $[\alpha, \beta)$, in the course of arithmetic encoding, so that rescaling is not possible. Suppose also that $[\alpha, \beta) \subseteq [1/4, 3/4)$; so, we have $1/4 \le \alpha < 1/2 < \beta \le 3/4$.

Now, the eventually-to-be-discovered dfwld r in $[\alpha, \beta)$, whose binary expansion constitutes the rest of the code (added on to the code already generated) is either in $[\alpha, 1/2)$ or in $[1/2, \beta)$. In the former case, $r = (.01\ldots)_2$ because $\alpha \ge 1/4$; in the latter, $r = (.10\ldots)_2$ because $\beta \le 3/4$. The point is that the first and second bits of the binary expansion of r are different. Therefore, if you know one, you know the other.

The transformation $x \to 2x - 1/2$ doubles the directed distance from x to $1/2$; call it the "doubling expansion around $1/2$" if you like. Further, if r is the dfwld in the final interval to be discovered by subdividing $[\alpha, \beta) \subseteq [1/4, 3/4)$, according to the source text, then $2r - 1/2$ will be the dfwld in final interval obtained by so subdividing $[2\alpha - 1/2, 2\beta - 1/2) \subseteq [0, 1)$. (Verify!) Inspecting the effect of this transformation on $r \in [1/4, 3/4)$ we see: if $r = (.01a_3a_4\ldots)_2$ then $2r - 1/2 = (.0a_3a_4\ldots)_2$, and if $r = (.10a_3a_4\ldots)_2$ then $2r - 1/2 = (.1a_3a_4\ldots)_2$. That is, the effect of this transformation on $r \in [1/4, 3/4)$ is to delete the second bit of its binary expansion. But that bit is the opposite of the first bit, which will be discovered the next time a rescaling occurs.

This leads to the following rules for using the underflow expansion.

1. Keep track of the *underflow count*, the number of times that the underflow expansion has been applied since the last rescaling.

2. When the current interval $[\alpha, \alpha + \ell)$ satisfies $1/4 \le \alpha < 1/2 < \alpha + \ell \le 3/4$, replace α by $2\alpha - 1/2$ and ℓ by 2ℓ, and add one to the underflow count.

3. Upon rescaling if the underflow count is k, add 01^k to the code stream if the rescaling transformation is $x \rightarrow 2x$, and 10^k to the code stream if the rescaling transformation is $x \rightarrow 2x - 1$; reset the underflow count to 0. [ξ^k means ξ iterated k times. When $k = 0$, this means the empty string.]

Let's try encoding $babc$, when $S = \{a, b, c, d\}$, in that order, with $f_a = .4$, $f_b = .3$, $f_c = .2$, and $f_d = .1$, using rescaling and the underflow expansion.

Next letter or rescale or underflow	α	ℓ	New code	Underflow count
	0	1		0
b	.4	.3		0
$x \rightarrow 2x - 1/2$	.3	.6		1
a	.3	.24		1
$x \rightarrow 2x - 1/2$	.1	.48		2
b	.292	.144		2
$x \rightarrow 2x$	.584	.288	011	0
$x \rightarrow 2x - 1$	.168	.576	1	0
c	.5712	.1152		0
$x \rightarrow 2x - 1$	.1424	.2304	1	0
$x \rightarrow 2x$	.2848	.4608	0	0

The code is, therefore, 0111101, the last 1 arising from the dfwld $1/2$ in the final interval, $[.2848, .7456)$.

Notice that if the rules for encoding with the underflow expansion are used for encoding $bacb$, we wind up with a final interval containing $1/2$, no code generated, and an underflow count of 4. In this section, we ignore final underflow counts and give the code as 1. In modified arithmetic encoding, we may wind up with $10^4 = 10000$, possibly followed by some more bits of local significance.

Notice that the underflow expansion prevents the current interval from getting arbitrarily small, in arithmetic encoding. But it does not prevent the endpoints of the current interval from having increasingly lengthy representations, if we require exact computation.

6.1.2 Decoding

We assume that the decoder has been supplied the code word for the source word and the length N of the source word. If the code word is 0, then the source word is $s_1 \cdots s_1 = s_1^N$. In all other cases, from the code word the decoder knows the number r, the dfwld in the interval $A(i_1, \ldots, i_N)$ corresponding to the sought-for source word.

As mentioned above, the decoder could, in principle, recover the source word $s_{i_1} \cdots s_{i_N}$ from r and N by computing all the intervals $A(j_1, \ldots, j_N)$, $1 \leq j_1, \ldots, j_N \leq m$, and deciding which of them contains r. A moment's thought shows that this would be senselessly inefficient. A much more sensible approach is to find the index i_1 such that $r \in A(i_1)$, then the index i_2 such that $r \in A(i_1, i_2)$, and so on, until $i_1, \ldots, i_N$—or, equivalently, $s_{i_1}, \ldots, s_{i_N}$—are found. The process neatly exploits the fact that the intervals $A(i_1), A(i_1, i_2), \ldots$ are *nested*, i.e., each is a subinterval of the preceding.

Once we have found $A(i_1, \ldots, i_k)$, we are looking for the unique $j \in \{1, \ldots, m\}$ such that r lies in the subinterval $A(i_1, \ldots, i_k, j)$ of $A(i_1, \ldots, i_k)$. This j will be i_{k+1}. How do we go about finding i_{k+1}? This is where we have to refer to the method by which the interval $A(i_1, \ldots, i_k)$ is sliced up. Suppose $A(i_1, \ldots, i_k) = [\alpha, \alpha + \ell)$. The endpoints of the intervals at the next level are α, $\alpha + f_1\ell, \ldots, \alpha + (\sum_{i<j} f_i)\ell, \ldots, \alpha + (\sum_{i<m} f_i)\ell$, $\alpha + \ell$. Therefore, i_{k+1} will be the index satisfying

$$\alpha + \left(\sum_{i < i_{k+1}} f_i \right)\ell \leq r < \alpha + \left(\sum_{i \leq i_{k+1}} f_i \right)\ell$$

or, equivalently,

$$\sum_{i < i_{k+1}} f_i \leq \frac{r - \alpha}{\ell} < \sum_{i \leq i_{k+1}} f_i.$$

So, to sum up: having found $A(i_1, \ldots, i_k) = [\alpha, \alpha + \ell)$ containing r, find i_{k+1}, the largest index among those j such that $\sum_{i<j} f_i \leq \frac{r-\alpha}{\ell}$. Then

$$A(i_1, \ldots, i_k, i_{k+1}) = \left[\alpha + \left(\sum_{i < i_{k+1}} f_i \right)\ell, \alpha + \left(\sum_{i \leq i_{k+1}} f_i \right)\ell \right);$$

iterate the process until $i_1, \ldots, i_N$ have been found.

Let's try the procedure in the case $S = \{a, b, c, d\}$, $f_a = .4$, $f_b = .3$, $f_c = .2$, and $f_d = .1$ and the decoder is given the code word 1 and $N = 4$.[2] As before, we use the source letters themselves rather than indices. In the following table, α denotes the left-hand endpoint of the current interval and ℓ denotes its length; also, $r = 1/2 = (.1)_2$:

α	ℓ	$\dfrac{r - \alpha}{\ell}$	Next letter
0	1	$1/2 = .5$	b (since $.4 \leq .5 < .7$)
.4	.3	$\frac{.5-.4}{.3} = .3\overline{3}$	a (since $0 \leq .3\overline{3} < .4$)
.4	.12	$\frac{.5-.4}{.12} = \frac{5}{6} = .8\overline{3}$	c (since $.7 \leq .8\overline{3} < .9$)
$.4 + (.12)(.7)$ $= .484$	.024	$\frac{.5-.484}{.024} = \frac{2}{3}$	b

and the process is complete since $N = 4$.

Now, let's try it with the code 111000001 and $N = 4$. We know the right answer: *ccda*. And, this time, we will put into the mix the decoder's version

---

of rescaling and the underflow expansion. The rescaling and underflow transformations will be applied to r, of course. We start with $r = (.111000001)_2 = 449/512$

α	ℓ	r	$\dfrac{r - \alpha}{\ell}$	Next letter or rescale or underflow
0	1	$\frac{449}{512}$	$\frac{449}{512} \approx .88$	c
.7	.2	$\frac{449}{512}$	wait for rescaling	$x \to 2x - 1$
.4	.4	$\frac{193}{256}$	$\frac{\frac{193}{256} - .4}{.4} \approx .88$	c
$.68 = .4 + (.7)(.4)$	$.08 = (.4)(.2)$	$\frac{193}{256}$	wait	$x \to 2x - 1$
.36	.16	$\frac{65}{128}$	wait	$x \to 2x - 1/2$
.22	.32	$\frac{33}{64}$	$\frac{\frac{33}{64} - .22}{.32} \approx .92$	d
$.22 + (.9)(.32) = .508$	.032	$\frac{33}{64}$	wait	$x \to 2x - 1$
.016	.064	$\frac{1}{32}$	wait	$x \to 8x$
.128	.512	.25	$\frac{.25 - .128}{.512} \approx .24$	a

We apologize for cheating by combining three rescalings of the form $x \to 2x$ into one, $x \to 8x$, in the next to last line of the decoding table above.

There is a way to simplify the decoding process that combines rescaling with the discovery of the next source letter. This method superficially seems more efficient than the decoding method (or methods, if you count with and without rescaling as different) used so far, but it contains a fatal defect that limits its applicability. We will discuss this defect in Section 6.3. Meanwhile, for all its practical defects, this method may be of academic interest, so here it is.

Given $r = r(i_1, \ldots, i_N)$ and N, we generate two sequences, $r_0, r_1, r_2, \ldots$ and $j_1, j_2, \ldots$ (or, equivalently, $s_{j_1}, s_{j_2}, \ldots$), as follows:

1. Set $r_0 = r$.

2. For $1 \le k \le N$, having found $j_1, \ldots, j_{k-1}$ and r_{k-1}, the decoder finds j_k, the largest index such that $\sum_{j < j_k} f_j \le r_{k-1}$.

3. If $k = N$, the decoder is done. Otherwise, the decoder sets $r_k = f_{j_k}^{-1}(r_{k-1} - \sum_{j < j_k} f_j)$ and returns to 2, with k replaced by $k + 1$.

You are asked to show, in Exercise 6.1.4, that the sequence $j_1, \ldots, j_N$ is actually the sought-for sequence $i_1, \ldots, i_N$ such that $r = r(i_1, \ldots, i_N)$.

For example, let us decode the code words for *bacb* and for *ccda* that were found on page 144. As usual, we will use source letters instead of numeric indices.

Code: 1; $r = r_0 = .5$; $N = 4$. Decoding table:

k	r_k	next letter
0	.5	b (since $.4 \leq r_0 < .7$)
1	$(.3)^{-1}(.5-.4)$ $= 1/3 = .33\ldots$	a
2	$(.4)^{-1}(1/3-0)$ $= 5/6 = .833\ldots$	c
3	$(.2)^{-1}(\frac{5}{6}-.7)$ $= 2/3 = .66\ldots$	b

Thus we decode 1, with $N = 4$, as $bacb$.

Code: 111000001; $r = r_0 = \frac{449}{512} = 0.876953125$; $N = 4$. Table:

k	r_k	next letter
0	$449/512$	c
1	$(.2)^{-1}(449/512-.7)$ $= \frac{453}{512} = .884765625$	c
2	$(.2)^{-1}(\frac{453}{512}-.7)$ $= \frac{473}{512} = .923828125$	d
3	$(.1)^{-1}(\frac{473}{512}-.9)$ $= \frac{61}{256} = .23828125$	a

The source word decoded by this process is $ccda$, which is what it should have been.

Notice that we do not have to find the intervals $A(i_1, \ldots, i_k)$, $1 \leq k \leq N$, in this method. Calculating those mysterious r_k does the job.

Exercises 6.1

1. Suppose that $S = \{a, b, c, d\}$ and $f_a = .35$, $f_b = .3$, $f_c = .25$, and $f_d = .1$.

   (a) Encode $bbbb$, $abcd$, $dcba$, and $badd$ by the method of this section, assuming that the decoder will be given the source word length.

   (b) Decode 11, 010001, 10101, and 0101, assuming the source word lengths are all 4.

2. One of the disadvantages of the method of arithmetic coding described in this section is that the encoder must supply the decoder with the length N of the source word that has been encoded. Supplying information of a different type outside the main code stream is usually extremely inconvenient.

   What if the encoder were to supply just the code and not N? Then the decoder would know the dfwld r in $A(i_1, \ldots, i_N)$, where $s_{i_1} \cdots s_{i_N}$ is the source word encoded. The problem is that r may also be the dfwld in various intervals $A(i_1, \ldots, i_k)$, $k < N$; if r is the dfwld in $A(i_1, \ldots, i_k)$, then r is the dfwld in any interval $A(i_1, \ldots, i_k, \ldots, i_t)$, $t > k$, that r happens to lie in, because the intervals are getting smaller and the denominator of r is not getting any bigger.

Here is one way that the encoder could communicate the source word length N without special arrangements. Having found $A(i_1,\ldots,i_N)$ and r, the encoder finds the smallest value of k such that r is the dfwld in $A(i_1,\ldots,i_k)$, and then adds $N-k$ zeros to the code string. For example, on page 144, $bcba$ would be encoded 1000, while $ccda$ would be encoded 111000001, the same as before, because the dfwld in $A(ccd)$ is $\frac{225}{256} = (.11100001)_2$.

The decoder would proceed by the recovering-the-intervals method, with the additional burden of checking whether or not r is the dfwld in each successive interval. (Since rescaling amounts to shifting out the prefixes in which the interval endpoints agree, this checking is not too terrible; you apply rescaling to r until the current value of r is $1/2$ or zero.) Once r is the dfwld, then the decoder knows from the number of zeros remaining in the code how long the source word is, and can proceed to decode by any effective method.

Pretty clearly there are some awkwardnesses and inefficiencies to be dealt with in the implementation of such a coding method, and adding those zeros makes the code representatives of source words longer. Still, you might keep this in mind as a possible solution to the problem of specifying the source text length in arithmetic coding. Compare it with the trick employed in Section 6.4, for elegance. Now, some exercise!

(a) In Exercise 1(a) above, how would the encoding be different if the encoder communicates the source word length (4) by adding zeros to the code words?

(b) In the situation of Exercise 1, decode 011000, assuming that the source word length has been indicated by those three extra zeros, as described above.

3. In the situation of Exercise 1 above, suppose that decoding has to proceed before encoding is finished, and the encoder, by rescaling, is supplying as much of the code string as possible to the decoder, and well as the length of the source string read through so far. Suppose that the encoder informs the decoder that the first 3 bits of the code string are 010 and that seven source letters have been read (and no more bits of the code string are available beyond 010). What are those first seven source letters? (After attempting this problem, see Section 6.3.3.)

4. Suppose that $S = \{s_1,\ldots,s_m\}$ is a source alphabet with relative frequencies $f_1 \geq \cdots \geq f_m > 0$, and $s_{i_1}\cdots s_{i_N}$ is a source word. For $1 \leq k \leq N$, let $A(i_1,\ldots,i_k)$ have endpoints α_k and $\alpha_k + \ell_k$. (No rescaling in this problem!) Let $r \in A(i_1,\ldots,i_N)$. (If you want, take r to be the dfwld in $A(i_1,\ldots,i_N)$.) Let $r_0, r_1,\ldots$ and $j_1, j_2,\ldots$ be the sequences described at the end of Section 6.1.2; i.e., $r_0 = r$, j_k is the largest index among $1,\ldots,m$ satisfying $\sum_{j<j_k} f_j \leq r_{k-1}$, and $r_k = f_{j_k}^{-1}(r_{k-1} - \sum_{j<j_k} f_j)$, $k = 1,2,\ldots$. Prove the

validity of the method that involves the r_k, i.e., that $i_k = j_k$, $k = 1, 2, \ldots$, by showing that $r_k = \frac{r - \alpha_k}{\ell_k}$, $k = 1, 2, \ldots$. [Go by induction on k; note that $r_0 = \frac{r - \alpha_0}{\ell_0} = r$. Next show that $i_1 = j_1$ and $r_1 = \frac{r - \alpha_1}{\ell_1}$. In the induction step, assume that $i_{k-1} = j_{k-1}$, $r_{k-1} = \frac{r - \alpha_{k-1}}{\ell_{k-1}}$, and show that $i_k = j_k$ and that $r_k = \frac{r - \alpha_k}{\ell_k}$.]

5. If the first few source letters are all s_1, what will the partial code word provided by rescaling look like? Try it with $f_1 = .4$ and source text $W = s_1 s_1 s_1 s_1 \cdots$. Now you know the only case, alluded to in Section 6.1.1, in which the partial code word provided by rescaling is not necessarily an initial segment of the final code word: it is the case in which the source text W is s_1^N.

6.2 What's good about dfwld coding: the compression ratio

Suppose that $A = [\alpha, \beta)$ is an interval of length $\ell \leq 1$ and that t is an integer satisfying $2^{-t} \leq \ell < 2^{-(t-1)}$. The basis for method 1 on page 144 is the observation that A must contain some fraction of the form $x/2^t$, x an integer; if not, then A would be contained in an interval of the form $(\frac{x}{2^t}, \frac{x+1}{2^t})$ and would thus have length $< 2^{-t}$.

Therefore, the binary expansion through the last 1 of the dfwld in A (neglecting the possibility that the dfwld might be zero) is of length no greater than t. On the other hand, $\ell < 2^{-(t-1)}$ implies $t < \log_2(1/\ell) + 1$. These observations give rise to the following.

6.2.1 Theorem *Suppose that $S = \{s_1, \ldots, s_m\}$ is a source alphabet, and the source letters have relative frequencies $f_1, \ldots, f_m > 0$ in the source text. Then the average length of the code words for the source words of length N, derived by the arithmetic coding method of Section 6.1, is no greater than $NH(S) + 1$ where (as usual) $H(S) = -\sum_{j=1}^{m} f_j \log_2 f_j$.*

Proof: Let (as in Chapter 7) $f(i_1, \ldots, i_N)$ denote the relative frequency of $s_{i_1} \cdots s_{i_N}$ among all source words of length N, for $1 \leq i_1, \ldots, i_N \leq m$. Let $t(i_1, \ldots, i_N)$ denote the length of the arithmetic code word for $s_{i_1} \cdots s_{i_N}$; so the average length of the code words for the source words of length N is

$$\sum_{1 \leq i_1, \ldots, i_N \leq m} f(i_1, \ldots, i_N) t(i_1, \ldots, i_N).$$

By the observations preceding this theorem, and the fact that the interval $A(i_1, \ldots, i_N)$ has length $\prod_{j=1}^{N} f_{i_j}$, we have

$$\sum_{1 \le i_1,\ldots,i_N \le m} f(i_1,\ldots,i_N) t(i_1,\ldots,i_N)$$

$$< \sum_{\overline{i}} f(i_1,\ldots,i_N) \left(\log_2 \left(\prod_{j=1}^{N} f_{i_j} \right)^{-1} + 1 \right)$$

$$= \sum_{\overline{i}} f(i_1,\ldots,i_N) \left(\sum_{j=1}^{N} \log_2 f_{i_j}^{-1} \right) + \sum_{\overline{i}} f(i_1,\ldots,i_N)$$

$$= \sum_{i_1=1}^{m} \left(\sum_{1 \le i_2,\ldots,i_N \le m} f(i_1,\ldots,i_N) \right) \log_2 f_{i_1}^{-1}$$

$$+ \sum_{i_2=1}^{m} \left(\sum_{1 \le i_1,i_3,\ldots,i_N \le m} f(i_1,\ldots,i_N) \right) \log_2 f_{i_2}^{-1}$$

$$+ \cdots + \sum_{i_N=1}^{m} \left(\sum_{1 \le i_1,\ldots,i_{N-1} \le m} f(i_1,\ldots,i_N) \right) \log_2 f_{i_N}^{-1} + 1$$

$$= \sum_{i_1=1}^{m} f_{i_1} \log_2 f_{i_1}^{-1} + \sum_{i_2=1}^{m} f_{i_2} \log_2 f_{i_2}^{-1} + \cdots + \sum_{i_N=1}^{m} f_{i_N} \log_2 f_{i_N}^{-1} + 1$$

$$= NH(S) + 1. \qquad \square$$

The reader should scrutinize and ponder the equation

$$\sum_{1 \le i_2,\ldots,i_N \le m} f(i_1,\ldots,i_N) = f_{i_1},$$

one of N such used in the proof above. See Chapter 7.

Theorem 6.2.1 draws a conclusion about zeroth-order arithmetic encoding, and that conclusion is that the average number of bits per source letter achieved by pure dfwld encoding, as in Section 6.1, is vanishingly close to the Holy Grail, $H(S)$ (widely believed, although not proven, except for replacement methods by fixed encoding schemes, to be an absolute lower bound on the number of bits per source letter achievable by uniquely decodable zeroth-order coding methods; see comment 4 below). We will see a corresponding theorem about higher-order arithmetic encoding in Section 7.3. We prefer not to attempt a rigorous definition of "zeroth-order," but it might help the reader to understand the comments below if we mention the following. Zeroth-order statistical coding methods are distinguished by being based on the relative source letter frequencies alone, and not on more complicated statistical information like the relative frequencies of certain two-letter sequences, for instance. A perfect zeroth-order source is a source that emits source letters randomly and *independently*, with certain probabilities. Note that it is *not* assumed in Theorem 6.2.1 that the source is a perfect zeroth-order source; it is the coding method that is zeroth-order.

Comments

1. Perhaps it should be emphasized that "the source," envisaged by Shannon as a mysterious probabilistic finite state automaton (see Section 7.5), produces, or could produce, an infinite amount of "source text." The usual situation is that we do not know the structure of the source and can only carry out statistical studies of finite samples of source text. The relative frequencies $f_1, \ldots, f_m$ are really a priori probabilities that we would be able to calculate exactly if only we knew the structure of the source; usually we can only hope to approximate them by statistical study. Roughly speaking, the Law of Large Numbers says that our approximations will be good with high probability if our sample of the source text is large.

In Theorem 6.2.1, $f_1, \ldots, f_m$ are assumed to be exact, the true a priori probabilities of the various source letters being emitted by the source (at randomly chosen instants). However, analysis similar to that contained in the proof of Theorem 6.2.1 shows that if $f_1, \ldots, f_m$ are merely good approximations to the true relative source frequencies $\widehat{f}_1, \ldots, \widehat{f}_m$, in the sense that the ratios $\widehat{f}_i / f_i$ are all close to 1, and if these approximate relative source frequencies are used to subdivide intervals in zeroth-order arithmetic encoding of source words of length N, then the average number of bits per code representative of these source words will be no greater than $NH(S) + 1 + \epsilon$, where $\epsilon \to 0$ as $\max_i |1 - (\widehat{f}_i / f_i)| \to 0$, and $H(S) = -\sum_{i=1}^{m} \widehat{f}_i \log_2 \widehat{f}_i$, the true zeroth-order source entropy. Note also that $H(S)$ will be well approximated in this case by $-\sum_{i=1}^{m} f_i \log_2 f_i$.

2. It may be useful to think of the source words of length N that are being encoded arithmetically as blocks of N consecutive letters taken at random from the infinite source text referred to above. Because we are considering all possible source words of length N, the proof of Theorem 6.2.1 has to resort to an averaging process that may possibly obscure the important point that zeroth-order arithmetic coding with respect to the relative source frequencies $f_1, \ldots, f_m$ will encode any source word w in around $-\log_2 \ell(w)$ bits or less, where, for $w = s_{i_1} \cdots s_{i_N}$, $\ell(w) = \prod_{j=1}^{N} f_{i_j}$, the length of the "final interval" $A(w)$. Notice that if w is completely typical, and the source letters occur in w in exactly the proportions $f_1, \ldots, f_m$, then $\ell(w) = \prod_{i=1}^{m} f_i^{f_i N}$ and $-\log_2 \ell(w) = -N \sum_i f_i \log_2 f_i = NH(S)$. If you don't mind a certain amount of fudging, you could say that this observation establishes that zeroth-order arithmetic coding of text from a source S encodes "typical" source text in around $H(S)$ bits per source letter.

Bell, Cleary, Neal, and Witten [8, 84] call $-\log_2 \ell(w)$ the *entropy of the message* w, and they and Langdon and Rissanen [43, 44] interpret it as the number of bits that "ought" to be allocated to the encoding of w, if the given relative source frequencies are correct.

3. It may be worth emphasizing again that in Theorem 6.2.1 there are no assumptions about the nature of the source, beyond the existence of the rela-

tive source letter frequencies $f_1, \ldots, f_m$ and $f(i_1, \ldots, i_N), 1 \leq i_1, \ldots, i_N \leq m$. In particular, the source is not assumed to be a perfect zeroth-order source, emitting the source letters randomly and independently. The information given about the source (i.e., the relative source letter frequencies) is zeroth-order, but the source itself need not be.

4. Restated in statistical shopkeeper's terms, Theorem 6.2.1 says that dfwld arithmetic coding uses, on average, no more than $H(S) + 1/N$ bits per source letter, in the encoding of source words of length N. Now as has been mentioned from time to time in the information theory part of this book (see Chapter 2, the discussion following Theorem 4.3.7, and Section 4.6), the source entropy $H(S)$ bears the interpretation of being the average amount of information carried by an individual source letter in the source text, and the use of the base 2 in the logarithm involved in $H(S)$ means that information is being measured in *bits*. That is, the average source letter in the source text carries $H(S) = \sum_{j=1}^{m} f_j \log_2(1/f_j)$ bits of information.

Therefore, it should be impossible for any *lossless* (zeroth-order statistical) method of encoding to encode source text using, on the average, less than $H(S)$ bits per source letter. The content of Theorem 6.2.1 is the often repeated mantra that arithmetic coding is *optimal* among lossless zeroth-order statistical coding methods, that it takes you as close to the shrieking limit of compression theoretically achievable by such methods as you could wish. (In fact, the words "zeroth-order" often are omitted from this mantra.)

Let us digress briefly into controversy concerning those words "should be impossible" in the paragraph above. It is a widely held belief that "should be" can be confidently replaced by "is" in that statement (although, as noted, it would be reckless to omit "zeroth-order statistical"). This belief arises from faith in Shannon's quantification of *information*, which treats information as an incompressible fluid; you can compress text by squeezing out redundancy and unused space between nuggets of information, but you cannot (so the belief says) put a certain amount of information into a container (a code representation) that is too small for that amount of information, without losing information. Try pouring 12 ounces of your favorite liquid into a glass that holds only 10 ounces, and the intuitive idea becomes clear.

We are believers, also, but would like to point out that, so far as we know, it has never been rigorously proven that there cannot be a source S and a binary zeroth-order lossless statistical encoding method which encodes text from this source in fewer than $H(S)$ bits per source letter. For one thing, there is the problem of saying exactly what we mean by a zeroth-order statistical encoding method. Notice that the Noiseless Source Coding Theorem (5.4.3 and also 7.2.1) establishes the result for a particular class of such methods, namely, replacement via encoding scheme. Bell, Cleary, and Witten, in their canonical classic on lossless methods [8], attempt (p. 47) a proof of the Noiseless Source Coding Theorem in apparently greater generality, but their proof may suffer from some difficulties, located near the beginning of the attempt. In fact, Exercise 6.2.1 at the end of this section shows that what they seem to aim to prove there cannot be proven because it is not true—although this assertion can be debated on the grounds that some fudging is possible in the understanding, in Bell, Cleary, and Witten's proof, of what it means for a "set of codes" to represent "messages" (i.e., source words), and on the admission below of a hidden cost that is left out of account in Exercise 6.2.1. Perhaps a more devastating objection to their attempted proof, and a test for all such attempted proofs, is provided by the observation that it is quite normal for higher order Huffman encoding to encode in significantly fewer bits per source letter than $H(S)$ (see Section 7.1). Therefore, a proof based on assumptions about the encoding method that does not rule out higher order Huffman encoding must be fallacious.

Without giving away any answers, we note that it appears that Exercise 6.2.1 suggests that it is possible, in a certain extreme case, to encode source words of length N by dfwld arithmetic coding at an average cost per source letter of slightly less (where "slightly less" is a function of N) than $H(S)$ bits per source letter. This is enough to shoot down certain rash statements about the entropy being an absolute lower bound on the "average code word length," or at least to make us skeptical of such statements; but the result of the exercise, and of the preceding theorem, leave out of account a certain hidden cost of arithmetic coding as described in Section 6.1, namely, the necessity of supplying to the decoder the length N of the source word. This will require $1 + \lfloor \log_2 N \rfloor$ bits, and will therefore increase the average number of bits per source letter by about $(1 + \log_2 N)/N$. When you throw this into the average computed in Exercise 6.2.1, "slightly less" than $H(S)$ becomes "slightly more."

In Section 6.4 we will consider an algorithm for arithmetic coding that simulates pure dfwld encoding without passing the length N of the source word encoded. However, this algorithm operates with a new cost, an extraneous source symbol *EOF* for "end of file," to be used once at the end of the source text. This symbol has to be assigned a relative frequency, and the other source letters have to have their relative frequencies trimmed. We haven't done the arithmetic, but surely this new device costs enough to keep the average number of bits per source letter above the entropy. We are Shannonite believers, you see, but we like to keep track of what has been proven and what has not.

Exercises 6.2

1. Suppose that $m = 2^L$ for some positive integer L, and that $f_j = 2^{-L}$, $j = 1, \ldots, m$. Note that, by Theorem 5.4.3, since the f_j are integral powers of $1/2$, and since Huffman's algorithm always gives the smallest $\bar{\ell}$ that can be achieved with a prefix code, in this case the $\bar{\ell}$ produced by Huffman's algorithm will be H, the source entropy.

   (a) Compute H. If the source letters $s_1, \ldots, s_{2^L}$ are really the binary words of length L, what is the compression ratio achieved by Huffman's algorithm in this case?

(b) Suppose that the source is a perfect zeroth-order source, meaning that $f(i_1, \ldots, i_N) = \prod_{j=1}^{N} f_{i_j} = 2^{-NL}$ for all positive integers N and $1 \leq i_1, \ldots, i_N \leq m$, with $f(i_1, \ldots, i_N)$ denoting, as in the text, the relative frequency of the source word $s_{i_1} \cdots s_{i_N}$ among all source words of length N. Recall that $H(S^N) = N H(S)$ in this case.

Therefore, this assumption implies that encoding S^N by Huffman's algorithm will result in no compression at all by (a). The average code word length achieved by applying Huffman's algorithm to S^N will be NL; thus the average number of bits per original source letter in the encoding will be L.

Show that the average length of a code word representing a source word of length N, derived by the arithmetic coding method of Section 6.1, is $NL - 1 + 2^{-(NL-1)}$. (Get started by noting that the dfwld in $A(i_1, \ldots, i_N)$ is the left hand-endpoint. These left-hand endpoints are $\frac{j}{2^{NL}}, 0 \leq j \leq 2^{NL} - 1$. You will probably need to know that $\sum_{j=1}^{k} j 2^{j-1} = (k-1)2^k + 1$, $k = 1, 2, \cdots$. You can prove this by induction; or, differentiate both sides of $\sum_{j=0}^{k} x^j = \frac{x^{k+1}-1}{x-1}$ and plug in $x = 2$.)

Thus, arithmetic coding appears to beat Huffman in this case. Not by much, but then it is a severely intractable case. However, the appearance is deceptive because, as noted in the remarks at the end of this section, you have to pass the length N of the source text to the decoder along with the code text, and that adds around $\log_2 N$ bits to the total code package; this is not much compared with NL, but it is much bigger than $1 - 2^{-(NL-1)}$.

2. Suppose that $s_1, \ldots, s_m$ are binary words with the SPP and, in the class of files to be parsed by the s_j, the relative frequencies of $s_1, \ldots, s_m$ in the resulting source text are $f_1, \ldots, f_m$, respectively. Suppose that $\bar{L} = \sum_{j=1}^{m} f_j \operatorname{lgth}(s_j)$. Show that the typical compression ratio achieved in arithmetically encoding original files that translate into source words of length N is no less than $\bar{L}/[H(S) + N^{-1}(2 + \log_2 N)]$.

6.3 What's bad about dfwld coding and some ways to fix it

As promised, we look at some of the impractical features of pure dfwld arithmetic coding and make some suggestions. In Section 6.4 all of these impracticalities will be overcome, with the sacrifice of a certain purity.

6.3.1 Supplying the source word length

In the method of Section 6.1, the encoder must supply the decoder with the
length N of the source text, but how is this to be done? If the encoder sends the
binary representation of N at the beginning of the code stream, how is the decoder to know when that representation is finished and the code proper begins?
Some possibilities:

1. If it is known that there will never be more than M source letters in
any source text to be dealt with, then you can reserve $\lfloor \log_2 M \rfloor + 1$ bits at the
beginning of the code text for transmitting N. If no bound on the source text
length is known, you can still use this device by choosing M reasonably large
and reserving $\lfloor \log_2 M \rfloor + 2$ bits at the beginning of the code text; the last bit of
these is a "warning bit" which, if set at 1, warns the decoder that the expression
for N has overflowed the allotted space and will continue into the next block
of $\lfloor \log_2 M \rfloor + 2$ bits equipped with its warning bit; and so on. The code finally
commences when the last warning bit is zero.

The disadvantage of this solution to the problem of supplying N is that it
compounds the problem discussed in 6.3.3, below. Presumably the encoder will
keep a count of the source letters while encoding. If the encoder is to convey
the number N of source letters at the *beginning* of the code text, then there
will be a great delay; the decoder will not even get a peek at the partial code
word supplied by rescaling until the encoding is complete. This is all right in
those leisurely situations in which the encoded, compressed text is to be stored
away and decompressed later, but the other kind of situation is encountered with
increasing frequency.

We could convey N or a running count of the source letters encoded in some
other location outside the main code stream. However, providing companion
locations or parallel streams is inconvenient precisely in those situations when
we are in a hurry and hope to decode on the heels of encoding. We will have
more to say about this in 6.3.3.

2. The method of Exercise 6.1.2 smoothly communicates the source word
length by adding a certain number of zeros onto the code word. The method
imposes an extra burden of computation on both the encoder and the decoder,
but this disadvantage is not as important as the fact that this method appears
to be incompatible with solutions to the problem addressed in 6.3.3; how can
decoding proceed on the heels of encoding if the decoder does not know whether
a string of zeros in the code stream is part of the regular code word or part of the
extra zeros at the end? This problem could be dealt with by providing a marker
of, or a pointer to, the end of the regular code word, but this would again raise
the technical difficulty of supplying an extra location or stream of information
outside the main code stream.

3. The algorithm of Section 6.4, yet to come, eliminates the necessity of
counting the source letters, at the cost of introducing an extra source letter,
usually called *EOF*, for "end of file." This extra letter will be used once, to

mark the end of the source text.

The disadvantage in this trick is that compression will be somewhat less than optimal, not so much because of the extra bits required for *EOF* as because the original relative source frequencies will have to be trimmed a bit to make room for the small relative frequency to be assigned to *EOF*. However, the entropy of the modified source can be made as close as desired to the original entropy by making f_{EOF} sufficiently small (because $x \log_2 x$ is a continuous function of $x > 0$ and $x \log_2 x \to 0$ as $x \downarrow 0$). Therefore, by analysis similar to that in the proof of Theorem 6.2.1 and the remarks following, you can get within a cat's whisker of optimal lossless compression, on the average, for long source texts, using *EOF* and the algorithm of Section 6.4.

6.3.2 Computation

The arithmetic coding method of Section 6.1 requires exact computations, both in encoding and decoding. These are costly, especially the multiplications. The length of $A(i_1, \ldots, i_N)$ is $\prod_{j=1}^{N} f_{i_j}$, which on the average requires around N times the number of bits to store (never mind compute) as the average number of bits per number required to store the (rational) numbers $f_1, \ldots, f_m$.

Rescaling and the underflow expansion may appear to relieve the burden of exact computation. However, note that these operations involve multiplying the interval lengths by powers of two. Therefore, odd factors of the denominators of the f_j are never reduced by rescaling, and, if bigger than one, will cause the complexity of and storage space required for exact computations to grow inexorably, approximately linearly with the number of source letters. This observation inspires the first of three suggestions for lessening the burden of exact computation.

Replace the f_j by approximations which are dyadic fractions. For example, if $m = 4$ and $f_1 = .4$, $f_2 = .3$, $f_3 = .2$, and $f_4 = .1$, you could take $\tilde{f_1} = \frac{102}{256}$, $\tilde{f_2} = \frac{77}{256}$, $\tilde{f_3} = \frac{51}{256}$, and $\tilde{f_4} = \frac{26}{256}$, these being the closest (by most definitions of closeness) dyadic fractions with common denominator 256 to the actual values of f_1, f_2, f_3, and f_4. It may appear that replacing the f_j in this case by these approximations will actually increase the burden of computation, because the approximations are nastier-looking fractions than the original f_j, and this is indeed a consideration; we could make life easier if we replace the f_j by dyadic fraction approximations with denominator 8 or 16—but then our approximations would not be very close to the true relative frequencies, and that might affect compression deleteriously. [It can be shown, by analysis similar to that in the proof of Theorem 6.2.1, that as approximate relative frequencies tend to the true relative frequencies, the average number of bits per source letter in code resulting from arithmetic coding of source words of length N, using the approximate relative frequencies, will eventually be bounded above by $H(S) + \frac{1+\epsilon}{N}$, for any $\epsilon > 0$, where $H(S)$ is the true source entropy. Thus good approximations ensure good approaches to optimal lossless encoding; however, not much is known about the penalty to be paid for bad approximations.]

Encode blocks of source letters of a certain fixed length. After each block is encoded, the encoder starts over on the next block. The computational advantage is that the denominators of the rational numbers that give the interval endpoints and lengths cannot grow without bound, even if the relative source frequencies are not dyadic fractions, since the calculations start over periodically.[3]

But how will the decoder know where the code for one block ends and the code for the next block begins? If an efficient method of providing non-binary markers or a parallel pointer/counter stream outside the main code stream is ever devised, this might be a good place to use it. In the absence of any such technical convenience, we could use a modification of the method of Section 6.4, with the artificial source letter *EOB* for "end of block" to be inserted by the encoder into the source text at the end of each block. Of course, this device costs something in diminished compression. The longer the blocks, the less the cost of *EOB*, but the greater the cost of computation.

Use approximate arithmetic. This third suggestion for avoiding computational arthritis in arithmetic coding is the method actually used in the proposed implementation in Section 6.4. The interval $[0, 1)$ is replaced by an "interval" of consecutive integers, $\{0, \ldots, M - 1\}$, which we will continue to denote $[0, M)$, and in subdivisions of this interval the source words are allocated blocks of consecutive integers *approximately* as they would be allocated in pure dfwld arithmetic coding using the full interval of real numbers from 0 to M. Thus, if $f_1 = .4$, $f_2 = .3$, $f_3 = .2$, $f_4 = .1$, and $M = 16$, then $A(1) = A(s_1) = \{0, 1, 2, 3, 4, 5\} = [0, 6)$, $A(2) = A(s_2) = \{6, 7, 8, 9, 10\} = [6, 11)$, etc. Are you worried that subsequent subdivision will shrink the intervals to lengths less than one so that they may fail to contain any integers at all? Well may you worry! This unpleasant possibility is taken care of by starting with M sufficiently large, with respect to the relative source frequencies, and by rescaling and applying the underflow expansion.

This trick solves the problem of exact computation by simply doing away with exact computation. The disadvantage lies in the level of compression achievable. This disadvantage has been considered in [8, 33, 48], but there is room for further analysis. Some experimental results comparing pure dfwld

---

[3]You might observe that block arithmetic encoding amounts to using an encoding scheme for S^N, where N is the length of the source blocks. This encoding scheme definitely does not satisfy the prefix condition; for instance, the single digit 1 is the code word representative of some member of S^N, and is also the first digit of the code representatives of a great many others.

However, the luxury of instantaneous decoding available with a prefix-condition encoding scheme for S^N is illusory. If $|S|$ is fairly large, say $|S| = 256$, and N is fairly hefty, say $N = 10$, then an encoding scheme for S^N would have a huge number, $|S^N| = |S|^N$, of lines; $256^{10} = 2^{80}$ is an unmanageable number of registers necessary to store an encoding scheme. So a nice prefix-condition scheme for S^N is of no practical value in any case. You can think of the decoding process in block arithmetic coding as a clever and relatively efficient way of looking up code words in an encoding scheme without having actually to store the scheme.

arithmetic encoding involving exact computation with slapdash integer-interval methods of the type in Section 6.4 appears in [51].

Note that if M is a multiple, or, better yet, a power, of the common denominator of $f_1, \ldots, f_m$, then interval subdivision of the blocks of integers substituting for real intervals in this form of arithmetic coding is sometimes exact. In practice, M is usually taken to be a power of 2, $M = 2^K$. It might be a shrewd move in these cases to replace the f_j by dyadic fraction approximations with common denominator 2^k, with k being an integer divisor of K. (Of course, it is a luxury to know the f_j beforehand. In adaptive arithmetic coding, to be described in Chapter 8, the f_j are changing as the source is processed, and it is not convenient to repeatedly replace them by approximations.)

6.3.3 Must decoding wait until encoding is completed?

According to the description of dfwld arithmetic encoding in Section 6.1, in order that a source text be encoded and the code word for it subsequently decoded, the encoder has to read through the entire source text and compute the dfwld r associated with the source text, and the decoder has to wait until encoding is completed before beginning to decode.

Compare this train of events with the corresponding operation in the case of replacement encoding, in which each occurrence of a source letter is replaced by a binary word. In all forms of such encoding (including the adaptive varieties, which will be described in Chapter 8), encoding can begin as soon as scanning of the source text begins, and decoding can begin as soon as the beginnings of the code text are supplied to the decoder. Clearly there are situations in which it is highly desirable, or even indispensable, that encoding not wait upon the reading of the entire source text, nor decoding upon the delivery of the entire code text.

Can this apparent disadvantage of dfwld arithmetic coding be overcome? Well, yes; for instance, one could limit the delays by resorting to encoding of blocks of source letters of a pre-set length, as discussed above, at the cost of providing a parallel counter or pointer stream, or an extra source letter, EOB, which would be discarded by the decoder.

Is there any way to take advantage of the partial code words, prefixes of the final code word, supplied to the decoder by the encoder through rescaling? Yes: as noted at the end of Section 6.1.1, the decoder can deduce what the source text is right up to and including the last letter processed by the encoder if the decoder is supplied with the code text shifted out by rescaling together with the number of source letters processed so far. In Exercise 6.1.3 you were asked to struggle with the deduction process. Let us look now at what the decoder has to do.

Suppose that the code text supplied by rescaling after the scanning of the prefix W of the source text, say with N letters, is a binary word u. Recall from the introductory discussion of rescaling that u consists of the part of the initial segments of the binary expansions of the endpoints of the interval $A(W)$ where those binary expansions agree. [In case $A(W) = [1 - \epsilon, 1)$, this statement holds true if you take $1 = (.11 \cdots)_2$.] That is, the binary expansion of the lower endpoint of $A(W)$ looks like $(.u0 \cdots)_2$, and the binary expansion of the upper endpoint is $(.u1 \cdots)_2$. Therefore, provided no interval endpoints are dyadic fractions, $(.u1)_2$ is the dfwld in $A(W)$! That is, given u and N, the decoder needs only to tack 1 on the end of u and decode normally.

Thus, in Exercise 6.1.3, in which the relative source frequencies make it unlikely that any interval endpoint other than 0 or 1 is a dyadic fraction, you take 010, tack on a 1 to get 0101, and decode normally by any of the methods of Section 6.1.1. [Note $r = (.0101)_2 = 5/16$ and $N = 7$.] You should get *acdcaca*. You can check that this is correct by encoding *acdcaca*, with rescaling, to see if 010 is the partial code word provided by rescaling.

The problem of dyadic fraction interval endpoints is a nuisance, but can be overcome. As mentioned in the footnote on page 145, this would not be a problem if we made our intervals open on the left, closed on the right, and that is one way out. Even with intervals closed on the left, note that if $(.u1)_2$ is *not* the dfwld in $A(W)$, then either $(.u1)_2$ is in $A(W)$, in which case normal decoding of $u1$ gives the source word W, or $(.u1)_2$ is the upper endpoint of $A(W)$. This second possibility can be checked for by the decoder, and adjustments can be made.

Although we will not dwell upon it here, this process of decoding from knowing N and the partial code supplied by rescaling can be adapted so that it proceeds right "on the heels" of encoding, with the decoder's rescaling sweeping away old code and keeping the eager decoder one source letter behind the encoder.

The great impediment to our happiness with this method of eager decoding is the necessity of supplying the source letter count N corresponding to the partial code supplied by rescaling. If N is to be conveyed by some pointer/counter stream parallel to the regular code stream, compression is seriously reduced. Perhaps there are situations in which code can be delivered to the decoder in conformity with a certain rhythm, so that the decoder gets N by some sort of timing device; barring some such trick, this sort of decoding on the heels of encoding appears infeasible.

Another, more promising, path to allowing decoding to follow soon upon the start of encoding arises from the observation that in the decoding of Section 6.1, we need only decide in which of several large intervals $(r - \alpha)/\ell$ lies. We usually do not need to know $(r - \alpha)/\ell$ exactly, which means that we usually do not need to know r exactly.

What would happen if we replaced r by the approximation $\tilde{r}$ obtained by truncating the binary expansion of r somewhere – i.e., if we tried to proceed using just the (current) first few bits of the (current) code stream? The approximation $\tilde{r}$ will be a little less than r, so $(\tilde{r} - \alpha)/\ell$ will be a little less than $(r - \alpha)/\ell$. If the latter is exactly equal to the lower endpoint of the interval

$A(s_k) = [\sum_{j<k} f_j, \sum_{j\le k} f_j)$ in which it lies (so that we ought to decode s_k) then we are doomed: $(\tilde{r} - \alpha)/\ell$ will be in the next interval down, we will wrongly decode s_{k-1}, the next "current interval" will be wrong, and we will be in a world of trouble. The same catastrophe will occur if $(r - \alpha)/\ell$ is not equal to the lower endpoint of $A(s_k)$, but is very close to it, and $\tilde{r}$ is not close enough to r to put $(\tilde{r} - \alpha)/\ell$ in $A(s_k)$.

These catastrophes could be avoided at some cost in compression if we roughened the arithmetic coding process by putting some space – an "error zone" – between the intervals into which the current interval is subdivided. That is, the initial intervals $A(s_1), \ldots, A(s_m)$ would not cover $[0, 1)$, and subsequent subdivisions would be similar to the first. Then we can proceed to decode fearlessly, replacing r by $\tilde{r}$, provided we have figured out how far to take the binary expansion of r, to obtain $\tilde{r}$, so as to ensure that whenever $(r - \alpha)/\ell$ is in $A(s_k)$, $1 < k \le m$, then $(\tilde{r} - \alpha)/\ell$ will be greater than the upper endpoint of $A(s_{k-1})$.

This roughening is somehow, happily, built into the algorithm of Section 6.4, but not explicitly. The algorithm is a discrete simulation of pure dfwld arithmetic coding which corrects all the defects of the pure process that we have discussed here, at a controllable cost.

Exercises 6.3

1. Suppose that $S = \{a, b, c, d\}$, and $f_a = .35$, $f_b = .3$, $f_c = .25$, and $f_d = .1$, as in Exercise 6.1.1. Find the dyadic fractions $\widehat{f_a}, \widehat{f_b}, \widehat{f_c}$, and $\widehat{f_d}$ with common denominator 16, adding up to 1, such that $(\widehat{f_a}, \widehat{f_b}, \widehat{f_c}, \widehat{f_d})$ is as close as possible to (f_a, f_b, f_c, f_d). (Take "as close as possible" to mean that $\sum_{s \in S} |f_s - \widehat{f_s}|$ is minimized.)

Redo Exercise 6.1.1 using the $\widehat{f_s}$ as the relative source frequencies. Does rescaling do much to curb the growth of the denominators of the interval lengths?

2. $S = \{a, b, c, d\}$, $f_a = .4$, $f_b = .3$, $f_c = .2$, and $f_d = .1$. The encoder, rescaling whenever possible, passes to the decoder the following information, one line at a time (λ stands for the empty string):

Number of source letters processed	New bits added to the code stream by rescaling
1	λ
2	10
3	λ
4	1010
5	110
6	λ
7	λ

Decode on the run, on the heels of the encoding, as best you can. (Note that the code string, with $N = 7$, stands at 101010110, so you can always check your work by decoding 1010101101, with $N = 7$.)

6.4 Implementing arithmetic coding

In this section, some of the practical considerations for implementing arithmetic coding are examined. As with all the probabilistic methods presented, any model producing symbol probabilities can be used with arithmetic coding. The examples in this section use the simplest case of a fixed order-0 model.

So far, arithmetic coding has been presented with the understanding that the encoded stream is determined after the *entire* input stream is examined, although rescaling may produce a good part of the code as the source is processed. In practice, it is generally not feasible to maintain the precision required to compute the interval corresponding to the entire source stream, even with rescaling. In addition, in many applications transmission must begin before the entire stream has been coded.

Typically, arithmetic will be of limited precision; in fact, an approximate version of dfwld arithmetic coding can be implemented entirely with integer arithmetic (32 bits of precision is common). On many machines, integer operations are much faster than floating-point, and, in addition, portability considerations are simpler.

The scheme of this section will use the rescaling and incremental transmission described earlier: as soon as a digit in the binary representation of the final interval is determined, send that bit out as part of the encoded stream and then expand the interval. In order to prevent the current interval from becoming too small, the underflow expansion will be done in the case that the current interval is short but includes $1/2$ as an interior point.

The decoder will need a method to determine when all symbols have been recovered. If the source length cannot be provided "up front," then another method will be needed to terminate decoding. One possibility is to enlarge the symbol set S by adding a special end-of-file symbol, denoted by EOF. Of course, enlarging S has a price: the new symbol requires code space. In practice, this may be quite small; see Section 6.3 and also [34] for some discussion.

The algorithm that we will describe here for practical arithmetic coding is due to Witten, Neal, and Cleary [84]; in their honor, we will call it the WNC algorithm, for short. It is best thought of as a simulation of pure dfwld arithmetic encoding and decoding.

The main feature of the WNC algorithm is that the interval $[0, 1)$ will be replaced by a finite set of consecutive integers, $\{0, \ldots, M-1\}$, to be denoted $[0, M)$. The choice of M is critical, and will be discussed later. In practice, M is always a power of 2, but there is no harm in leaving M unspecified, in what follows.

The WNC encoding algorithm follows dfwld encoding exactly, with the reservations that computations are replaced by "integer arithmetic", best explained by example below; the rescaling and underflow expansions are obligatory, not optional; and the finish of the encoding process is not just "add 1 for

1/2, the dfwld in the final interval"—there is more to it than that.

The parallels and differences between the two processes are given in Table 6.1. In this table, $[L, H) = \{L, \ldots, H - 1\}$ will be the "current interval" in WNC decoding, where L and H are integers. As before, the "current interval" in dfwld encoding will be denoted $[\alpha, \beta)$. The operation "round down" will be denoted $\lfloor \cdot \rfloor$. Both dfwld and WNC assume an ordered source alphabet $s_1, \ldots, s_m$ with positive relative frequencies $f_1, \ldots, f_m$, usually (but not necessarily) in non-increasing order. In the WNC algorithm, one of the s_i, usually s_m, is EOF, with a small putative relative frequency obtained by taxing the relative frequencies of the real source letters.

Table 6.1: *Encoding comparison between dfwld and WNC methods.*

	Dfwld	WNC
Starting interval	$[0, 1)$	$[0, M) = \{0, \ldots, M - 1\}$
New current interval after reading s_k	$\alpha \leftarrow \alpha + (\sum_{i<k} f_i)(\beta - \alpha)$ $\beta \leftarrow \alpha + (\sum_{i\leq k} f_i)(\beta - \alpha)$	$L \leftarrow L + \lfloor (\sum_{i<k} f_i)(H - L) \rfloor$ $H \leftarrow L + \lfloor (\sum_{i\leq k} f_i)(H - L) \rfloor$
Rescaling	When $\beta \leq 1/2$: $\quad \alpha \leftarrow 2\alpha, \beta \leftarrow 2\beta$. When $1/2 \leq \alpha$: $\quad \alpha \leftarrow 2\alpha - 1, \beta \leftarrow 2\beta - 1$.	When $H \leq M/2$: $\quad L \leftarrow 2L, H \leftarrow 2H$. When $M/2 \leq L$: $\quad L \leftarrow 2L - M, H \leftarrow 2H - M$.
Underflow expansion	When $1/4 \leq \alpha < 1/2 < \beta \leq 3/4$: $\quad \alpha \leftarrow 2\alpha - 1/2$, $\quad \beta \leftarrow 2\beta - 1/2$.	When $M/4 \leq L < M/2 < H \leq 3M/4$: $\quad L \leftarrow 2L - \lfloor M/2 \rfloor$, $\quad H \leftarrow 2H - \lfloor M/2 \rfloor$. (Usually M is even, so $\lfloor M/2 \rfloor = M/2$.)
Ending encoding	At the end, carry out rescaling until $\alpha < 1/2 < \beta$, then add 1 to the code stream	At the end, having read EOF, carry out rescaling and the underflow expansion until neither can be carried out. At this point, either $L < M/4 < M/2 < H$ or $M/4 \leq L < M/2 < 3M/4 < H$. In the former case add 01^{k+1} to the code stream, where k is the underflow count; in the latter case add 10^{k+1} to the code stream.

6.4.1 Example (WNC encoding) (a) $S = \{a, b, EOF\}$, $f_a = 6/10$, $f_b = 3/10$, $f_{EOF} = 1/10$; $M = 16$. We encode $abaEOF$, using a table similar to those in Section 6.1.

Next letter or rescale or underflow	L	H	New code	Underflow count
	0	16		0
a	0	$0+\lfloor(.6)16\rfloor=9$		0
b	$0+\lfloor(.6)9\rfloor=5$	$0+\lfloor(.9)9\rfloor=8$		0
$x\to 2x$	10	16	0	0
$x\to 2x-16$	4	16	1	0
a	4	$4+\lfloor(.6)12\rfloor=11$		0
$x\to 2x-8$	0	14		1
EOF	$0+\lfloor(.9)14\rfloor=12$	14		1
$x\to 2x-16$	8	12	10	0
$x\to 2x-16$	0	8	1	0
$x\to 2x$	0	16	0	0

Thus the code for $abaEOF$ is 01101001. The last "01" is added because the final interval is $[0,16)$; $L=0 < M/4 = 4 < M/2 = 8 < H = 16$. See the last part of Table 6.1.

(b) $S=\{a,b,c,EOF\}$, $f_a=.4$, $f_b=.3$, $f_c=.2$, and $f_{EOF}=.1$. This time, we take $M=32$. We encode $bacbEOF$.

Next letter or rescale or underflow	L	H	New code	Underflow count
	0	32		0
b	12	22		0
$x\to 2x-16$	8	28		1
a	8	$8+\lfloor(20).4\rfloor=16$		1
$x\to 2x$	16	32	01	0
$x\to 2x-32$	0	32	1	0
c	22	28		0
$x\to 2x-32$	12	24	1	0
$x\to 2x-16$	8	32		1
b	$8+\lfloor(.4)24\rfloor=17$	$8+\lfloor.7(24)\rfloor=24$		1
$x\to 2x-32$	2	16	10	0
$x\to 2x$	4	32	0	0
EOF	29	32		0
$x\to 2x-32$	26	32	1	0
$x\to 2x-32$	20	32	1	0
$x\to 2x-32$	8	32	1	0

The output code is: 011110011110.

6.4.2 The WNC algorithm for encoding An ordered source alphabet $s_1,\ldots,$ s_m, including a special symbol EOF, and corresponding relative frequencies $f_1,\ldots,f_m$ are given. Also, a (large) positive integer M has been chosen. The following applies to encoding a source word in which EOF occurs once, at the end. For $j=1,\ldots,m+1$, let $F_j=\sum_{i<j}f_i$.

1. A current interval $[L, H)$ is initialized as $[0, M) = \{0, \ldots, M - 1\}$ and maintained at each step. Also, an underflow count is initialized at 0 and maintained to the end of the file.

2. (Underflow condition) If the current interval $[L, H)$ satisfies $M/4 \leq L < M/2 < H \leq 3M/4$, replace the current interval by $[2L - \lfloor M/2 \rfloor, 2H - \lfloor M/2 \rfloor)$ and add 1 to the underflow count. [If M is even, the round down signs may be deleted from the preceding.]

3. (Shift condition) If the current interval $[L, H)$ satisfies $H \leq M/2$, replace the current interval by $[2L, 2H)$ and output 01^k, $k =$ underflow count, to the code stream. If $M/2 \leq 1$, replace the current interval by $[2L - M, 2H - M)$ and output 10^k to the code stream. In either case, reset the underflow count to 0.

4. If none of the conditions in 2 or 3 hold, look at the next source letter (indicated by a pointer). If it is s_j, assign $L \leftarrow L + \lfloor F_j(H - L) \rfloor$, $H \leftarrow L + \lfloor F_{j+1}(H - L) \rfloor$ and move the pointer forward, unless $s_j = EOF$.

5. Repeat 2-4 until EOF has been encountered and none of the conditions in 2 or 3 hold. If, at this point, $L < M/4 < M/2 < H$, output 01^{k+1} to the code stream, where k is the underflow count. Otherwise, output 10^{k+1}. The encoding is now finished.

Decoding WNC decoding differs significantly from pure dfwld decoding in that the decoder does not use the entire code stream to decode, but rather just the (current) first $N = \lceil \log_2 M \rceil$ bits of the code stream. These appear in a register called v (for value), and change as decoding proceeds, as code bits from the right are shifted into v and (one of the first two) code bits on the left (start) of v are deleted. (We will make the process clear below.) It is this dependence on only the first few bits of the code stream, in decoding, together with the use of integer arithmetic, which makes WNC encoding and decoding so practical and fast.

The decoder tracks the encoding process. This means that a current interval $[L, H)$ is maintained and whenever any of the conditions in 2 and 3 of the algorithm in 6.4.2 hold, the appropriate expansion brings about a shift into and out of v. It is not necessary to keep an underflow count, but the underflow expansion brings about an unusual shift into the v register:

$$\boxed{a_1 a_2 \ldots a_N} \, a_{N+1} a_{N+2} \cdots \rightarrow \boxed{a_1 a_3 \ldots a_N a_{N+1}} \, a_{N+2} \cdots$$

That is, the second bit in v is deleted and a new code bit is shifted in. Meanwhile, the rescaling expansions in 3 of 6.4.2 both bring about

$$\boxed{a_1 a_2 \ldots a_N} \, a_{N+1} a_{N+2} \cdots \rightarrow \boxed{a_2 \ldots a_N a_{N+1}} \, a_{N+2} \cdots$$

(In all of the above, $a_1 a_2 \ldots a_N a_{N+1} a_{N+2}$ are the bits of the current code stream and the box represents the v register.)

Decoding of source letters occurs when, finally, no condition in 2 or 3 of 6.4.2 holds. In order to decode, the different subintervals $[L_j, H_j)$ of $[L, H)$

corresponding to the source letters $s_1, \ldots, s_m$ have to be computed, and v has to be viewed as the binary expansion of a non-negative integer; whichever interval $[L_j, H_j)$ the value v falls in determines which s_j is next decoded, and $[L_j, H_j)$ replaces $[L, H]$ as the current interval. When EOF is decoded, decoding stops.

One very slight inconvenience of the use of integer arithmetic and the round down operation is that it is not determinable invariably which interval $\lfloor L_j, H_j)$ v lies in simply by calculating $\lfloor \frac{v-L}{H-L} \rfloor$. We could be off by one interval if we try to use the "$\frac{r-\alpha}{\ell}$" method of 6.1. There are ways of avoiding the full burden (see the discussion at the end of "Implementation and performance issues", below), but, in what follows, we will calculate the subintervals $[L_j, H_j)$ corresponding to the letters s_j, $j = 1, \ldots, m$, whenever the time to decode has arrived.

6.4.3 Example (decoding) (a) $S = \{a, b, EOF\}$, $f_a = 6/10$, $f_b = 3/10$, $f_{EOF} = 1/10$, $M = 16$, and the code is 01101001; $N = 4$.

v	L	H	$[L_a, H_a)$	$[L_b, H_b)$	$[L_{EOF}, H_{EOF})$	Decode or ...
$(0110)_2 = 6$	0	16	$[0, 9)$	$[9, 14)$	$[14, 16)$	a
0110	0	9	$[0, 5)$	$[5, 8)$	$[8, 9)$	b
0110	5	8				$x \rightarrow 2x$
1101	10	16				$x \rightarrow 2x - 16$
$(1010)_2 = 10$	4	16	$[4, 11)$	$[11, 14)$	$[14, 16)$	a
1010	4	11				$x \rightarrow 2x - 8$
$(1100)_2 = 12$	0	14	$[0, 8)$	$[8, 12)$	$[12, 14)$	EOF

The decoded source message is: $abaEOF$.

(b) $S = \{a, b, c, EOF\}$, $f_a = .4$, $f_b = .3$, $f_c = .2$, $f_{EOF} = .1$, $M = 32$, and the code is 011110011110. $[N = 5]$.

v	L	H	$[L_a, H_a)$	$[L_b, H_b)$	$[L_c, H_c)$	$[L_{EOF}, H_{EOF})$	Decode or ...
$(01111)_2 = 15$	0	32	$[0, 12)$	$[12, 22)$	$[22, 28)$	$[28, 32)$	b
01111	12	22					$x \rightarrow 2x - 16$
$(01110)_2 = 14$	8	28	$[8, 16)$	$[16, 22)$	$[22, 26)$	$[26, 28)$	a
01110	8	16					$x \rightarrow 2x$
11100	16	32					$x \rightarrow 2x - 32$
$(11001)_2 = 25$	0	32	$[0, 12)$	$[12, 22)$	$[22, 28)$	$[28, 32)$	c
11001	22	28					$x \rightarrow 2x - 32$
10011	12	24					$x \rightarrow 2x - 16$
$(10111)_2 = 23$	8	32	$[8, 17)$	$[17, 24)$	$[24, 29)$	$[29, 32)$	b
10111	17	24					$x \rightarrow 2x - 32$
01111	2	16					$x \rightarrow 2x$
$(11110)_2 = 30$	4	32	$[4, 15)$	$[15, 23)$	$[23, 29)$	$[29, 32)$	EOF

Decoded source message: $bacbEOF$. We leave the formulation of the WNC decoding algorithm as an exercise.

Implementation and performance issues

Some issues related to implementation on a machine have been discussed in previous sections. In this section, some additional programming considerations are examined. The section concludes with a few notes on performance issues.

The first question regarding WNC encoding/decoding we need to deal with is: how large does M have to be, and why? The larger M is, the more accurate is the simulation of pure dfwld arithmetic coding provided by the WNC algorithms (on the grounds that long intervals of consecutive integers are more divisible than short ones, and so better simulate the continuum), and thus the closer we are to the Holy Grail, encoding losslessly in $H(S)$ bits per source letter. On the other hand, letting M be simply enormous is computationally impractical.

There is a more prosaic consideration concerning the size of M, other than compression performance: during encoding, when the current interval $[L, H)$ is subdivided into subintervals $[L_j, H_j)$, $j = 1, \ldots, m$, it must never happen that $L_j = H_j$ for some j. For if $L_j = H_j$ then $[L_j, H_j)$ is empty, and if the next letter is s_j, the encoding process will proceed to crash, or enter an infinite rescaling loop.

Given $S = \{s_1, \ldots, s_m\}$ and positive relative frequencies $f_1, \ldots, f_m$, let $F_j = \sum_{i<j} f_i$, $j = 1, \ldots, m+1$. The disaster we have to avoid is $L + \lfloor F_j(H-L) \rfloor = L + \lfloor F_{j+1}(H-L) \rfloor$, i.e., $\lfloor F_j(H-L) \rfloor = \lfloor F_{j+1}(H-L) \rfloor$, for some j, when we have a current interval $[L, H)$ satisfying none of the conditions in 2 or 3 of 6.4.2. (That is, the time has come to read the next source letter and replace $[L, H)$ by the subinterval $[L_j, H_j)$ corresponding to that source letter.)

Notice that $F_{m+1} = 1 > F_m = 1 - f_m$, and $H - L$ is a positive integer, so $\lfloor F_m(H-L) \rfloor < H - L = \lfloor F_{m+1}(H-L) \rfloor$. Therefore, the disaster we fear will never occur with $j = m$. Therefore, to avoid calamity it suffices that $f_j(H - L) \geq 1$ (why?) for $j = 1, \ldots, m-1$.

If none of the conditions in 2 and 3 of 6.4.2 holds, then either $L < M/4 < M/2 < H$ or $L < M/2 < 3M/4 < H$. Let us suppose that M is divisible by 4. Then we see that if the current interval $[L, H)$ is about to be subdivided, it must be that $H - L \geq M/4 + 2$. Putting this together with the condition in the paragraph preceding we obtain the following.

6.4.4 *Given $S = \{s_1, \ldots, s_m\}$ and positive relative frequencies $f_1, \ldots, f_m$, let $f_{\min} = \min[f_1, \ldots, f_{m-1}]$. The WNC encoding algorithm, applied to source text over S, will not crash due to an empty interval if the integer M is divisible by 4 and $M \geq (4/f_{\min}) - 8$.*

In the situation of 6.4.1(a), for example, $f_{\min} = .3$ and we could have proceeded with $M = 8$. In 6.4.1(b), $f_{\min} = .2$ and we could have proceeded with $M = 12$. Of course, the compression achievable with these smaller values of M will probably not match that achievable with the larger values, 16 and 32, over the long haul, but it might be interesting to experiment and find out roughly how much we lose with the smaller values.

The test of sufficiency of M given in 6.4.4 becomes important in adaptive arithmetic coding, in which the relative frequencies vary—see Section 8.3. In every case it is good policy to make *EOF* the last in the order of source letters, with the smallest relative frequency.

As mentioned earlier, the arithmetic is easiest to follow in the case that the interval length M satisfies $M = 2^N$ for some $N \in \mathbb{N}$, which is a natural choice for implementing on a machine. The values for the interval (and the value of v used in decoding) can be maintained in N bits. In addition, a number of the steps in the scheme can be managed as simple bitwise operations.

To be precise, if the current interval is to be maintained in N bits, then the register for the right endpoint will contain $H - 1$ rather than H. It turns out that this is the natural choice if bitwise operations are to be used. The shift condition in step 3 of the algorithm 6.4.2 is a simple test on the leftmost (called the *most significant bit* or MSB) of the N bits of L and $H - 1$: if $\text{MSB}(L) = \text{MSB}(H-1)$ then shift these registers left, sending the MSB to the output and giving new values for L and $H - 1$. The shift on L doubles the value represented in the lower $N - 1$ bits, which is what is desired. However, the shift on the right endpoint is performed on $H - 1$ and hence 1 must be added to the result.

The underflow condition in step 2 of the algorithm also corresponds to simple bitwise operations. Underflow occurs when the two leftmost bits of L and $H - 1$ are '01' and '10', respectively. In the expansion, the second bit is deleted in each of L and $H - 1$, the last $N - 2$ bits are shifted one space left, a 0 is the new last bit of the L register, and a 1 is the new last bit of $H - 1$. (Also, the underflow count is incremented.)

An example is the best way to understand the process. The encoding of 6.4.1(a) is repeated in Table 6.2. The format of the table has changed somewhat, in order to better illustrate the bitwise operations. The calculations haven't changed, but the current interval is maintained in N-bit registers as L and $H - 1$, with the understanding that this corresponds to the interval $[L, H)$. The left endpoint of the corresponding subinterval for each symbol is listed as L_a, L_b, and L_{EOF}, respectively. The result, of course, is the same as before. The algorithm gives '01101001' as the encoded stream.

For illustration, the table lists all of the left endpoints L_a, L_b, and L_{EOF}. This is more arithmetic than is required and would be expensive if S has many symbols. Only the endpoints corresponding to the current input symbol are needed (and can be calculated from number 4 of 6.4.2).

In decoding, only the subinterval corresponding to the value is needed. In the decoding examples we calculated all the subintervals $[L_i, H_i)$ corresponding to different letters, when it came time to decode, but there is a way to avoid this calculation. Suppose that the relative frequencies f_i are rational numbers, say $f_i = c_i/C$, c_i and C positive integers, $i = 1, \ldots, m$. [We use the letters c and C here to suggest the word "counts", in anticipation of adaptive arithmetic coding. See Chapter 8.] Let $C_0 = 0$ and $C_i = \sum_{j \le i} c_j$, $i = 1, \ldots, m$. Thus $C_m = C$ and $c_i/C = F_{i+1}$, $i = 0, \ldots, m$, with the F_j as defined previously. Now the amount

Table 6.2: *The encoding of Example 6.4.1 (a).*

Symbol	L	$H-1$	L_a	L_b	L_{EOF}	Output
			Current interval			
start	0000	1111	0000	1001	1101	
a	0000	1000	0000	0101	1000	
b	$\underline{0}101$	$\underline{0}111$	expand[a] $x \mapsto 2x$			0
	$\underline{1}010$	$\underline{1}111$	expand[b] $x \mapsto 2(x - M/2)$			1
	0100	1111	0100	1011	1110	
a	$0\underline{1}00$	$1\underline{0}10$	expand[c] $x \mapsto 2(x - M/4)$			underflow
	0000	1101	0000	1000	1100	
EOF	$\underline{1}100$	$\underline{1}101$	expand[b] $x \mapsto 2(x - M/2)$			10
	$\underline{1}000$	$\underline{1}011$	expand[b] $x \mapsto 2(x - M/2)$			1
	$\underline{0}000$	$\underline{0}111$	expand[a] $x \mapsto 2x$			0
	0000	1111				

[a]Using notation from the C programming language, the expansion is $L \ll 1$ and $(H-1) \ll 1 \mid 1$.
[b]The expansion is the same as above, but the leftmost bit must be discarded.
[c]This can be written as the bitwise operations $(L \& (M/4 - 1)) \ll 1$ and $((H-1) \wedge (3M/4)) \ll 1 \mid 1$.

of arithmetic can be minimized by scaling the value v back to the subintervals $[C_{i-1}, C_i)$ in order to find the current output symbol. To see how this works, consider the stage in decoding where $L \le v < H$ and we wish to find i so that $L_i \le v < H_i$ (giving output symbol s_i). From the formulas in 6.4.2, this means

$$\left\lfloor \frac{C_{i-1}}{C}(H-L) \right\rfloor \le v - L < \left\lfloor \frac{C_i}{C}(H-L) \right\rfloor$$

and hence

$$C_{i-1}(H-L) < (v - L + 1)C \le C_i(H-L).$$

Every term is an integer, and it follows that

$$C_{i-1} \le w = \left\lfloor \frac{(v-L+1)C - 1}{H-L} \right\rfloor < C_i. \qquad (6.1)$$

The steps can be reversed, showing that the scaled value w satisfies $C_{i-1} \le w < C_i$ if and only if $L_i \le v < L_{i+1}$. It is w (along with the cumulative counts, in the case of adaptive coding; see Section 8.3) which can be used by the decoder to find the current symbol.

As an example of the use of w to find the current symbol, consider the second step in the decoding example in 6.4.3 (a) where $v = 6$ and the current interval is $[L, H) = [0, 9)$. We can take $C_1 = 6$, $C_2 = 9$, and $C = C_3 = 10$. Our calculation gives

$$w = \left\lfloor \frac{(v-L+1)C - 1}{H-L} \right\rfloor = \left\lfloor \frac{(6-0+1)10 - 1}{9-0} \right\rfloor = \left\lfloor \frac{69}{9} \right\rfloor = 7.$$

Since $C_1 = 6 \leq w = 7 < 9 = C_2$, it follows that the current symbol is $s_2 = b$. At this stage, the current interval corresponding to 'b' must be calculated and then decoding continues.

Precision

If, for some reason, we require the M in the WNC algorithm to be small, we may allow rough and arbitrary approximation of the relative source frequencies. If we allow the $f_{\min}$ in 6.4.4 to be as large as $1/m = 1/|S|$, then, by the analysis preceding 6.4.4, the rather minimal condition $|S| \leq M/4 + 2$ and M divisible by 4 will guarantee that the algorithm can proceed. This will not leave much room to accurately reflect probabilities, however, and in practice M may be much larger than $4|S|$. Performance considerations will place an upper bound on the number of bits which can be required for the calculations. Letting $f_i = \frac{c_i}{C}$ as above, $i = 1, \ldots, m$, clearly $f_{\min} \geq 1/C$ and therefore the subintervals $[L_i, H_i)$ will be nonempty if $C \leq M/4 + 2$. With some programming care, the intermediate calculations can be done if $CM \leq 2^p$, where p is the number of bits of precision. Hence, if $M = 2^m$, it suffices to require that $C \leq 2^c$, where

$$\log_2 |S| \leq c \leq m - 2 \quad \text{and} \quad c + m \leq p. \tag{6.2}$$

Today, $p = 32$ is common, and $m = 16$ and $c = 14$ may be a natural choice. For large symbol sets, conditions (6.2) could be rather unpleasant even with larger p; see Moffat, Neal, and Witten [48] for an improved coder which is more flexible.

As an alternate viewpoint (especially in the case that p is small), one could consider that m and c are given, and then choose S appropriately. Finally, note that the number of underflow bits is not bounded by the algorithm. However, since any pending underflow bits are all the same, only a count need be maintained.

Performance

For a given model, Huffman coding is "best possible" among probabilistic methods which replace source symbols by an integral number of bits. However, it is not optimal in the sense of "entropy." As the simplest example, consider an alphabet with two symbols. Regardless of the probabilities, Huffman will assign a single bit to each of the symbols, giving no compression.

On the other hand, arithmetic coding is optimal, and it can do better than Huffman. In particular, arithmetic coding on the two-symbol alphabet can yield compression, in contrast to Huffman. It is important to note that implementations of arithmetic coding (such as that presented here) will be somewhat less than optimal, due to the integer arithmetic and other compromises. Since Huffman is nearly optimal in many cases [23], the choice between Huffman and arithmetic is not as simple as the theory might suggest, although it appears safe to say that, in practice, arithmetic coding usually gives better compression.

Implementing arithmetic coding is not much more difficult than Huffman coding, but execution speed has been a serious concern. In the case of fixed probabilities (static or semi-static coding), Huffman will be significantly faster. In the adaptive case, however, maintaining the Huffman codes is expensive in time and memory. An optimized arithmetic coding implementation can be faster, use less memory, and give better compression [48–51].

There has been considerable work to reduce the number of multiplications required. Using approximate probabilities can permit replacement of multiplications by simple shift operations. The QM-coder (a binary arithmetic coder) used in JPEG image compression schemes [57] is an attempt to maximize performance with such methods.

JPEG provides an example of another consideration in choosing a compression method. As discussed in Chapter 10, lossy JPEG schemes get most of their compression using a transform method, and then Huffman or arithmetic coding is used on the output. The QM-coder used in the arithmetic mode may be covered by patent, according to the Independent JPEG Group (IJG) [25]. The free JPEG software implements the Huffman portion of the specification. Arithmetic coding may offer some additional compression, but the IJG writes "Since arithmetic coding provides only a marginal gain over the unpatented Huffman mode, it is unlikely that very many [JPEG] implementations will support it."

Exercises 6.4

1. A stream from $S = \{a, b, c, EOF, d\}$ is to be encoded using the statistics $f_a = 3/7$ and $1/7 = f_b = f_c = f_{EOF} = f_d$.

   (a) Encode 'db', followed by EOF, using (integer) arithmetic coding with $M = 2^4$ (i.e., the registers for L and $H - 1$ are 4-bit). Keep the symbols in the order listed when assigning subintervals.

   (b) Given that the encoder and decoder have agreed on the algorithm, what information must be passed in order for the decoder to recover the source string?

   (c) Decode the result of (a), showing the details. Minimize the arithmetic by using 6.1 to find the symbol. (Let $C = 7$.)

2. It is sometimes possible to reduce the size of the output file at the last stage of the encoding process. Consider encoding b followed by EOF, using the arrangements from Exercise 1.

   (a) Show that the algorithm gives '1000001'.

   (b) Let's try to save a few bits at the last stage. Assume that the decoder understands that any encoded string ends with an infinite number of zeros. In our problem, the final interval includes 0, and if there are no underflow bits pending, then two bits could be saved at the final stage. In every case (even if there are underflow bits), sending a single 1 suffices. (Why?)

Suppose the algorithm is modified so that the encoder never outputs until it has a maximum number of equal consecutive bits (and the decoder understands that the encoded string ends with all zeros). Show that these changes give '1' as the encoded string for this example. Are there any "problems" with these changes?

3. In Exercise 1, show that it is not possible to scale the symbol counts and choose c to satisfy 6.2. However, after expansions, the current interval $[L, H)$ is always wide enough to assign symbols to distinct subintervals.

4. Verify that the bitwise operations in the footnotes to the table on page 174 agree with the expansions described in the algorithm.

5. For a given S, the minimal condition $|S| \leq M/4 + 2$ guarantees that the algorithm can proceed. Name two important advantages in choosing M larger than this lower bound.

6. Langdon and Rissanen [44] describe a method called *bit stuffing* to handle the "carry-over problem" (which corresponds to the underflow case discussed in this chapter). The current interval length is maintained in a fixed-width register, but normalized so that it represents lengths in $[1/2, 1)$; i.e., the register is shifted so that the (implied) binary point is followed by a 1). The code string is shifted by the same amount in order to maintain alignment.

Table 6.3 illustrates the process with a binary arithmetic code.[4] The source alphabet is $S = \{0, 1\}$, and the first portion of the string to be encoded is '0100010'. The probabilities depend on the context s, and $P(1 \mid s)$ denotes the probability of 1 in context s. The symbol '0' is "more probable" in this example, and is assigned to the left subinterval at each stage. Following the notation of [44], the code string at each stage is $C = C(s)$, and the (normalized) length is contained in $A = A(s)$. It is understood that the length $A(s1)$ is obtained by truncating the value $P(1 \mid s)A(s)$, and then $A(s0) = A(s) - A(s1)$.[5]

The lines containing "shift" indicate the number of left shifts to be performed on A and C so that the most significant digit of A is 1 (and the result is shown in the following line of the table). This allows the working part of C to be maintained in a fixed-length register (4 bits in the table), but additions can cause *carry-over* into the digits preceding the binary point. In the example, the last 0 input leads to a string of ones in C. If the next input symbol is 1, then a carry-over may propagate up this long string of 1s, converting each to 0, and terminating only when it reaches a 0.

---

[4]See also Langdon's paper [43] for additional notes and extended examples of a similar algorithm.

[5]In this example, the probabilities are of the form $1/2^j$, and so the calculation is particularly simple (and amounts to shifting $A(s)$ right by j bits). These probabilities may be approximations to the actual values and result in a multiplication-free scheme. See [44] for a discussion on how these estimates should be chosen.

Table 6.3: *Binary arithmetic coding with bit stuffing.*

string s	input	$P(1\mid s)$	$C(s)$	$A(s)$	$A(s0)$	$A(s1)$
null	0	$1/2^2$	.0000	1	.1100	.0100
0	1	$1/2$	.0000	.1100	.0110	.0110
01			.0110	.0110	shift 1	
01	0	$1/2^2$	0.1100	.1100	.1001	.0011
010	0	$1/2^2$	0.1100	.1001	.0111	.0010
0100			0.1100	.0111	shift 1	
0100	0	$1/2^3$	01.1000	.1110	.1101	.0001
01000	1	$1/2$	01.1000	.1101	.0111	.0110
010001			01.1111	.0110	shift 1	
010001	0	$1/2$	011.1110	.1100	.0110	.0110
0100010			011.1110	.0110	shift 1, bit stuff	
0100010			01110.1100	.1100		

In order to limit the number of bits that can be affected by carry-over, an extra 0 is "stuffed" into C (shown in the last line of the table). This extra zero blocks the propagation of a carry: the digits which appear to the left of this stuffed bit no longer participate in the arithmetic, and can be sent out.

(a) Assume that the probability at the last line of the table is $P(1\mid s) = 1/2^2$. If the next input symbol is '1', show that the carry propagates into the stuffed bit.

(b) The string $s = 01000101$ (i.e., the stream in the table along with the digit from (a)) is encoded. Recall that any number in $[C(s), C(s) + A(s))$ determines a valid encoded stream. Show that '01111011' is the best representative.[6]

The decoder must also manage the stuffed bit. After receiving a predetermined number of consecutive 1s (three, in this example), the decoder examines the next bit (the stuffed bit). If the bit is 0, then no carry at the next stage has propagated into the current digits, and the stuffed bit is ignored. If the bit is 1, then it is added to the current value of the codeword.

(c) Show the decoding details for the encoded stream '01111011'.

(d) What is the cost of bit stuffing?

---

[6]Langdon and Rissanen do not include the leading 0 (apparently since the decoder can manage this case); it has been retained here since it simplifies the discussion.

6.5 Notes

A complete implementation of the scheme by Witten, Neal, and Cleary appears in their well-known paper [84]. This article is the basis for corresponding material in *Text Compression* [8]. Appendix C gives addresses where the original and optimized versions may be found. A separate implementation of the same coding scheme can be found in the book by Nelson and Gailly [53]. Portability has been a goal in this code, but it should be noted that the sources make a few optimizations that assume certain widths on data types. Moffat, Neal, and Witten [48] provide a number of improvements to the earlier version.

The QM-coder used in JPEG is a descendent of the IBM Q-coder, developed out of work on compressing bilevel images. A description of the Q-coder can be found in a series of articles in the *IBM Journal of Research and Development*, November 1988. Rabbani and Jones [59] present a short section on the Q-coder, and [57] contains a discussion of the QM-coder.

Binary coders (such as the Q-coder) are an important special case in arithmetic coding. Alphabets with more than two symbols can be managed by encoding the current bit according to a suitable context, although performance may be unsatisfactory [34, 49, 50]. Howard and Vitter [34, 35] discuss modeling and coding methods to improve the speed while preserving most of the compression.

Ross Williams' thesis [82] contains a lengthy survey of text compression. The section on arithmetic coding is short, but includes an interesting alternate view of the scheme and a few notes on the basic ideas in the Q-coder development. The internet newsgroups comp.compression (established by Williams in 1991) and comp.compression.research can be good sources of information, although, like many newsgroups, there is a considerable amount of noise to filter. Jean-loup Gailly coordinates the FAQ (Frequently Asked Questions) for these newsgroups, which is a good source for introductory material, pointers to source code, references, and other information (see Appendix C).

Chapter 7

Higher-order Modeling

In understanding probabilistic or statistical coding methods, it is useful to think of the coding process or apparatus as divided into two autonomous packages, the *coder* and the *model*. The model, or statistical processor, passes information about the statistical nature of the source text to the coder, which then uses this information to encode the source text efficiently.

Chapters 5 and 6 were about different kinds of coder, and the model was very rudimentary; supposedly a statistical study of the source text was conducted before encoding to estimate the relative source frequencies, $f_1, \ldots, f_m$, which are supplied to the coder once and for all (for that source text). In this chapter and the next we will look at two different kinds of statistical processor more complicated than the plain old non-adaptive, zeroth-order model presumed heretofore. In Chapter 8, we take up adaptive methods. Here, we study higher-order non-adaptive methods. As noted in Chapter 8, the two sorts of model can be crossed to produce hybrid processors, higher-order adaptive models.

All of these different models can be used with either the Huffman or the arithmetic coder (or with other statistical coders, such as those based on Shannon's or Fano's methods). As long as the coder knows the current values of $f_1, \ldots, f_m$, the coder knows how to process the source text. In principle, the coder need not be tailored to fit the statistical processor. In practice, it may increase the efficiency if the coder is modified to mix better with the model, in the necessary exchange of information between them. We will present the higher-order models (and, in the next chapter, the adaptive statistical processor) in alliance with the Huffman and arithmetic coders, and speak as though there were such things as "kth-order Huffman encoding," or "adaptive arithmetic encoding," because it is convenient to do so, and because we think it might be easier for somebody learning about these models for the first time to do so in connection with the coders, so that they can see how the full package works. But we wish to emphasize, for reasons of academic purity, that, as the title of this chapter indicates, "higher-order" is a quality of the *model*, or statistical processor, and can be considered separately from any particular coding method. Perhaps it is time to reveal what higher-order modeling is all about.

Suppose $S = \{s_1, \ldots, s_m\}$, a source alphabet, and an integer $k \geq 0$ are given. In kth-order encoding, we assume that the relative source frequencies $f(i_1, \ldots, i_{k+1})$ of the words $s_{i_1} \cdots s_{i_{k+1}}$, among all source words of length $k+1$, are given.

In the cases $k = 0, 1, 2$, we write $f(i) = f_i$ (the relative frequency of s_i much used in the preceding chapters), $f(i, j) = f_{ij}$, the so-called *digram frequency* of $s_i s_j$ and $f(i, j, k) = f_{ijk}$, the *trigram frequency* of $s_i s_j s_k$.

If the relative frequencies $f(i_1, \ldots, i_{k+1})$ are given, then so are all the relative frequencies $f(i_1, \ldots, i_t)$, $1 \le i_1, \ldots, i_t \le m$, $1 \le t \le k+1$. For instance, $f(i_1, \ldots, i_k) = \sum_{j=1}^{m} f(i_1, \ldots, i_k, j)$. Notice that, for instance, when $k = 1$, $[f_{ij}] = F$, an $m \times m$ matrix of non-negative numbers, is an acceptable matrix of digram frequencies if and only if for each i, the ith row sum and the ith column sum of F are equal, and $1 = \sum_i \sum_j f_{ij}$.

7.1 Higher-order Huffman encoding

One way to use the hard-won knowledge of the relative frequencies $f(i_1, \ldots, i_{k+1})$ would be to treat S^{k+1} as the source alphabet and to produce an encoding scheme using Huffman's algorithm. This encoding scheme would have m^{k+1} lines.

In kth-order Huffman encoding, $k \ge 1$, we have, instead of one big scheme, rather a lot of little schemes, m^k of them, in fact, each with m lines, so the total hidden cost of kth-order encoding is about the same as that of zeroth-order encoding using the huge source alphabet S^{k+1}. Let us call each source word $s_{i_1} \cdots s_{i_k}$ of length k a *kth-order context*. For each such context, and $1 \le j \le m$, let

$$P(s_j \mid s_{i_1} \cdots s_{i_k}) = \frac{f(i_1, \ldots, i_k, j)}{f(i_1, \ldots, i_k)},$$

the conditional probability that, if you have just scanned the word $s_{i_1} \cdots s_{i_k}$ in the source text, the next letter will be s_j. The m^k encoding schemes come about by applying Huffman's algorithm to $S = \{s_1, \ldots, s_m\}$ equipped with the conditional relative frequencies $P(s_1 \mid s_{i_1} \cdots s_{i_k}), \ldots, P(s_m \mid s_{i_1} \cdots s_{i_k})$, for each context $s_{i_1} \cdots s_{i_k}$. Thus there is one scheme per context, which makes m^k of them, and each is an encoding scheme for S, and so has m lines.

Once you have all these schemes, how do you encode source text? Each occurrence of the letter s_j is encoded with the code word for s_j in the scheme associated with the context $s_{i_1} \cdots s_{i_k}$, the k-letter word immediately preceding that occurrence of s_j. Thus different occurrences of s_j may well be encoded differently. How, then, will the decoder be able to recognize the code for that occurrence of s_j, following $s_{i_1} \cdots s_{i_k}$? Very simple: the decoder has decoded the code text preceding the code for that occurrence of s_j, so the decoder knows that it is "in context $s_{i_1} \cdots s_{i_k}$"; the decoder proceeds to scan the code text with reference to the encoding scheme associated with the correct context.

The discerning reader will have detected that there is a problem with those first k letters in the source text, which are not preceded by a k-letter context. No problem—decide on some prefix-condition "starter scheme" for S and use it for those first k letters. (Of course, the decoder will have to be told what the starter scheme is.) It seems reasonable to use the Huffman scheme based on the relative frequencies $f_1, \ldots, f_m$ of the source letters, calculable as follows:

$$f_j = \sum_{1 \le i_1, \ldots, i_k \le m} f(i_1, \ldots, i_k, j).$$

7.1.1 Example $k = 1$, $S = \{s_1, s_2, s_3, s_4\}$, and

$$[f(i,j)] = [f_{ij}] = \begin{bmatrix} .16 & .10 & .10 & .04 \\ .08 & .17 & .04 & .01 \\ .14 & .01 & .01 & .04 \\ .02 & .02 & .05 & .01 \end{bmatrix}.$$

(How are these f_{ij} found? *Sampling*. But these particular f_{ij} were just made up.) We find (how?) that the relative source frequencies of s_1, s_2, s_3, s_4 are $f_1 = .4$, $f_2 = .3$, $f_3 = .2$, $f_4 = .1$, so we take the following as *starter scheme*:

$$s_1 \to 0, \quad s_2 \to 10, \quad s_3 \to 110, \quad s_4 \to 111.$$

Now we compute the context schemes. For context s_i, we are supposed to assign to $s_1, \ldots, s_4$ the conditional relative frequencies $f_{i1}/f_i, \ldots, f_{i4}/f_i$; since these are proportional to $f_{i1}, \ldots, f_{i4}$, we use these to form the Huffman tree. Similarly, in general, in kth-order encoding, the Huffman tree for context $s_{i_1} \cdots s_{i_k}$ is formable with the assignment of the $f(i_1, \ldots, i_k, j)$ to the s_j; it is not necessary to compute $P(s_j \mid s_{i_1} \cdots s_{i_k}) = f(i_1, \ldots, i_k, j)/f(i_1, \ldots, i_k)$.

Context s_1:

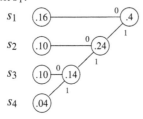

This gives the same scheme as the starter scheme.

Context s_2:

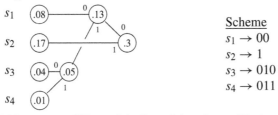

Scheme
$s_1 \to 00$
$s_2 \to 1$
$s_3 \to 010$
$s_4 \to 011$

(Of course, a different labeling of the edges will give a different scheme, but with the same code word lengths.)

Context s_3:

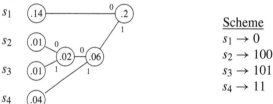

Scheme

$s_1 \to 0$

$s_2 \to 100$

$s_3 \to 101$

$s_4 \to 11$

Context s_4:

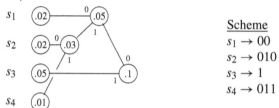

Scheme

$s_1 \to 00$

$s_2 \to 010$

$s_3 \to 1$

$s_4 \to 011$

So, for instance, with the four context schemes and the starter scheme at our disposal, the source text $s_2 s_1 s_1 s_2 s_1 s_4 s_3 s_3 s_1$ is encoded 10000100011111010. Check the encoding, and also check that the decoder can recover the source string from the code string, if supplied with the starter and the context schemes.

7.1.2 Computing the compression ratio Again, $S = \{s_1, \ldots, s_m\}$ and the relative frequencies $f(i_1, \ldots, i_{k+1})$ of the words in S^{k+1} are given. For a context $s_{i_1} \cdots s_{i_k}$ (assuming $k \geq 1$) and $1 \leq j \leq m$, let $\ell(i_1, \ldots, i_k, j)$ be the length of the code word for s_j in the encoding scheme for the context. The average length of a code word replacing a source letter (neglecting the starter scheme, the effect of which would be negligible with a large source text) is, by elementary considerations (see Section 1.8)

$$\bar{\ell}^{(k)} = \sum_{1 \leq i_1, \ldots, i_k \leq m} f(i_1, \ldots, i_k) \sum_{j=1}^{m} P(s_j \mid s_{i_1} \cdots s_{i_k}) \ell(i_1, \ldots, i_k, j)$$

$$= \sum_{1 \leq i_1, \ldots, i_{k+1} \leq m} f(i_1, \ldots, i_{k+1}) \ell(i_1, \ldots, i_{k+1}). \qquad \text{(Verify!)}$$

Thus, for instance, in Example 7.1.1, we have

$$[\ell(i, j)] = [\ell_{ij}] = \begin{bmatrix} 1 & 2 & 3 & 3 \\ 2 & 1 & 3 & 3 \\ 1 & 3 & 3 & 2 \\ 2 & 3 & 1 & 3 \end{bmatrix},$$

and $\bar{\ell}^{(1)} = \sum_{1 \leq i, j \leq 4} f_{ij} \ell_{ij} = 1.72$. (Verify!) By comparison, applying Huffman's algorithm to S, with $f_1 = .4$, $f_2 = .3$, $f_3 = .2$, $f_4 = .1$, gives $\bar{\ell} = \bar{\ell}^{(0)} = 1.9$. If you apply Huffman's algorithm to S^2, with $s_i s_j$ assigned relative frequency f_{ij}, you will get $\bar{\ell} = \bar{\ell}^{(0)}(S^2) = 3.53$. Thus the compression ratio achievable by this method, $2\bar{L}/3.53$, assuming the s_j themselves are binary words with average length $\bar{L}$, is less than that achieved by first-order encoding, $\bar{L}/1.72$.

Here is an academic question of practical importance. Given $f(i_1,\ldots,i_{k+1})$, $1 \le i_1,\ldots,i_{k+1} \le m$, let $\bar{\ell}^{(k)}$ be as defined above, and let $\bar{\ell}(S^{k+1})$ be the average code word length achieved by Huffman's algorithm applied to S^{k+1} as the source alphabet equipped with the relative frequencies $f(i_1,\ldots,i_{k+1})$. Is it always the case that $\bar{\ell}^{(k)} \le \frac{k}{k+1}\bar{\ell}(S^{k+1})$? In other words, is the compression achieved by kth-order Huffman encoding always at least as good as the compression achieved by zeroth-order encoding, treating S^{k+1} as the source alphabet? (Notice that when $k=0$, these two are the same.) It is somewhat surprising that the answer is: not always. See Exercise 7.1.4. Notice that the situation in that exercise is rather extreme. The next question is: under what conditions do we have $\bar{\ell}^{(k)} \le \frac{1}{k+1}\bar{\ell}(S^{k+1})$? It is a large question that probably does not have a snappy answer given the current state of our knowledge and terminology, but its obvious practical importance makes it worth looking into.

Here is another question of practical importance: in case $k \ge 1$, is it necessarily the case that $\bar{\ell}^{(k)} \le \bar{\ell}^{(k-1)}$? Or, can increasing the order sometimes give you worse compression? We suspect that $\bar{\ell}^{(k)} \le \bar{\ell}^{(k-1)}$ always holds, but we have no proof.

Exercises 7.1

1. Suppose $S = \{s_1, s_2, s_3, s_4\}$, with $s_1 = 000$, $s_2 = 001$, $s_3 = 01$, $s_4 = 1$, and digram frequencies $f(i,j) = f_{ij}$ given in

$$[f_{ij}] = \begin{bmatrix} .2 & .04 & .06 & .05 \\ .05 & .17 & .08 & .02 \\ .07 & .08 & .06 & .02 \\ .03 & .03 & .03 & .01 \end{bmatrix}.$$

(a) Find the single-letter relative frequencies f_1, f_2, f_3, f_4, and the compression ratio achieved if Huffman's algorithm is applied to S.

(b) Find the compression ratio achieved if Huffman's algorithm is applied to S^2 (with the relative frequencies f_{ij} given above, of course).

(c) Give the four context schemes for first-order encoding of this source and encode the source string $s_2s_2s_1s_3s_1s_1s_1s_3s_2s_3s_3s_1s_4$. (Use the scheme associated with (a) for the first letter, s_2. There are different correct schemes for the starter and the contexts, so, if doing this exercise as part of a problem set, clearly label your schemes.)

(d) Find the compression ratio achieved by first-order encoding of this source alphabet. (This does not mean the compression ratio achieved in part (c) on that small segment of source text, but in general, on the average, on very large "typical" blocks of source text.)

2. Suppose someone were to examine the source text of problem 1 and to discover the single-letter source frequencies, f_1, f_2, f_3, and f_4, but to remain ignorant of the digram frequencies f_{ij}. Suppose this person applies Huffman's algorithm to S^2, assuming the relative frequency of $s_i s_j$ among all two-letter source words to be $f_i f_j$.

   (a) What compression ratio would this person believe they have achieved, given their assumption about the digram frequencies?

   (b) What compression ratio would they actually have achieved?

3. The lazy but earnest person of problem 2 also tries first-order Huffman encoding of the source text of problem 1, again assuming that the relative frequency of $s_i s_j$ is $f_i f_j$.

   (a) What compression ratio does the encoder believe has been achieved by this method?

   (b) What is the actual compression ratio achieved?

4. Let $S = \{s_1, s_2, s_3\}$ and suppose the digram frequencies are given by

$$[f_{ij}] = \begin{bmatrix} .7 & .05 & .05 \\ .02 & .04 & .04 \\ .08 & .01 & .01 \end{bmatrix}.$$

Recalling the notation of this section, compute $\bar{\ell}^{(0)}$, $\bar{\ell}^{(1)}$, and $\bar{\ell}(S^2)$ for this source alphabet. Observe that $\bar{\ell}^{(1)} > \bar{\ell}(S^2)/2$.

7.2 The Shannon bound for higher-order encoding

Again, suppose that $k \geq 1$ and the "$(k+1)$-gram" frequencies $f(i_1, \ldots, i_{k+1})$ are given. Recall that the k-gram frequencies are then known: $f(i_1, \ldots, i_k) = \sum_{j=1}^{m} f(i_1, \ldots, i_k, j) = \sum_{j=1}^{m} f(j, i_1, \ldots, i_k)$.

In the preceding section we looked at kth-order Huffman encoding. Clearly other kth-order replacement scheme strategies are possible; you need only supply a prefix-condition scheme for encoding $s_1, \ldots, s_m$ for each context $s_{i_1} \cdots s_{i_k}$. Let, for some such association of schemes to contexts, $\ell(i_1, \ldots, i_k, j)$ be the length of the code word for s_j in the scheme corresponding to context $s_{i_1} \cdots s_{i_k}$, and

$$\bar{\ell}(i_1, \ldots, i_k) = \sum_{j=1}^{m} P(s_j \mid s_{i_1} \cdots s_{i_k}) \ell(i_1, \ldots, i_k, j)$$

$$= \sum_{j=1}^{m} \frac{f(i_1, \ldots, i_k, j)}{f(i_1, \ldots, i_k)} \ell(i_1, \ldots, i_k, j).$$

Then the average length of a code word replacing a source letter, using whatever our context schemes are, is

$$\bar{\ell} = \sum_{1 \leq i_1, \ldots, i_k \leq m} f(i_1, \ldots, i_k) \bar{\ell}(i_1, \ldots, i_k)$$

$$= \sum_{1 \leq i_1, \ldots, i_{k+1} \leq m} f(i_1, \ldots, i_{k+1}) \ell(i_1, \ldots, i_{k+1}),$$

as in the preceding section, where the method was kth-order Huffman encoding.

Since Huffman's algorithm gives the minimal $\bar{\ell}(i_1, \ldots, i_k)$ for each context $s_i \cdots s_{i_k}$, among prefix-condition schemes associable to that context, it follows from the preceding that $\bar{\ell}^{(k)}$, the value of $\bar{\ell}$ for kth-order Huffman encoding, will be the smallest kth-order $\bar{\ell}$ achievable. Therefore, in thinking about bounds on compression achievable with kth-order replacement schemes, we may as well stick with kth-order Huffman encoding. Henceforward, $\ell(i_1, \ldots, i_k, j)$ and $\bar{\ell}^{(k)}$ will be as in Section 7.1.

Let

$$H(i_1, \ldots, i_k) = -\sum_{j=1}^{m} P(s_j \mid s_{i_1} \cdots s_{i_k}) \log_2 P(s_j \mid s_{i_1} \cdots s_{i_k})$$

$$= -\sum_{j=1}^{m} \frac{f(i_1, \ldots, i_k, j)}{f(i_1, \ldots, i_k)} \log_2 \frac{f(i_1, \ldots, i_k, j)}{f(i_1, \ldots, i_k)},$$

the "entropy of the source in context $s_{i_1} \cdots s_{i_k}$." We define the kth-order entropy of $S = \{s_1, \ldots, s_m\}$ to be

$$H^{(k)}(S) = H^{(k)} = \sum_{1 \leq i_1, \ldots, i_k \leq m} f(i_1, \ldots, i_k) H(i_1, \ldots, i_k).$$

Plugging the full gory expression for $H(i_1, \ldots, i_k)$ into the expression for $H^{(k)}$, thrashing about and doing what comes naturally with logarithms, one finds that

$$H^{(k)}(S) = H(S^{k+1}) - H(S^k),$$

where

$$H(S^{k+1}) = -\sum_{1 \leq i_1, \ldots, i_{k+1} \leq m} f(i_1, \ldots, i_{k+1}) \log_2 f(i_1, \ldots, i_{k+1})$$

is the plain old zeroth-order entropy of S^{k+1}.

We have, for each context $s_{i_1} \cdots s_{i_k}$,

$$H(i_1, \ldots, i_k) \leq \bar{\ell}(i_1, \ldots, i_k) < H(i_1, \ldots, i_k) + 1,$$

by the Noiseless Coding Theorem, plus the fact that the average code word length obtainable by Huffman's algorithm is the best (smallest) obtainable with a prefix code.

7.2.1 Theorem *The average code word length $\bar{\ell}^{(k)}(S)$ achieved by kth-order Huffman encoding applied to a source alphabet S satisfies $H^{(k)}(S) \leq \bar{\ell}^{(k)}(S) < H^{(k)}(S) + 1$.*

Proof: By the preceding remarks,

$$
\begin{aligned}
H^{(k)}(S) &= \sum_{1 \leq i_1, \dots, i_k \leq m} f(i_1, \dots, i_k) H(i_1, \dots, i_k) \\
&\leq \sum_{1 \leq i_1, \dots, i_k \leq m} f(i_1, \dots, i_k) \bar{\ell}(i_1, \dots, i_k) = \bar{\ell}^{(k)}(S) \\
&< \sum_{1 \leq i_1, \dots, i_k \leq m} f(i_1, \dots, i_k)(H(i_1, \dots, i_k) + 1) \\
&= H^{(k)}(S) + \sum_{i_1, \dots, i_k} f(i_1, \dots, i_k) = H^{(k)}(S) + 1. \qquad \square
\end{aligned}
$$

7.2.2 Corollary *If the s_j are binary words with average length $\bar{L}$, then the compression ratio $\bar{L}/\bar{\ell}^{(k)}(S)$ achieved by kth-order Huffman encoding applied to S satisfies $\bar{L}/(H^{(k)}(S) + 1) < \bar{L}/\bar{\ell}^{(k)} \leq \bar{L}/H^{(k)}$.*

As mentioned in the last section, we do not know whether or not $\bar{\ell}^{(k)}$ always decreases as k increases. If this were the case, then increasing the order repays your effort with a better compression ratio. However, when $m = 256$, as is often the case, it is a lot of trouble to increase the order, and actual case studies with $k = 0, 1, 2, 3$ show a discouragingly small improvement in the compression ratio going from $k = 1$ to $k = 2$, and a minuscule improvement obtainable by taking $k = 3$.

This sort of experimental observation agrees with the behavior predicted by the theory developed by Claude Shannon [63, 65]. In practice it is impossible to let k get very large, much less go to infinity. And, in fact, there is a theoretical obstacle to letting k go to infinity: we would have to have an infinitely long source text, given our notion of how the relative frequencies $f(i_1, \dots, i_{k+1})$ are obtained. Shannon gets around this difficulty by envisioning "the source" as a probabilistic finite state automaton, a system of *states*; as time pulses on discretely, the current state changes (or not) at each pulse, and source letters are emitted. What the next state will be and which letter is emitted are both random variables depending on the current state—that is, the different possibilities have their probabilities, and those probabilities vary with the current state. Thus there is a hypothetically endless string of source letters emitted, with statistical properties, including the probabilities $f(i_1, \dots, i_{k+1})$, for each k, determined by the nature of the source automaton.

Is every source "language" correctly (whatever that means) *modeled* by some such source automaton? This is a far deeper question than we will ever answer, although we will have a bit more to say about it in Section 7.4. For now, let us assume that our source is one of these Shannon automata. Shannon showed that the kth-order entropies $H^{(k)}$ tend to a limit, let us call it $H^{(\infty)}$, which Shannon called the *entropy of the source*. Thus the Shannon bound $\bar{L}/H^{(k)}$ on the compression ratio achieved with kth-order Huffman encoding tends to a limit $\bar{L}/H^{(\infty)}$ if $H^{(\infty)} > 0$. Consequently, when $H^{(\infty)} > 0$, the compression ratio $\bar{L}/\bar{\ell}^{(k)}$ cannot be increased without bound by taking k larger and larger. The experimental case studies mentioned above, with $\bar{\ell}^{(2)}$ not much smaller than $\bar{\ell}^{(1)}$ and $\bar{\ell}^{(3)}$ *very* close to $\bar{\ell}^{(2)}$, are very much in accord with the picture suggested by Shannon's results and Theorem 7.2.1 of the compression ratio coming to a screeching halt at some unbreachable limit, as k increases.

This is an instance of difficult mathematics confirming intuition. If we require lossless compression, meaning that the original file shall always be completely recoverable from its encoded version, then surely there should be some natural limits, depending on the nature of the original file, to how much compression can be realized. However, it is important to realize that the Shannon bounds on the compression ratio, of the form $\bar{L}/H^{(k)}$, $k = 0, 1, \ldots, \infty$, apply to the replacement-by-encoding-scheme methods discussed in this chapter. As we have seen in the last chapter, these bounds can be beaten by other methods in some cases. So the natural bound to the compression ratio, even given a Shannon automaton-type source, may not be the Shannon bound $\bar{L}/H^{(\infty)}$.

One last remark about the Shannon bounds $\bar{L}/H^{(k)}$: Shannon asserts, but does not show, that the $H^{(k)}$ non-increase with k in case the source is a probabilistic finite state automaton. Therefore, the Shannon bounds on the compression ratio, $\bar{L}/H^{(k)}$, are going in the right direction (up!) as k increases, even though we do not know about the actual compression ratios, $\bar{L}/\bar{\ell}^{(k)}$, achievable by kth-order Huffman encoding.

We finish this section with an elementary verification of Shannon's assertion about the monotonicity of the $H^{(k)}$, without assuming anything about the nature of the source.

7.2.3 Theorem *Suppose that $k \geq 1$ and the $(k+1)$-gram frequencies $f(i_1, \ldots, i_{k+1})$, $1 \leq i_1, \ldots, i_{k+1} \leq m$, for an m-letter source S are known. Then*

$$H^{(k)}(S) \leq H^{(k-1)}(S).$$

Proof: Observe that $h(x) = -x \log_2 x$ has negative second derivative on $(0, \infty)$ and so is concave on $[0, \infty)$. (Note: $h(0) = 0$, by convention.) Therefore, for $\lambda_1, \ldots, \lambda_r \geq 0$ with $\sum \lambda_i = 1$, and $x_1, \ldots, x_r \geq 0$, $\sum_i \lambda_i h(x_i) \leq h(\sum_i \lambda_i x_i)$. Therefore,

$$H^{(k)}(S) = \sum_{1 \leq i_1, \ldots, i_k \leq m} f(i_1, \ldots, i_k) \sum_{j=1}^{m} h\left(\frac{f(i_1, \ldots, i_k, j)}{f(i_1, \ldots, i_k)}\right)$$

$$= \sum_{1 \leq i_2, \ldots, i_k, j \leq m} \sum_{i_1=1}^{m} f(i_1, \ldots, i_k) h\left(\frac{f(i_1, \ldots, i_k, j)}{f(i_1, \ldots, i_k)}\right)$$

$$= \sum_{1 \le i_2,\ldots,i_k,j \le m} f(i_2,\ldots,i_k) \sum_{i_1=1}^{m} \frac{f(i_1,\ldots,i_k)}{f(i_2,\ldots,i_k)} h\left(\frac{f(i_1,\ldots,i_k,j)}{f(i_1,\ldots,i_k)}\right)$$

$$\le \sum_{1 \le i_2,\ldots,i_k,j \le m} f(i_2,\ldots,i_k) h\left(\sum_{i_1=1}^{m} \frac{f(i_1,\ldots,i_k,j)}{f(i_2,\ldots,i_k)}\right)$$

$$= \sum_{1 \le i_2,\ldots,i_k,j \le m} f(i_2,\ldots,i_k) h\left(\frac{f(i_2,\ldots,i_k,j)}{f(i_2,\ldots,i_k)}\right)$$

$$= \sum_{1 \le r_1,\ldots,r_k \le m} f(r_1,\ldots,r_{k-1}) h\left(\frac{f(r_1,\ldots,r_k)}{f(r_1,\ldots,r_{k-1})}\right) = H^{(k-1)}(S)$$

(When $k = 1$, the sums over $i_2,\ldots,i_k, j$ are just sums over j.) □

7.2.4 Corollary With k and S as above,

$$(k+1)H^{(k)}(S) \le H(S^{k+1}) \le \frac{k+1}{k} H(S^k) \le (k+1)H(S).$$

The left-hand inequality, and its proof below, are due to Shannon [63].

Proof:

$$H(S^{k+1}) = (H(S^{k+1}) - H(S^k))$$
$$+ (H(S^k) - H(S^{k-1})) + \cdots + (H(S) - 0)$$
$$= H^{(k)}(S) + H^{(k-1)}(S) + \cdots + H^{(0)}(S)$$
$$\ge (k+1)H^{(k)}(S),$$

by the theorem above. Therefore, also,

$$H(S^{k+1}) \ge (k+1)H^{(k)}(S) = (k+1)(H(S^{k+1}) - H(S^k))$$

implies $H(S^{k+1}) \le \frac{k+1}{k} H(S^k)$. Hence,

$$H(S^{k+1}) \le \frac{k+1}{k} H(S^k) \le \frac{k+1}{k} \frac{k}{k-1} H(S^{k-1})$$

$$\le \cdots \le \frac{k+1}{k} \frac{k}{k-1} \cdots \frac{2}{1} H(S) = (k+1)H(S). \quad □$$

Exercises 7.2

1. Compute $H^{(0)}$ and $H^{(1)}$ for the source of Exercise 7.1.1, and the Shannon bounds $\bar{L}/H^{(0)}$ and $\bar{L}/H^{(1)}$.

2. Compute $H^{(0)}$ and $H^{(1)}$ for the source of Exercise 7.1.4.

3. Show that, for $k \ge 1$, $H^{(k)}(S) = H(S^{k+1}) - H(S^k)$, as asserted in this section.

4. Show that if the s_j occur randomly and independently in the source text, with relative frequencies $f_1,\ldots,f_m$, so that, for each k, $f(i_1,\ldots,i_k) =$

$\prod_{j=1}^{k} f_{i_j}$, $1 \le i_1, \ldots, i_k \le m$, then $H^{(k)}(S) = H^{(0)}(S) = H(S) = -\sum_{j=1}^{m} f_j \log_2 f_j$ for every $k \ge 0$.

5. Notice that, in the proof of Theorem 7.2.3, h is *strictly* concave. This implies that if $\lambda_1, \ldots, \lambda_r > 0$, $\sum \lambda_i = 1$, and $x_1, \ldots, x_r \ge 0$, then $\sum \lambda_i h(x_i) < h(\sum \lambda_i x_i)$ unless all the x_i are equal.

   Find a necessary and sufficient condition on the k- and $(k+1)$-gram frequencies for a source S for the equality $H^{(k)}(S) = H^{(k-1)}(S)$ to hold.

6. What does Corollary 7.2.4 say about the Shannon bounds on the compression ratios for zeroth-order Huffman replacement, kth-order Huffman replacement, and Huffman replacement using S^{k+1} or S^k as the source alphabet?

7.3 Higher-order arithmetic coding

For k a positive integer, kth-order encoding of any sort departs from the "$(k+1)$-gram" relative frequencies, $f(i_1, \ldots, i_{k+1})$, $1 \le i_1, \ldots, i_{k+1} \le m$, where $f(i_1, \ldots, i_{k+1})$ is the relative frequency of the source string $s_{i_1} \cdots s_{i_{k+1}}$ among all blocks of $k+1$ consecutive letters in the source text. Supposing that we know what these $(k+1)$-gram relative frequencies are, how would we proceed to take advantage of this knowledge in arithmetic coding?

The main idea is that, for $t \ge k$, the intervals $A(i_1, \ldots, i_t, j)$, $1 \le j \le m$, for the source words $s_{i_1} \cdots s_{i_t} s_j$, are obtained by subdividing $A(i_1, \ldots, i_t)$ into subintervals of lengths proportional to the numbers $f(i_{t-k+1}, \ldots, i_t, j)$, $j = 1, \ldots, m$. The actual probabilities associated to the s_j after $s_{i_1} \cdots s_{i_t}$ are $f(i_{t-k+1}, \ldots, i_t, j)/f(i_{t-k+1}, \ldots, i_t)$, where the k-gram frequencies $f(j_1, \ldots, j_k)$ are given by $f(j_1, \ldots, j_k) = \sum_{i=1}^{m} f(i, j_1, \ldots, j_k) = \sum_{i=1}^{m} f(j_1, \ldots, j_k, i)$. These probabilities are just constant multiples of the $f(i_{t-k+1}, \ldots, i_t, j)$ but, unlike the case of higher-order Huffman encoding, in which absolute probabilities are not important, you will have to use these probabilities in calculations.

For instance, let us return to Example 7.1.1, in which $S = \{s_1, s_2, s_3, s_4\}$, and the digram frequencies are given by

$$[f(i, j)] = [f_{ij}] = \begin{bmatrix} .16 & .10 & .10 & .04 \\ .08 & .17 & .04 & .01 \\ .14 & .01 & .01 & .04 \\ .02 & .02 & .05 & .01 \end{bmatrix}.$$

The single letter frequencies are $f_1 = .4$, $f_2 = .3$, $f_3 = .2$, and $f_4 = .1$. We will use these to start first-order encoding, so the first intervals are $A(1) = [0, .4)$, $A(2) = [.4, .7)$, $A(3) = [.7, .9)$, and $A(4) = [.9, 1)$. From there on we have context; every letter after the first has a predecessor. So, for instance, $A(2, 1) = A(s_2 s_1) = [.4, .48)$; the length is .08 because the probability of an s_1 when we are in first-order context s_2 is $\frac{.08}{.3}$, and this is multiplied by the length .3 of the interval $A(2)$. Similarly, $A(2, 1, 3) = A(s_2 s_1 s_3) = [.452, .472)$. (Because we are in context s_1, you look at the first row of the matrix of digram frequencies. The left-hand endpoint of $A(2, 1, 3)$ is $.4 + \frac{.16+.10}{.4}(.08) = .452$, and the length is $(.08)\frac{.10}{.4} = .02$.)

Notice that it is not necessary or advisable to put the source letters in order of non-increasing context probability in subdividing successive intervals. For instance, in the situation above, $A(4, 3)$ may as well be the third interval from the left in $A(4)$, not the first, even though f_{43} is the largest of f_{41}, f_{42}, f_{43}, and f_{44}.

Decoding in the manner of Section 6.1 proceeds as in that section, except that, after the first k letters have been decoded, the decoder has to keep account of the changing context probabilities. For example, with $k = 1$ and the digram relative frequencies as above, given 01111 and $N = 3$ the decoder could proceed as follows $[r = (.01111)_2 = 15/32]$:

Next letter	α	ℓ	$\frac{r-\alpha}{\ell}$
	0	1	15/32
s_2	.4	.3	about .23
s_1	.4	.08	about .86
s_3			

Thus, 01111 would be decoded, correctly, as $s_2 s_1 s_3$. (Check that $(.01111)_2$ is the dfwld in $[.452, .472) = A(2, 1, 3)$, found earlier.) Finding α and ℓ on the third line of the table above, after s_1 has been decoded, follows the same procedure as in encoding. It is notable that in the second line $\frac{r-\alpha}{\ell} \approx .23$ is compared, not to 0 and .4, but to 0 and $\frac{.08}{.3} \approx .27$, to decode s_1 on the next line. Similarly, on that line $\frac{r-\alpha}{\ell} \approx .86$ is compared to $\frac{.26}{.4} = .65$ and to $\frac{.36}{.4} = .9$, to decode s_3.

The methods of Section 6.4 can be adapted to higher-order situations—as in the examples just exhibited, it is a matter of keeping account of the context and adjusting the current source letter probabilities according to the context.

It is a bit of trouble to take—is it worth it? Zeroth-order dfwld arithmetic coding encodes in close to the zeroth-order source entropy bits per source letter; will kth-order arithmetic encoding do the job in something like the kth-order source entropy bits per letter? (See Section 7.2 for the definition of the kth-order source entropy, $H^{(k)}(S)$, and note Theorem 7.2.3 which says that $H^{(k)}(S)$ is non-increasing with k.) Happily, the answer is yes.

7.3.1 Theorem *The average length of the code representatives of source words of length N over a source alphabet S, if the code words are derived by kth-order dfwld arithmetic coding, where $k < N$ and the first k letters of each source word are processed in zeroth-order fashion, is no greater than $1 + kH(S) + (N - k)H^{(k)}(S)$.*

Thus the average number of bits per source letter with kth-order arithmetic coding applied to source words of length N is no greater than $\frac{1}{N} + \frac{k}{N}H(S) +$

$(1 - \frac{k}{N})H^{(k)}(S))$, which is quite close to $H^{(k)}(S)$, for large N.

Theorem 7.3.1 is proved by the same sort of thrashing around as in the proof of Theorem 6.2.1. The main constituents of the proof are the observations that the dfwld in an interval of length ℓ has a binary expansion of no more than $\log_2(\ell^{-1}) + 1$ bits, and that the length of the interval $A(s_{i_1} \cdots s_{i_N})$ derived by the kth-order subdivision procedure is

$$\left(\prod_{j=1}^{k} f_{i_j}\right) \frac{f(i_1, \ldots, i_{k+1})}{f(i_1, \ldots, i_k)} \frac{f(i_2, \ldots, i_{k+2})}{f(i_2, \ldots, i_{k+1})} \cdots \frac{f(i_{N-k}, \ldots, i_N)}{f(i_{N-k}, \ldots, i_{N-1})}.$$

We omit the details. See Exercise 2 below.

Exercises 7.3

1. Suppose $S = \{s_1, s_2, s_3, s_4\}$ and the digram frequencies are as in Example 7.1.1 (and again in this section).

   (a) Encode $s_2 s_2 s_2 s_2$, $s_1 s_2 s_3 s_4$, $s_4 s_3 s_2 s_1$, and $s_2 s_1 s_4 s_4$ by first-order dfwld arithmetic coding.

   (b) Decode 11, 01001, 10101, and 0101, assuming that the source word lengths are all 4, and that the encoding method was first-order dfwld arithmetic coding.

2. Prove Theorem 7.3.1. [The hard part will be the following: there are numbers $C(j_1, \ldots, j_{k+1})$, $1 \leq j_1, \ldots, j_{k+1} \leq m$, which satisfy

$$\sum_{1 \leq i_1, \ldots, i_N \leq m} f(i_1, \ldots, i_N) \sum_{r=1}^{N-k} g(i_r, \ldots, i_{r+k})$$

$$= \sum_{1 \leq j_1, \ldots, j_{k+1} \leq m} C(j_1, \ldots, j_{k+1}) g(j_1, \ldots, j_{k+1}),$$

where $f(i_1, \ldots, i_N)$ is the relative frequency of $s_{i_1} \cdots s_{i_N}$ among source words of length N, and g could be anything, but will be given by $g(j_1, \ldots, j_{k+1}) = \log_2 \frac{f(j_1, \ldots, j_k)}{f(j_1, \ldots, j_{k+1})}$ in the proof; the C's are given by $C(j_1, \ldots, j_{k+1}) = (N - k)f(j_1, \ldots, j_{k+1})$. Once you see this, the proof is straightforward.]

7.4 Statistical models, statistics, and the possibly unknowable truth

Statistical parameters such as the probabilities $f(i_1, \ldots, i_{k+1})$ are estimated, in this case by taking sample means, through real statistics collected from some messy reality and are then used to talk about, or to do calculations concerning, that reality. It is always the case that these parameters are used in conjunction with some sort of mental picture of what that messy reality is like. We dignify this mental picture with the term "statistical model."

In many cases the statistical model need not be spelled out. For instance, consider the parameter "average number of children per household in the United States." What's the model? There are children, there are households, and each child belongs to a household; we know all about this from our daily experience—no need to make a fuss about the picture of the reality to which the parameter applies or from which it is estimated.

Here is an example that shows that sometimes a fuss is in order. There is a probability distribution called the Poisson distribution which applies to simple statistical models concerning the number of occurrences of some specific event during specified time intervals. For instance, the Poisson distribution is used to talk about the number of cars passing a certain point on a certain road between, say, 2 and 3 PM every Tuesday, or the number of cesium atoms, in a certain hunk of cesium, that will decay in a month. You need one statistical parameter to use the Poisson distribution: it is the average number of occurrences of the event during the time interval. Clearly this parameter can be estimated by observation.

Let us take the time interval to be the 24 hours from midnight to midnight, and the event to be: the sun rises. Observation clearly suggests that the average number of occurrences of this event per time interval is one. Plugging this into the Poisson distribution, one finds that the probability that the sun will not rise tomorrow is $1/e$, the probability that it will rise exactly once is $1/e$, and, therefore, the probability that it will rise 2 or more times is $1 - (2/e)$, about $1/4$.

Have we just been lucky all these millenia? How do we resolve the disparity between our experience of the sun rising with probabilities calculated using the Poisson distribution? The resolution seems clear—shrug and dismiss the calculations on the grounds that the rising of the sun every day does not fall into the class of phenomena that even approximately conform to a statistical model for which the Poisson distribution is valid. We know it doesn't because the conclusions we get from the assumption that it does are absurd. We can leave it to the philosophers to sort out the a priori reasons why we should never have bothered applying the Poisson distribution to the rising of the sun.

Unfortunately, in dealing with a source text we are not on such familiar ground as we are with the rising of the sun. In attempting the replacement encoding that is the subject of this chapter, we estimate certain statistical parameters, the relative frequencies $f(i_1, \ldots, i_{k+1})$, and proceed essentially on the faith that we are taking advantage of a good statistical model of the source to achieve compression. The bottom line for us is the compression ratio.

Whatever the true nature of the source, there is associated to kth-order encoding a particular statistical model of the probabilistic finite state automaton type. For $k = 0$ there is only one state. At each pulse of time a letter is emitted, with each s_j having probability f_j of being emitted, and the system stays in that state.

For $k \geq 1$ there is one state for each context. When we are in context $s_{i_1} \cdots s_{i_k}$, we are also in that state. During the next pulse of time, a letter s_j is emitted with probability $P(s_j \mid s_{i_1} \cdots s_{i_k})$ and we move to state $s_{i_2} \cdots s_{i_k} s_j$. For example, the following state diagram, with the discs representing states and the labels on the arrows being the probabilities, depicts a statistical model for the situation of Exercise 7.1.4:

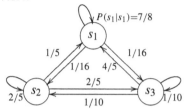

(We leave it to the reader to ponder whether or not the source diagrammed just above really will produce source text exhibiting the digram frequencies in Exercise 7.1.4. Since $f_{ij} = f_i P(s_j \mid s_i)$, and the $P(s_j \mid s_i)$ are given in the diagram, it suffices to verify that the single letter sequences in the source text will be $f_1 = .8$, $f_2 = f_3 = .1$.)

The practical question is, will kth-order Huffman or arithmetic encoding achieve good compression? This is not really the same as asking if the kth-order model is a good model of the source. For example, if the zeroth-order model is a perfect model of the source and the relative source frequencies are approximately equal, then no spectacular compression is possible by the methods of this chapter, and probably not by any methods. Good compression is achievable when the relative frequencies are decidedly unequal.

But you might feel that the question of "goodness" of the model is important, because you feel that if the kth-order model is "good," then, say, kth-order Huffman encoding, while it may not give very good compression, and even, as we have seen in the last chapter, may be slightly less compressive than arithmetic coding, will still give almost as good lossless compression as can be had. Perhaps this feeling is valid in the majority of cases; we would need some definitions of "goodness," and "as good as can be had" to analyze the situation. Meanwhile, here is an extreme and disturbing example that shows that goodness of the kth-order model does not always give as good compression as can be had.

7.4.1 Example $S = \{a, b, c, d\}$ and the source text is $\overline{abcd} = abcdabcd\cdots$. That is, the source text consists of the word $abcd$ repeated over and over.

If we knew or noticed that the source text is of this nature, then we could achieve great compression by what is called "run-length" encoding. If the particular chunk of source text that we want to encode is $(abcd)^N$, we could leave a note saying "repeat $abcd$ N times." The amount of space such a message would occupy would be no greater than $constant + \log_2 N$ bits, so the "local" compression ratio achieved, if a, b, c, d are binary words with average length $\bar{L}$, would be no less than $4N\bar{L}/(const + \log_2 N) \to \infty$ as $N \to \infty$. (This does not beat the Shannon bound $\bar{L}/H^{(\infty)}$, however, because $H^{(\infty)} = 0$ in this case. See Exercise 7.4.3.)

Suppose we do not notice the nature of the source and attempt kth-order Huffman encoding. With $k = 0$ we have $f_a = f_b = f_c = f_d = 1/4$, and Huffman's algorithm replaces each source letter with a 2-bit binary word. Let s_1, s_2, s_3, s_4 be auxiliary names for a, b, c, d. When $k \geq 1$, only four contexts $s_{i_1} \cdots s_{i_k}$ have positive probability (1/4 in each case) and for each of these we have $P(s_j \mid s_{i_1} \cdots s_{i_k}) = 1$ for one value of j, and $P(s_j \mid s_{i_1} \cdots s_{i_k}) = 0$ for other values of j. You carry out kth-order encoding by ignoring the contexts with zero probability. The result is that every source letter gets replaced by one bit. Thus the compression ratio achieved is $\bar{L}/1 = \bar{L}$ for each $k \geq 1$. This is better than the zeroth-order compression ratio, but a far cry from what is possible. Yet, the first-order model of this source is correct. See the comment in Exercise 2 below.

Exercises 7.4

1. Give the state diagram of the first-order model of the source in Example 7.1.1.

2. Give the state diagram of the first-order model of the source in Example 7.4.1. (Disconcertingly, this first-order model can be regarded as a perfect model of the source—it produces exactly the right source string, although not necessarily starting with 'a'. Yet first-order Huffman encoding does not achieve compression as good as can be obtained.)

3. Show that for the source of Example 7.4.1, $H^{(0)}(S) = 2$ and $H^{(k)}(S) = 0$, for $k \geq 1$. [Thus $H^{(\infty)}(S) = 0$. Perhaps disturbing examples like Example 7.4.1 are only possible with sources of zero entropy.]

4. Generating text according to some statistical model can be a mildly amusing experiment, and it also gives some insights into the model itself. The basic goal in compression is to remove redundancy, so the output of a good model/coder is typically rather random. We want to reverse the process, sending random data to a specific model to generate text. As Bell, Cleary, and Witten [8] remark, the output of this reverse process is a rough indication of how well the model captured information about the original source.

   Single-letter frequencies from a 133,000-character sample of English appear in Table 4.1. Text generated according to these probabilities is not likely to be mistaken for lucid prose, as the model knows very little about the structure of English. Higher-order models may do better, where the necessary statistics are gathered from a sample of "typical" text.

   As an example, models were built from Kate Chopin's *The Awakening*.[1] An order-1 model produced the following text when given random input:

   > asthe thetas tol t dinfrer the Yo Do smp thle s slawhee pss,
   > tepimuneatage le indave tha cars atuxpre ad merong? d ur atinsinth g

---

[1] The electronic source for the book was obtained from Project Gutenberg, available via ftp://uiarchive.cso.uiuc.edu/pub/etext. Thanks to Judith Boss and Michael Hart.

> teres runs l ie t ther Mrenorend t fff mbendit'sa aldrea ke Shintimal
> "Alesunghed thaf y, He," ongthagn buid co. fouterokiste singr. fod,

Moving to higher-order models will capture more of the structure of the source. An order-3 model, given the same random data, produced:

> assione mult-walking, hous the bodes, to site scoverselestillier from the
> for might. The eart bruthould Celeter, ange brouse, of him. They was
> made theight opened the of her tunear bathe mid notion habited. Mrs.
> She fun andled sumed a vel even stremoiself the was the looke hang!

Choose your favorite software tool and write a program that builds an order-k $(k > 0)$ model from a given source, and then uses that model to generate characters from random input.[2]

7.5 Probabilistic finite state source automata

The first-order state diagrams introduced in the preceding section are special cases (actually, slightly degenerate special cases) of diagrams that we will call probabilistic finite state source automata, or pfssa's, for short. These were introduced by Shannon in "A mathematical theory of communication" [63], although he gave them no special name. They are sources; they produce source text. It appears that Shannon entertained the belief that human language, produced by a single hypothetical human, could be well approximated – perhaps *simulated* would be a better word – by a large pfssa, or even that human language production could be described exactly by a very large pfssa.

A pfssa is a finite *directed graph* (i.e., a diagram consisting of *nodes* and *arrows* or *arcs* among the nodes, with each arc having a starting node at one end and a not necessarily different finishing node at the arrowhead end) in which each arc e is furnished with two labels. One label, $g(e)$, is a positive probability, and the other label, $s(e)$, is a letter from a fixed source alphabet S. The labeling must satisfy two requirements: every source letter must appear somewhere, and for each node, the sum of the probability labels on the arcs leaving that node is one.

7.5.1 Example The following diagram shows a pfssa, with three nodes, S_1, S_2, S_3, over a source alphabet $S = \{a, b, c, d\}$.

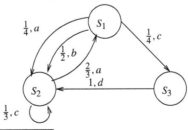

---

[2]The authors used *awk*, which offers associative arrays and other magical features. The script contained roughly 80 lines of *awk* code, and required approximately 5 megabytes of memory for the order-3 example.

The nodes of a pfssa are called *states*, which is why we named the nodes by indexing the letter S. (It is just our bad luck that the words "source" and "state" start with the same letter.) Pfssa's differ from probabilistic finite state automata, pfsa's, only in the presence of the letter labels on the arcs. (Also, sometimes pfsa's come equipped with a designated "starting state" which may or may not ever be revisited. In this text, there will be no such starting states.) A pfssa produces source text in the following fashion: time is discretized, and an imaginary entity, the source gremlin, is moving among the states of the pfssa, making one move per pulse of time. When the gremlin makes a move, the gremlin chooses an arc leaving the state in which it currently resides probabilistically, with the different arcs having the chance of being chosen indicated by their probability labels. Whichever arc is chosen, the letter label of that arc is emitted; it is the next letter of the source text. Notice that the source text produced is hypothetically a two-way infinite string, with no beginning and no end. Because the text is produced probabilistically, there are typically many (a very great many!) different possible two-way infinite strings that could be produced by a given psffa. When we are reading a particular piece of source text, what we have on our hands is a finite substring of one of the possible two-way infinite strings of source letters producible by the pfssa.

Note that the first-order diagrams in Section 7.4 have no letter labels on the arcs. However, they can be considered pfssa's because the states are identified with the letters of a source text; in this identification, each letter stands for a first-order context, but there is no harm in using the same letter as indicating the "next letter", as well. That is, each arc is considered to be labeled with the letter with which its destination node is identified.

The mathematics of pfsa's had been pretty well worked out by the late 1940's, when Shannon created information theory, and he made good use of that mathematics in working out the essentials of pfssa's. What follows is a brief account of some of that mathematics, and some of those essentials.

For two states, S, S' in a pfsa $\mathcal{D}$, the *transition probability* from S to S', denoted $q(S, S')$, is the sum of the probability labels on the arcs going from S to S'. Thus $q(S, S')$ is interpretable as the probability that the pfsa gremlin will next be in S', if currently residing in S.

If the states of $\mathcal{D}$ are ordered, $S_1, \ldots, S_n$, then we will abbreviate $q(S_i, S_j) = q_{ij}$. Thus, for the pfssa given in 7.5.1, with states S_1, S_2, S_3, the matrix $Q = [q_{ij}]$ of transition probabilities is

$$Q = \begin{bmatrix} 0 & \frac{3}{4} & \frac{1}{4} \\ \frac{2}{3} & \frac{1}{3} & 0 \\ 0 & 1 & 0 \end{bmatrix}$$

Now, let us consider the proposition that there are *state probabilities* $P(S)$, the probability that, at a randomly selected pulse of time, at the end of that pulse the pfsa gremlin will be residing in state S. The problem with this "definition," aside from the fact that is presumes existence, is in the phrase "a randomly selected pulse of time." There is no probability assignment to the integers which gives each integer (i.e., each pulse of time) the same probability.

The standard way of dealing with this difficulty is rather daunting: the set of all possible two-sided sequences of states in which the gremlin might be residing is made into a probability measure space in such a way that for each state S and for any two different sequence places – i.e. pulses of time – the measures of the two sets of sequences with S appearing in those places are the same; call this common value $P(S)$. (See [66] for an account of how the measure is defined, in slightly different circumstances.)

Here is a more facile approach that bypasses the philosophical problems in defining the probabilities $P(S)$ and leads directly to their computation. If it were possible to pick a pulse of time at random, then the next pulse would be selected with equal randomness. Thus, if the state probabilities exist, they must satisfy

$$P(S') = \sum_S P(S)q(S, S')$$

for each state S' of $\mathcal{D}$, with the sum above taken over the set of states of $\mathcal{D}$. This leads to the following.

7.5.2 Definition If a pfsa $\mathcal{D}$ has states $S_1, \ldots, S_n$ with transition probabilities $Q = [q(S_i, S_j)] = [q_{ij}]$, and if the linear system $Q^T \underline{p} = \underline{p}$ has a unique solution among the probability vectors

$$\underline{p} = \begin{pmatrix} p_1 \\ \vdots \\ p_n \end{pmatrix},$$

then (and only then) $p_i = P(S_i)$ will be called the state probability of (or, the probability of the state) S_i, $i = 1, \ldots, n$.

7.5.3 For example, for the pfssa of Example 7.5.1, regarded as a pfsa, it is straightforward to verify that the homogeneous linear system with matrix of coefficients

$$Q^T - I = \begin{bmatrix} -1 & \frac{2}{3} & 0 \\ \frac{3}{4} & -\frac{2}{3} & 1 \\ \frac{1}{4} & 0 & -1 \end{bmatrix}$$

has a unique solution among the probability vectors, namely $\begin{pmatrix} p_1 \\ p_2 \\ p_3 \end{pmatrix} = \begin{pmatrix} 4/11 \\ 6/11 \\ 1/11 \end{pmatrix}$.

On the other hand, for the pfsa

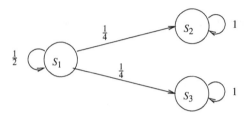

the system

$$Q^T \underline{p} = \begin{pmatrix} \frac{1}{2} & 0 & 0 \\ \frac{1}{4} & 1 & 0 \\ \frac{1}{4} & 0 & 1 \end{pmatrix} \begin{pmatrix} p_1 \\ p_2 \\ p_3 \end{pmatrix} = \begin{pmatrix} p_1 \\ p_2 \\ p_3 \end{pmatrix}$$

has an infinite number of solutions among the probability vectors: $\begin{pmatrix} 0 \\ t \\ 1-t \end{pmatrix}$ is a

solution for any $t \in [0, 1]$.

A directed graph (digraph) is *strongly connected* if for any ordered pair (u, v) of different nodes in the graph there is a directed walk (i.e., a walk along arcs in the directions of the arcs) in the digraph from u to v. For example, the underlying digraph of the pfssa in Example 7.5.1 is strongly connected, while the pfsa in 7.5.3 is not strongly connected—there is no way to walk from S_2 to S_3 in the digraph. The following is the great truth about strongly connected pfsa's upon which much else is based.

7.5.4 Suppose that $\mathcal{D}$ is a strongly connected pfsa with transition probabilities $Q = [q_{ij}]$. Then $Q^T \underline{p} = \underline{p}$ has a unique solution among the probability vectors $\underline{p}$.

See [21] for a proof of this, and a characterization of those $\mathcal{D}$ for which the conclusion of 7.5.4 holds. (It is not necessary that $\mathcal{D}$ be strongly connected.)

Henceforward, our pfssa's will be assumed to be strongly connected, with states $S_1, \ldots, S_n$, transition probabilities q_{ij}, $1 \le i, j \le n$, and state probabilities $p_i = P(S_i)$. Let the source alphabet be $S = \{s_1, \ldots, s_n\}$. Then for $1 \le i_1, \ldots, i_k, \le m$, $1 \le k$, the relative frequency of the k-gram $s_{i_1} \ldots s_{i_k}$ in the source text is given by

$$f(i_1, \ldots, i_k) = \sum_{j=1}^{n} p_j \sum_{e_1, \ldots, e_k} \prod_{t=1}^{k} g(e_t)$$

in which the inner sum is taken over those sequences $e_1, \ldots, e_k$ on a directed walk starting from S_j whose letter labels are $s_{i_1}, \ldots, s_{i_k}$, in that order. (Recall that $g(e)$ is the probability on the arc e.) Thus, for instance, for the pfssa of 7.5.1, in view of 7.5.3, using the letters instead of indices to indicate the k-gram, the single letter frequencies are given by $f(a) = \frac{4}{11}\frac{1}{4} + \frac{6}{11}\frac{2}{3} = \frac{5}{11}$, $f(b) = \frac{4}{11}\frac{1}{2} = \frac{2}{11}$, $f(c) = \frac{4}{11}\frac{1}{4} + \frac{6}{11}\frac{1}{3} = \frac{3}{11}$, $f(d) = \frac{1}{11} \cdot 1 = \frac{1}{11}$. Meanwhile, examples of digram frequencies are $f(ac) = \frac{4}{11}\frac{1}{4}\frac{1}{3} + \frac{6}{11}\frac{2}{3} \cdot \frac{1}{4} = \frac{4}{33}$, and $f(ad) = 0$.

With the relative k-gram frequencies at our disposal, for all $k = 1, 2, \ldots$, we have the kth-order entropies $H^{(k)}(\mathcal{D})$ of the text produced by a strongly connected pfssa $\mathcal{D}$, as in section 7.2, and $H^{(\infty)}(\mathcal{D}) = \lim_{k \to \infty} H^{(k)}(\mathcal{D})$. One of Shannon's great theorems is that $H^{(\infty)}(\mathcal{D})$ is directly computable if the structure of $\mathcal{D}$ is known: it is the average, over the states, of the ordinary, zeroth-order entropies of the zeroth-order sources you would get by turning each arc into a loop, returning to the state it comes from.

7.5.5 Theorem (Shannon [63]) *Suppose that $\mathcal{D}$ is a strongly connected pfssa with states $S_1, \ldots, S_n$, with state probabilities $P(S_i) = p_i, i = 1, \ldots, n$. Suppose that the source alphabet is $S = \{s_1, \ldots, s_m\}$, and for each $i \in \{1, \ldots, m\}$, $j \in \{1, \ldots, n\}$, $h_{ij} = P(s_j \mid S_i)$ is the sum of the probability labels on arcs leaving S_i with letter label s_j. Then*

$$H^{(\infty)}(\mathcal{D}) = -\sum_{i=1}^{n} p_i \sum_{j=1}^{m} h_{ij} \log h_{ij}.$$

For example, if $\mathcal{D}$ is the pfssa in 7.5.1, then

$$H^{(\infty)}(\mathcal{D}) = \frac{4}{11}[\frac{1}{4}\log 4 + \frac{1}{2}\log 2 + \frac{1}{4}\log 4]$$
$$+ \frac{6}{11}[\frac{2}{3}\log\frac{3}{2} + \frac{1}{3}\log 3] + \frac{1}{11} \cdot 0.$$

Simulating a source with a pfssa

Suppose we have some source $\mathcal{D}$ emitting text, with source alphabet $S = \{s_1, \ldots, s_m\}$. We do not presume anything about the inner workings of the source; it need not be a pfssa. We suppose that we have somehow (perhaps by sampling the text) obtained the relative $(k+1)$-gram frequencies $f(i_1, \ldots, i_{k+1}) = f(s_{i_1} \ldots s_{i_{k+1}}), 1 \le i_1, \ldots, i_{k+1} \le m$, of blocks of $k+1$ consecutive letters in the source text, for some $k \ge 0$.

The *kth-order simulant* $\mathcal{D}^{(k)}$ of the given source is the pfssa whose states are (or, are identified with) those kth order contexts $s_{i_1} \ldots s_{i_k}$ which have positive relative frequency $f(i_1, \ldots, i_k) = \sum_{j=1}^{m} f(i_1, \ldots, i_k, j)$. When $k = 0$ there is only one state, and the arcs are loops with (probability, letter) labels $(f_i, s_i), i = 1, \ldots, m$. When $k > 0$, for each context $s_{i_1} \ldots s_{i_k}$ with positive relative frequency there is an arc to the state of each context $s_{i_2} \ldots s_{i_k} s_j$ such that $f(i_1, \ldots, i_k, j) > 0$, with probability label $\frac{f(i_1, \ldots, i_k, j)}{f(i_1, \ldots, i_k)}$ and letter label s_j. [When $k = 1$ this might be a bit misleading. In this case there is an arc from state (context) s_i to state s_j, whenever $f(i, j) > 0$, with probability label $f(i, j)/f_i$ and letter label s_j.] As noted previously, in the special case $k = 1$ it is convenient to let the name of the destination state serve as the letter label on the arc. In fact, a convenience of the same type applies to any of these simulant pfssa's, when $k > 0$; the letter label on an arc into a state corresponding to a context will be understood to be the last letter of that context.

The first-order simulant of the pfssa in 7.5.1 has four states, corresponding to the four letters a, b, c, d. Using the convention above about letter labels, this simulant is

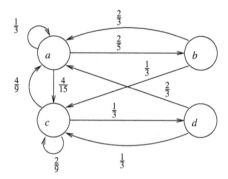

A good deal of calculation went into making this simulant. For example, to calculate the probability on the arc from context a to context c (which arc has letter label c, it is understood), we calculated

$$f(ac) = \frac{6}{11}\frac{2}{3}\frac{1}{4} + \frac{4}{11}\frac{1}{4}\frac{1}{3} = \frac{4}{33}$$

and then $f(ac)/f(a) = (4/33)/(5/11) = 4/15$.

The second-order simulant for the pfssa of 7.5.1 will have 10 states, not 16, because six of the second-order contexts, namely ad, bb, bd, cb, db and dd, never occur in the text generated by that pfssa. In that second-order simulant there are, for instance, arcs from the state ab to the states ba and bc, and only to those. There are arcs from bc to ca and cc, but not to cd, because $f(bcd) = 0$ (i.e., bcd does not appear in the source text). The arc from state ab to state bc has probability label $f(abc)/f(ab) = (2/33)/(2/11) = 1/3$ and letter label c.

We are just using the pfssa of 7.5.1 as an example – there doesn't seem to be any good reason to replace a perfectly good pfssa as in 7.5.1 by a simulant that is more complicated than the original. Simulants are for sources of which the inner workings are unknown, but the statistics of the source text output are available.

Shannon had simulants in mind as pfssa simulations of natural language production, using either the alphabet of the language, plus some punctuation marks, as a source alphabet, or, more promisingly, a significant subset of the set of words of the language. If one took as the source alphabet only the 1,000 most commonly used English words, and compiled statistics by running through a huge assortment of modern texts, the third-order simulant obtainable might have as many as $(1000)^3 =$ one billion states. Higher order simulants will be impossibly huge. So it appears that while simulants might be of interest in the theory of pfssa's and their uses in simulating sources, they are not really practical for playing around with artificial language production.

Simulants are too complicated for natural language simulation in one way, and not complicated enough in another. We think that a rich area of exploration would be opened by letting time play a more dynamic role in the operation of pfssa's: let the probability labels on the arcs be allowed to vary from pulse to pulse. So far as we know, no work has been done on this obvious idea—and we shall do none here. We return to the subject of simulation by pfssa's and finish this section by giving some facts and raising some questions.

Suppose that $\mathcal{D}^{(k)}$ is the kth-order simulant of a source $\mathcal{D}$ with source alphabet $S = \{s_1, \ldots, s_m\}$. Proofs of 7.5.7–7.5.9 can be found in [21]; a proof of 7.5.6 will appear soon.

7.5.6 If $\mathcal{D}$ is a "true source" then $\mathcal{D}^{(k)}$ is strongly connected. More precisely, $\mathcal{D}^{(k)}$ is strongly connected if the $(k+1)$-gram frequencies used to construct it satisfy

(i) $\sum_{j=1}^{m} f(i_1, \ldots, i_k, j) = \sum_{j=1}^{m} f(j, i_1, \ldots, i_k)$ for each $i_1, \ldots, i_k \in \{1, \ldots, m\}$; and

(ii) for no proper subset S' of S is it the case that $f(i_1, \ldots, i_{k+1}) > 0$ only for $(i_1, \ldots, i_{k+1}) \in (S')^{k+1} \cup (S \setminus S')^{k+1}$.

7.5.7 If $\mathcal{D}^{(k)}$ is strongly connected, the state probabilities for $\mathcal{D}^{(k)}$ are what you would expect: $P\left("s_{i_1} \ldots s_{i_k}"\right) = f(i_1, \ldots, i_k)$.

7.5.8 If $\mathcal{D}^{(k)}$ is strongly connected, the relative $(k+1)$-gram frequencies in the source text produced by $\mathcal{D}^{(k)}$ are the same as those in the source text produced by $\mathcal{D}$. Therefore, $H^{(j)}(\mathcal{D}) = H^{(j)}(\mathcal{D}^{(k)}), 0 \leq j \leq k$.

7.5.9 $H^{(\infty)}(\mathcal{D}^{(k)}) = H^{(k)}(\mathcal{D}^{(k)})$.

Two sources with the same source alphabet are *equivalent* if the relative k-gram frequencies in the texts produced by the two sources are the same, for every k. The great theoretical goal, given a source of unknown structure, is to produce a pfssa which is equivalent to the given source. Let us call a source which satisfies (i) and (ii) of 7.5.6 for every $k \geq 1$ a *true source*.

7.5.10 *Is every true source equivalent to a strongly connected pfssa?* If not, how can the source texts produced by strongly connected pfssa's be distinguished from source texts produced by true sources that are not equivalent to pfssa's?

7.5.11 *If $\mathcal{D}$ is a true source with kth-order simulant $\mathcal{D}^{(k)}$, and if $H^{(\infty)}(\mathcal{D}) = H^{(k)}(\mathcal{D})$, are $\mathcal{D}$ and $\mathcal{D}^{(k)}$ necessarily equivalent?*

Exercises 7.5

1. Find the state probabilities, single letter, digram, and trigram relative frequencies, and the zeroth-, first-, and second-order simulants of the following pfssa.

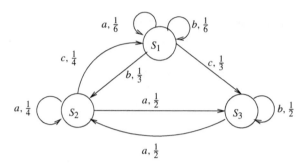

2. Consider the following pfssa $\mathcal{D}$:

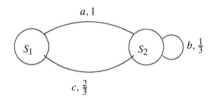

Notice that for $k \geq 2$, a k-gram in the text generated by this pfssa must end in one of ab, ac, bb, bc, ca. For a non-empty word $u \in \{a, b, c\}^+$, let u' denote the word obtained by deleting the first letter of u (on the left). Thus, if $\mathrm{lgth}(u) = 1$, then $u' = \lambda$, the empty word. We will use this notation to talk about kth-order contexts in the text generated by $\mathcal{D}$, and the kth order simulants of $\mathcal{D}$. For instance, for $k \geq 3$, a state uab in $\mathcal{D}^{(k)}$ ($\mathrm{lgth}(u) = k - 2$) has arcs to the states $u'abb$ and $u'abc$.

(a) For $k \geq 3$, describe the states in $\mathcal{D}^{(k)}$ receiving arcs from the states of the form uab, uac, ubb, ubc, and uca. Give the probability and letter labels on these arcs.

(b) By 7.5.6 and 7.5.7, $H^{(k)}(\mathcal{D}) = H^{(k)}(\mathcal{D}^{(k)}) = H^{(\infty)}(\mathcal{D}^{(k)})$. Use this to show that the sequence $(H^{(k)}(\mathcal{D}))_{k \geq 0}$ is strictly decreasing, and that, therefore, $\mathcal{D}$ is not equivalent to any of its simulants.

Chapter 8

Adaptive Methods

The methods to be described in this chapter are, as in Chapters 5 and 6, methods for lossless encoding of source text over a fixed source alphabet $S = \{s_1, \ldots, s_m\}$. The new feature is that no statistical study of the source text need be done beforehand. The encoder starts right in encoding the source text, initially according to some convention, but with the intention of modifying the encoding procedure as more and more source text is scanned. In the cases of adaptive Huffman (Section 8.1) and adaptive arithmetic (Section 8.3) encoding, the encoder keeps statistics, in the form of *counts* of source letters, as the encoding proceeds, and modifies the encoding according to those statistics. In interval and in recency rank encoding (Section 8.4), encoding of a source letter also depends on the statistical nature of the source text preceding the letter, but the statistics gathering is not boundless; the encoding rules do not change, but rather cleverly take into account the recent statistical history of the source text.

With the encoding procedure varying according to the statistical nature of the source text (unlike higher-order encoding in which the encoding procedure varies according to the syntactic nature of the source text), you might wonder, how is decoding going to work? No problem, in principle; notice that after the decoder has managed to decode an initial segment of the source text, the decoder then knows as much about the statistical nature of that segment as did the encoder, having come that far in the encoding process. If the decoder knows the rules and conventions under which the encoder proceeded, then the decoder will know how to decode the next letter. Thus, besides the code stream, the decoder will have to be supplied with the details of how the encoder started, and of how the encoder will proceed in each situation. The hidden cost suffered here is fixed, independent of the length of the source text, and is, therefore, usually essentially negligible; indeed, the required understanding between encoder and decoder can be "built in" in any implementation of an adaptive method and need not be resupplied for each instance of source text.

A note on classification: the "dictionary" methods to be described in Chapter 9 are adaptive, but so different from the methods of this chapter that they deserve a chapter to themselves. Some would argue that interval and recency rank encoding are really dictionary methods; we leave debate on the matter to those who enjoy debate.

8.1 Adaptive Huffman encoding

In adaptive Huffman encoding, counts of the source letters are kept as scanning of the source text proceeds, and the Huffman tree and corresponding encoding scheme change accordingly. The counts are proportional to the relative frequencies of the source letters in the segment of source text processed so far. (Or, almost proportional—in many cases it is convenient to start with the source letter counts set at some small positive value, like one, which makes the ostensible relative frequencies slightly different from the true relative frequencies. The difference diminishes as you go deeper into the source text. It is possible to start with all counts at zero. The drawback is that the presence of zero node weights complicates the tree management technique due to Knuth, to be described in the next section.) Thus it makes sense to use the counts as weights on the leaf nodes—the nodes associated with the source letters—of the Huffman tree.

Unfortunately, different Huffman trees can be constructed from the same leaf node weights because of choices that sometimes have to be made when two or more nodes carry the same weight. Also, different encoding schemes are associable with the same Huffman tree, because of different choices that can be made in labeling the edges. Therefore, to do adaptive Huffman encoding and decoding, we suffer the annoying necessity of adopting explicit conventions governing how we start and how the choices are to be made in drawing the Huffman trees and labeling their edges. We will also need conventions governing how to go from one Huffman tree to the next after a count has been incremented.

The conventions of this section, to be described below, are not really practical, but will (we think) serve better than "the real thing" as an introduction to adaptive Huffman coding. The real thing, i.e., the actual conventions used in practice are a bit trickier to describe; they are well adapted for computer implementations, but not so easy to walk through in doing pencil and paper exercises. We describe the real conventions, the Knuth-Gallager method of tree management, in Section 8.2.

In this section we shall draw our Huffman trees horizontally, as we did in Chapter 5. The weights in the leaf nodes will be non-increasing as you scan from top to bottom. Each leaf node establishes a *level* in the tree; all nodes will be on one of these levels. When two nodes are merged, i.e., when they become siblings, the parent node will be on the higher of the two levels of the sibling nodes. (Thus the root node is guaranteed to be on the level of the highest leaf, not that this is any great advantage.) Here's an example:

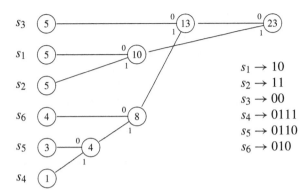

$$s_1 \rightarrow 10$$
$$s_2 \rightarrow 11$$
$$s_3 \rightarrow 00$$
$$s_4 \rightarrow 0111$$
$$s_5 \rightarrow 0110$$
$$s_6 \rightarrow 010$$

This example illustrates two other conventions we will observe: (1) in the labeling of the edges (or "branches") of the tree, the edge going from parent to highest sibling is labeled zero, and the lower edge is labeled 1; (2) we "merge from below in case of ties." For instance, in the construction of the tree above, when it came to pass that the smallest weight of an unmerged node was 5, there were three such nodes to choose among; we chose to merge the two lowest in level.

It remains to specify how to get started, and how the tree will be modified when a count is incremented.

Start: All letters start with a count of 1, and the initial ranking of the source letters, from top to bottom, will be $s_1, \ldots, s_m$. Thus the initial Huffman tree for source text over $S = \{s_1, s_2, s_3, s_4, s_5, s_6\}$ will look like this:

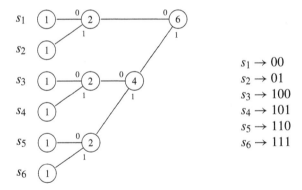

$$s_1 \rightarrow 00$$
$$s_2 \rightarrow 01$$
$$s_3 \rightarrow 100$$
$$s_4 \rightarrow 101$$
$$s_5 \rightarrow 110$$
$$s_6 \rightarrow 111$$

Update after count incrementation: The assignment of leaf nodes to source letters is redone so that the weights are once again in non-increasing order, and the source letter whose count has gone up by one is now attached to the lowest possible node consistent with this monotonicity; the ranking of the other letters is unchanged, but for the possible promotion of that one letter. For instance, in the first example, if any of s_4, s_5, or s_6 is the next letter scanned, the weight in the corresponding leaf node is increased by one and the letter-to-leaf assignments remain the same. In this particular example, incrementing any one of these letter counts does not affect the tree structure, and the encoding scheme remains the same. If s_2 is scanned, the next tree has its top three leaf nodes assigned to s_2, s_3, and s_1, in that order, with weights 6, 5, and 5, respectively. If s_1 is scanned, the order of the top 3 letters is s_1, s_3, s_2; if s_3 is scanned, the order stays as it is.

Let's try some encoding! We will take $S = \{s_1, s_2, s_3, s_4, s_5, s_6\}$ and source text starting $s_3 s_3 s_2 s_6 s_1 s_1 s_2 s_6 s_5 s_1 s_3 s_3 s_5 s_6 s_1 s_2 s_2 s_2$. We leave the full encoding of these 18 letters as Exercise 8.1.1, but this is how the code will start:

$$\underline{100}\ \underline{00}\ \underline{110}\ \underline{111} \cdots$$
$$\ \ \ s_3\ \ \ s_3\ \ \ s_2\ \ \ s_6$$

The Huffman tree after the first two letters (s_3's) are scanned looks like this:

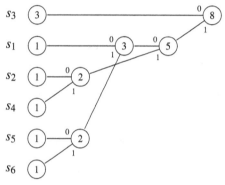

(Are you surprised that the tree comes out like this? Recall the convention that when there is a choice of unmerged nodes with smallest possible weights, merge the two at the lowest possible level.)

Now let's try decoding. With $S = \{s_1, s_2, s_3, s_4, s_5, s_6\}$, and all the conventions mentioned above, suppose the decoder is faced with

$$01100011001011111100\ldots$$

The decoder knows the starting scheme; scanning left to right, the decoder recognizes 01, which is the code word in the starting scheme for s_2. Now the decoder knows what the next Huffman tree will be:

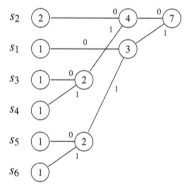

Having recorded s_2, the decoder resumes scanning and soon recognizes 10, the code word in the current scheme, derived from the tree above, for s_1. The decoder records s_1, makes a new tree, with s_1 now having a count of 2, and forges on.

8.1.1 Compression and readjustment

Keeping letter counts as the source text is scanned amounts to estimating the relative source frequencies by sampling. Therefore, the Huffman tree and associated encoding scheme derived from the letter counts are expected to settle down eventually to the fixed tree and scheme that might have arisen from counting the letters in a large sample of source text before encoding. Because adaptive Huffman encoding is a bit of trouble, and decoding is slow, you might think it is worth the trouble to do a statistical study of the source text first, and proceed with a fixed encoding scheme.

But adaptive Huffman encoding offers an advantage over plain zeroth-order Huffman encoding that can be quite important in situations when the nature of the source text might change. Let us illustrate by taking an extreme case, in which the source text consists of the letter a repeated 10,000 times, then the letter b repeated 10,000 times, then the letter c repeated 10,000 times, and finally the letter d repeated 10,000 times. A thorough statistical study of the source text will reveal that the relative source frequencies are all $1/4$. Plain zeroth-order Huffman encoding will represent a, b, c, and d by the four binary words of length 2. (Of course, first-order Huffman encoding will do very well in this example, but we are trying to compare zeroth-order adaptive and nonadaptive Huffman encoding.)

Now, adaptive Huffman encoding will start well, and, whatever conventions are in force, will encode almost all of those a's by single digits. However, the b's will be encoded with 2 bits each, and then the c's and d's with 3 bits each, squandering the early advantage over static Huffman encoding. Obviously the problem is that when the nature of the source text changes, one or more of the letters may have built up such hefty counts that it takes a long time for the other letters to catch up. To put it another way, the new statistical nature of the source text is obscured by the statistics from the earlier part of the source text.

This defect has a perfectly straightforward remedy, first proposed, we think, by Gallager [23]. The trick is to periodically multiply all the counts by some fraction, like $1/16$, and round down. (If the counts are stored as binary numbers, this operation is particularly easy if the fraction is an integral power of $1/2$.) The beauty of this trick is that if the source text is not in the process of changing its statistical nature, no harm is done, because the new counts after multiplication and rounding are approximately proportional to the old counts; and if the source text is changing, the pile of statistics from earlier text has been reduced from a mountain to a molehill, and it will not take so long for the statistical lineaments of the current text to emerge in the letter counts.

Of course, the decoder must know when and by how much the counts are to be scaled down. Also, if zero counts are not allowed, then steps might have to be taken to deal with occasional rounding down to zero. But this is a mere annoying detail.

8.1.2 Higher-order adaptive Huffman encoding

For $k \geq 1$, kth-order adaptive Huffman encoding proceeds as you might suppose: for each kth-order context $s_{i_1} \cdots s_{i_k}$ the encoder keeps counts of the source letters occurring in that context (i.e., the scanning of $s_{i_1} \cdots s_{i_k} s_j$ causes the count of s_j in context $s_{i_1} \cdots s_{i_k}$ to increase by 1). Huffman trees are maintained for each context. The encoder and decoder have to agree on conventions for tree formation and maintenance (updating) and also on how to get started; pretty obviously these starting rules will be more extensive than in the zeroth-order case. It seems reasonable to agree to encode the first k letters (before any context is established) according to the zeroth-order adaptive Huffman conventions, whatever they may be.

For example, let us take $S = \{s_1, \ldots, s_6\}$ and the source text we looked at before,

$$s_3 s_3 s_2 s_6 s_1 s_1 s_2 s_6 s_5 s_1 s_3 s_3 s_5 s_6 s_1 s_2 s_2 s_2 \cdots \qquad (*)$$

With tree formation and updating conventions as before, and with all single letter counts starting at 1 (as before) for zeroth-order encoding, let us stipulate that all counts start at 0 in the context schemes; thus, each Huffman tree, in each context, starts off like so:

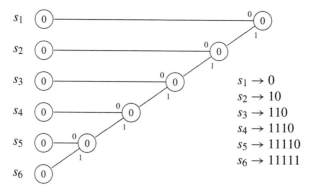

$$s_1 \rightarrow 0$$
$$s_2 \rightarrow 10$$
$$s_3 \rightarrow 110$$
$$s_4 \rightarrow 1110$$
$$s_5 \rightarrow 11110$$
$$s_6 \rightarrow 11111$$

Note that we are allowing zero counts in this instance.

If we attempt second-order adaptive Huffman encoding of $(*)$, with the various conventions and stipulations, the encoding starts

$$\underbrace{100}_{s_3} \, \underbrace{00}_{s_3} \, \underbrace{10}_{s_2} \, \underbrace{11111}_{s_6} \cdots$$

The first two code words are just as before, in the zeroth-order case. After that, the encoding scheme will be stuck on the scheme above, the starting scheme for all contexts, until a context repeats. In (∗) the first context to repeat is $s_2 s_6$. However, since s_1 followed $s_2 s_6$ the first time $s_2 s_6$ occurred, the encoding scheme for context $s_2 s_6$ will not have changed, and the s_5 that follows its second occurrence will be encoded 11110. (Verify!) By contrast, by the second occurrence of context $s_1 s_2$, the tree for that context is as follows:

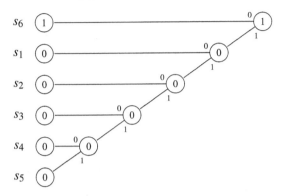

Thus the s_2 following the second occurrence of $s_1 s_2$ is encoded 110.

If the source alphabet is quite large, say $m = |S| = 256$, for instance, then maintaining the context trees for $k > 0$ can be expensive. There are m^2 contexts of order 2, and each context tree has m leaf nodes, and that is really a lot of stuff to maintain and to hunt through. Of course, the same volume of stuff has to be kept in *static* (non-adaptive) higher-order Huffman encoding. One way to lessen the burden in static kth-order Huffman encoding is to store the array of $(k+1)$-gram relative frequencies in some sort of alphabetized order so that the line $f(i_1, \ldots, i_k, 1), \ldots, f(i_1, \ldots, i_k, m)$ for context $s_{i_1} \cdots s_{i_k}$ can be quickly found, and then the tree and scheme for that context can be quickly constructed (and stored, if desired) from those relative frequencies. The savings achieved are not great; it does not take much more space to store the static context schemes than it does to store the $(k+1)$-gram relative frequencies.

In adaptive higher-order Huffman encoding we do not have the simplicity of fixed encoding schemes for each context, and it turns out that something similar to the storage procedure suggested above is useful. Letter counts in contexts replace the $(k+1)$-gram relative frequencies. As we shall see in the next section, lists of counts with a few frills and pointers added can function as easily updated Huffman trees/schemes.

There is a further savings possible with adaptive kth-order Huffman (and with adaptive kth-order arithmetic) coding that takes advantage of the fact that some contexts may occur very infrequently. The trick is not to reserve space and maintain counts for a particular context until that context actually occurs. For more detail on the implementation of adaptive higher-order encoding, we refer the interested reader to Bell, Cleary, and Witten [8].

Exercises 8.1

1. Complete the zeroth-order adaptive Huffman encoding of the source text labeled (∗), above, under the conventions of this section.

2. Give the first-order adaptive Huffman encoding of the source text labeled (∗), under the conventions of this section. (In particular: in each context the letter counts start at 0.)

3. Complete the decoding of $0110001100101111100\ldots$, with $S = \{s_1, s_2, s_3, s_4, s_5, s_6\}$, under the conventions for zeroth-order adaptive Huffman coding in this section.

4. Decode the fragment of code in problem 3 under the conventions for first-order adaptive Huffman coding in this section (with S as above).

8.2 Maintaining the tree in adaptive Huffman encoding: the method of Knuth and Gallager

The problem with the version of adaptive Huffman encoding presented in Section 8.1 is in the updating of the Huffman tree, after a count is incremented. Sometimes the tree does not change at all, and sometimes it changes drastically. It can change drastically even if the order of assignment of source letters to leaf nodes does not change. There does not seem to be any easy way to see what the next tree will look like; you have to construct the full tree at each stage, after each source letter, and that is a lot of trouble, especially if $m = |S|$ is large.

In practice, the Huffman tree at each stage is a sequence of registers or locations, each representing a node. The contents of each register will be (a) the weight of the node, (b) pointers to the sibling children of that node, or an indication of which source letter the node represents, should it be a leaf, (c) a pointer to the parent of the node, unless the node is the root node, and (d) optionally, an indication, in case the node has sibling children, as to which edge to them is labeled 0, and which is labeled 1. (We will see later why this feature could be optional.)

A tree stored in this way can be used for decoding more easily than the encoding scheme associated with the tree. In order to decode binary code text, start at the root node of the tree; having scanned the first bit, go to the register associated with the sibling child of the root node indicated by that bit, whether 0 or 1. Continue in this way until you arrive at a leaf node; decode the segment of code just scanned as the source letter assigned to that leaf, update the tree, return to the root node, and resume scanning.

Thus, in any Huffman encoding, whether adaptive or not, the Huffman tree can serve efficiently as the Huffman encoding scheme. Now, keeping in mind the sequence-of-registers form of the Huffman tree in storage, we will look at the efficient tree-updating algorithm due to Knuth [40], based on a mathematical result of Gallager [23]. First, we need to understand Gallager's result.

The tree resulting from an application of Huffman's algorithm to an initial assignment of weights to the leaf nodes belongs to a special class of diagrams called *binary trees*.[1] It is easy to see by induction on m that a binary tree with m leaf nodes has a total of $2m - 1$ nodes altogether, and thus $2m - 2$ nodes other than the root. (See Exercise 8.2.3.) Gallager's result is the following.

8.2.1 Theorem *Suppose that T is a binary tree with m leaf nodes, $m \geq 2$, with each node u assigned a non-negative weight $\mathrm{wt}(u)$. Suppose that each parent is weighted with the sum of the weights of its children. Then T is a Huffman tree (meaning T is obtainable by some instance of Huffman's algorithm applied to the leaf nodes, with their weights), if and only if the $2m - 2$ non-root nodes of T can be arranged in a sequence $u_1, u_2, \ldots, u_{2m-2}$ with the properties that*

(a) $\mathrm{wt}(u_1) \leq \mathrm{wt}(u_2) \leq \cdots \leq \mathrm{wt}(u_{2m-2})$ *and*

(b) *for* $1 \leq k \leq m - 1$, u_{2k-1} *and* u_{2k} *are siblings, in* T.

We leave the proof of this theorem as an exercise (see Exercise 8.2.4) for those interested.

Both Gallager [23] and Knuth [40] propose to manage the Huffman tree at each stage in adaptive Huffman encoding by ordering the nodes $u_1, \ldots, u_{2m-1}$ so that the weight on u_k is non-decreasing with k and so that u_{2k-1} and u_{2k} are siblings in the tree, for $1 \leq k \leq m - 1$. (u_{2m-1} will be the root node.) This arrangement allows updating after a count increment in a leaf node in, at worst, a constant times m operations (where m is the number of leaf nodes), while redoing the whole tree from scratch requires a constant times m^2 operations. Plus, you never catch a break redoing the whole tree; it always takes the same number of operations, whereas applying the method of Knuth and Gallager sometimes updates the tree in very few operations.

We refer to *the* method of Knuth *and* Gallager because the two methods are essentially the same, a fact that may not be evident from their descriptions; but the fact is that they always result in the same updated tree (except, possibly, for the leaf labels) by essentially the same steps. Gallager's, the first on the scene, is the slower and more awkward to carry out. Indeed, you can think of Gallager's method as a somewhat inefficient way of carrying out Knuth's algorithm, although the historical truth is that Knuth's algorithm is a clever improvement of Gallager's method. We shall give an account of Gallager's method here as a historical curiosity, and to warm the reader up for Knuth's algorithm, but the reader keen on applications may skip the subsection on Gallager's method.

---

[1]The technical definition of binary trees is easy but unmemorable: a binary tree is an acyclic connected graph in which exactly one node has degree 2 and all the other nodes have degrees 3 or 1. You can think of a binary tree as the finite diagram obtained by starting at the root node, drawing two edges or branches to two sibling children of the root node, and continuing in this way, deciding for each new node whether it will have two children or remain a childless leaf. The finiteness requirement says that you cannot continue forever allowing some node or other to have children.

In describing the method, we will identify the nodes of the tree with the registers or locations in which information about the nodes are stored. Both versions of the method use *node interchange*, which results in entire subtrees of the current tree being lopped off and regrafted in new positions. It works like this: suppose u_i and u_j are distinct nodes in a node-weighted binary tree. To exchange these nodes, switch the node weights and the names of associated source letters, in case one or both of u_i, u_j are leaf nodes, between the locations. Leave the "forward" pointers to the parents as they are, in locations u_i and u_j. So, effectively, the nodes have exchanged parents; but they keep their children and other descendants, and to effect this you have to go to the registers of the children of u_i (and of u_j), if any, and change their forward pointers so that they point to u_j (resp. u_i). Also, the backward pointers, if any, in u_i and u_j have to switch. (Of course, the names of the nodes are switched as well, so that what was once referred to as u_i is now u_j, but this has nothing to do with what is going on in the registers.)

For example, consider the Huffman tree below, lifted from Knuth's paper. Edges will substitute for pointers and the weights are indicated inside the nodes, as usual. The names of the nodes are beside the nodes, as well as the names of the associated source letters, in the case of leaf nodes.

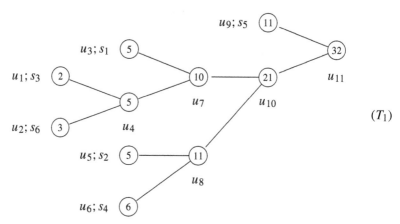

(T_1)

Verify that the weight stored at u_k is a non-decreasing function of k, and that successive nodes $u_{2k-1}, u_{2k}, 1 \le k \le 5$, in the list $u_1, \ldots, u_{10}$ are siblings.

Now, the interchange of u_4 and u_5 results in the tree (T_2). This also is a Huffman tree, with nodes in the "Gallager order" described in Theorem 8.2.1. It is worth noting that if you start with one Huffman tree with nodes in Gallager order and interchange two nodes, if the two nodes have the same weight then the result will be a Huffman tree with nodes in Gallager order. If the two nodes do not have the same weight then the resulting tree could be a Huffman tree, but, in any case, the nodes will not be in Gallager order; requirement (a) will be no longer satisfied.

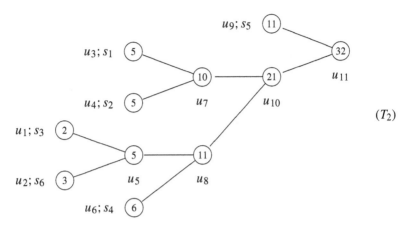

(T_2)

Note also the "subtree regrafting" feature of this interchange. The subtree originally rooted at u_4 has been snipped off and regrafted into the tree, now with u_5 as the root. The same is true of the subtree originally rooted at u_5— it is rerooted at u_4—but this rerooting is not very dramatic since that subtree consists of a single node.

8.2.1 Gallager's method

We suppose that the nodes $u_1, \ldots, u_{2m-1}$ of the Huffman tree are stored in Gallager order; for $1 \leq k \leq m-1$, we refer to u_{2k-1}, u_{2k} as a sibling pair.

Suppose a count wt(u) in a leaf node u is to be incremented. Change its weight to wt$(u) + 1$ and inspect the next sibling pair up from the one u is in. If the smaller weight in this sibling pair, say on the node v, is smaller than wt$(u) + 1$ (which will be the case if and only if wt$(u) =$ wt(v)), interchange u and v. In case there is a choice, interchange u with the sibling with larger index. Now again compare wt$(u) + 1$ with the weights in the next sibling pair up (from v). Interchange if wt$(u) + 1$ is larger than one of these. Continue interchanging until the incremented weight, wt$(u) + 1$, is, in fact, smaller than both weights in the sibling pair beyond the node on which that weight now resides, or until there is no sibling pair beyond that node (which will happen if and only if the current residence of wt$(u) + 1$ is a child of the root node).

Suppose the incremented weight is now at node $\tilde{u}$. Next increment the weight on the parent of $\tilde{u}$ by one. If the parent of $\tilde{u}$ is the root node, you are done. Otherwise, proceed with the parent of $\tilde{u}$ as with u, interchanging until its incremented weight is no greater than that on either node in the next sibling pair up. Then increment its parent—and so on.

For example, suppose the current Huffman tree is T_1, above, and s_6 is scanned. The count in u_2 is increased to 4. The next sibling pair up from u_2 is u_3, u_4, each with weight 5, so we leave location u_2 alone (for now; it will soon have its forward pointers changed) and increment the weight in u_4 by 1; it is now 6. This is greater than the weight in u_5, in the next sibling pair up, so nodes u_4 and u_5 are interchanged. (So, in particular, nodes u_1 and u_2 are now the children of u_5.) Now the weight 6 in u_5 is less than each weight in u_7 and u_8, corresponding to the next sibling pair up, so increment the weight in u_8 (the parent of u_5) by 1 up to 12. This is greater than the weight in u_9, so exchange u_8 and u_9. Finally, increment the root node. The result of all this:

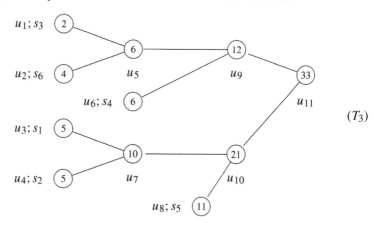

(T_3)

It is heartily recommended that the reader work through the steps in this example.

8.2.2 Knuth's algorithm

The main differences between the methods of Knuth and Gallager are that Knuth's calls for node interchange *before* any counts are incremented, and there is no emphasis at all on sibling pairs. (Nor did there need to be, really, in Gallager's method.) All node interchanges are of nodes of equal weight, so there need be no interchange of weights between locations. Interchanges involve only changes of pointers and of source letter identities of leaf nodes.

We start with the nodes in Gallager order, $u_1, \ldots, u_{2m-1}$. Suppose that the count on a leaf node u, currently with count $\text{wt}(u)$, is to be incremented. Look up the list from u and find the node $\tilde{u}$ furthest up the list with the same weight, $\text{wt}(u)$, as u, and interchange u and $\tilde{u}$. [This $\tilde{u}$ will be the same node discovered in the first stage of Gallager's method.] Now go to the parent of $\tilde{u}$. If it is the root node, go to the incrementation phase, described below. Otherwise, find the node the furthest up the list from the parent of $\tilde{u}$ with the same weight as the parent of $\tilde{u}$, interchange the parent of $\tilde{u}$ with that node, proceed to the next parent, and so on.

After all the node interchange is finished, the leaf node now corresponding to the source letter scanned, and each node on the unique path from that node to the root node, have their weights increased by 1, and the update is complete.

For example, let us again suppose that T_1 is the current state of the Huffman tree, and that s_6 is scanned. The current count 3 in u_2, corresponding to s_6, is the only 3 appearing, so we move to the next parent node, u_4; u_5 is the location with greatest index (5) bearing the same weight as u_4 (5, by coincidence), so nodes u_4 and u_5 are interchanged. The current picture is then T_2, above.

The next parent to be processed is the node u_8 with weight 11. The highest indexed node with weight 11 is currently u_9, so u_8 and u_9 are interchanged. The next parent is then the root node, so we are done. Now the counts in locations u_2, u_5, u_9, and u_{11} are increased by 1, and we are ready to scan the next letter. Notice that the tree arrived at after incrementation is T_3, the tree arrived at by Gallager's method, and that all the interchanges that took place were between the same pairs of nodes, in the same order, as in Gallager's method. It is always thus, if all weights on the nodes are positive. We leave verification of this statement to the reader.

There is a slight problem with Knuth's algorithm as described when zero counts are allowed on the leaf nodes; it can happen that the u_i with largest index i, with the same weight as the node v you are currently looking at, is u_{2m-1}, the root node. In this eventuality, simply interchange v with u_{2m-2} and proceed to the incrementation stage. By an accident of logic and reference, because of the business about sibling pairs, this provision is built into Gallager's method.

You may be wondering about labeling of the edges of the Huffman tree with 0 or 1, in the Knuth-Gallager procedure. The easiest rule is that if a parent u has children u_{2k-1}, u_{2k}, in the current Gallager ordering of the nodes, then let the edge from u to u_{2k} be labeled 0, and the edge from u to u_{2k-1} be labeled 1. In practice, it might be convenient to have a couple of bits in the register corresponding to u reserved for signaling this labeling, even though it can be figured out from the order of the sibling children in the current Gallager ordering of the nodes.

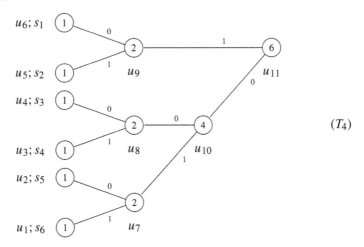

(T_4)

The Knuth-Gallager procedure requires some sort of initialization convention, because whether the initial counts are set at 1 or 0, there are many choices as to the Gallager order of the nodes at the outset. In the exercise set to follow,

$S = \{s_1, \ldots, s_6\}$, the initial counts are set at 1, and the initial Gallager ordering is indicated by (T_4). Please note that by the edge-labeling convention mentioned above, the edge from u_{11} to u_{10} will be labeled 0.

Exercises 8.2

1. With the initialization and the edge labeling conventions mentioned above, do encoding Exercise 8.1.1 with Knuth's algorithm governing the tree updating.

2. Similarly, do decoding Exercise 8.1.3 with Knuth's algorithm.

3. (a) Show that in a binary tree, there must be at least two leaf nodes that are siblings. [If a binary tree is defined as a tree formed by a certain process, this proposition is evident; the last two children formed will be siblings and leaf nodes.

    Here is another proof. Take a node a greatest distance from the root node. It and its sibling must be leaf nodes. Why?]

   (b) Show that a binary tree with m leaf nodes has $2m - 1$ nodes, total. [Use (a) and go by induction on m.]

4. Prove Theorem 8.2.1. You might follow the following program.

   (a) If T is a binary tree generated by applying Huffman's algorithm to the non-negatively weighted leaf nodes, then the two smallest node weights appear on sibling leaf nodes, by appeal to the procedure of formation. By a similar appeal, the tree obtained by deleting those two leaf nodes, thereby making their parent a leaf, is also a Huffman tree.

   The proof that the nodes of T can be put in Gallager order is now straightforward, by induction on the number of leaf nodes of T.

   (b) Suppose T is a binary tree with non-negatively weighted nodes, with each parent weighted with the sum of the weights of its children. Suppose the nodes of T can be put in Gallager order, $u_1, u_2, \ldots, u_{2m-1}$. If all the weights on nodes are positive, then u_1 and u_2, siblings, must be leaf nodes. (Why?) If zero appears as a weight, then it is possible that one of u_1, u_2 is a parent, say of $u_{2k-1}, u_{2k}, k \geq 2$, but only if the weights on u_{2k-1} and u_{2k} are both zero (verify this assertion, under the assumptions), in which case all the weights on $u_1, \ldots, u_{2k}$ are zero. Switch the sibling pairs u_1, u_2 and u_{2k-1}, u_{2k}, in the ordering; the new ordering is still a Gallager ordering. Switch again, if necessary, and continue switching until the first two nodes in the ordering are sibling leaf nodes. (How can you be sure that all this switching will come to an end with the desired result?) Now consider the tree T' obtained from T by deleting u_1, u_2. Draw the conclusion that T is a Huffman tree, by induction on the number of leaf nodes. (The important formalities are left to you.)

8.3 Adaptive arithmetic coding

If you understand the general procedure in arithmetic coding, the main idea in adaptive arithmetic coding (including higher-order adaptive arithmetic coding) is quite straightforward. Counts of the source letters are maintained in both the encoding and in the decoding; in the case of higher-order adaptive arithmetic coding, counts are maintained in the different contexts. Whatever the current interval is, it is subdivided into subintervals, corresponding to the source letters, with lengths proportional (or, in the case of integer intervals, approximately proportional) to the current source letter counts. It is simplest to maintain the order $s_1, \ldots, s_m$ of the source letters, as in higher-order arithmetic coding. That is, there is usually no good reason to rearrange the source letters, and thus the order of the next subintervals, so that the counts of the letters and the lengths of the subintervals are in non-increasing order.

Explicitly, if $S = \{s_1, \ldots, s_m\}$, if the current interval $A = A(s_{i_1} \cdots s_{i_k})$ (possibly the result of rescaling) has left-hand endpoint α and length ℓ, and if $s_1, \ldots, s_m$ have counts $c_1, \ldots, c_m$, respectively, then the endpoints of the subintervals $A(s_{i_1} \cdots s_{i_k} s_1), \ldots, A(s_{i_1} \cdots s_{i_k} s_m)$ will be α, $\alpha + \frac{c_1}{C}\ell$, $\alpha + \frac{c_1 + c_2}{C}\ell, \ldots,$ $\alpha + \frac{c_1 + \cdots + c_{m-1}}{C}\ell, \alpha + \ell$, where $C = \sum_{i=1}^{m} c_i$.

Let's try an example of adaptive dfwld encoding in the fashion of Section 6.1. We take $S = \{a, b, c, d\}$, initial counts all 1, and we will encode $bbca$. At the outset, the intervals $A(a)$, $A(b)$, $A(c)$, and $A(d)$ are, respectively, $[0, 25)$, $[.25, .5)$, $[.5, .75)$, and $[.75, 1)$. After the first b is scanned, the counts of a, b, c, d become $1, 2, 1, 1$, respectively, so $A(b) = [.25, .5)$ is broken into $[.25, .3)$, $[.3, .4)$, $[.4, .45)$, and $[.45, .5)$. When the next b is scanned, we are in interval $A(bb) = [.3, .4)$, and rescaling is possible. The resulting interval will be divided into subintervals with lengths in the proportions $1, 3, 1, 1$, which will make the arithmetic annoying. Here is the encoding table, similar to these in Section 6.1, with an extra column for the counts. As usual, α is the left-hand endpoint and ℓ is the length of the current interval.

Counts of a, b, c, d	Next letter or rescale	α	ℓ	Code so far
$1, 1, 1, 1$		0	1	
$1, 1, 1, 1$	b	$.25$	$.25$	
$1, 2, 1, 1$	b	$.3$	$.1$	
$1, 3, 1, 1$	rescale	$.2$	$.4$	01
$1, 3, 1, 1$	c	$7/15 = .2 + (.4)\frac{4}{6}$	$1/15 = (.4)\frac{1}{6}$	01
$1, 3, 2, 1$	a	$7/15$	$\frac{1}{105} = \frac{1}{15} \cdot \frac{1}{7}$	01

The last interval is $[7/15, 7/15 + \frac{1}{105}) = [\frac{7}{15}, \frac{10}{21})$, the dfwld in which is $(.01111)_2$, easily obtained by carrying out the binary expansions of the endpoints until they disagree. We add the bits of the binary expansion of the dfwld to the "code so far" to obtain 0101111 as the code for *bbca*.

Perhaps we are making too much of this. The point is that in both the encoding and the decoding, however they are carried out, the role of the relative source frequencies f_j (or, in the higher-order cases, of the conditional context probabilities $f(i_1,\ldots,i_k,j)/f(i_1,\ldots,i_k))$ is taken over by the constantly changing ratios c_j/C. (In the higher-order cases, these are relative to the context.) Whatever you did in Section 6.1, you are doing the same in adaptive arithmetic coding, with the constantly updated c_j/C playing the roles of the f_j.

The same is true in the more practical setting of Section 6.4, but here problems arise. Suppose we replace the continuum $[0,1)$ and exact computation by an integer interval $[0,M) = \{0,1,\ldots,M-1\}$, and approximate computation. If the relative source frequencies $f_1,\ldots,f_m$ are known ahead of time, M can be chosen so that, with "underflow condition" rescaling, every letter will always have a non-empty interval as a subinterval of the current interval. But with adaptive arithmetic coding, we know not what horrors await. M is chosen ahead of time, and it may well be that for some j, c_j/C will fall so low that the subinterval allocated to s_j in the current interval vanishes. Another, related danger is that when Gallager's idea of allowing for statistical change in the source text by multiplying the counts by a fraction and rounding down (see Section 8.1) is carried out, some source letter will get count zero, and thus be allocated an empty subinterval in the current interval.

One simple way to deal with both problems is to modify Gallager's method by rounding up instead of rounding down, so that each source letter gets a positive count, and to agree, between the encoder and the decoder, that the fractionalizing procedure will be repeated, if necessary, until the new count sum C satisfies $C \leq M/4 + 2$. Recall from Section 6.4 that this inequality guarantees that every source letter will be allocated a non-empty integer interval, in the subdivision of the current interval in the algorithm of that section.

Adaptive Huffman and adaptive arithmetic encoding have been regarded as approximately equivalent in effectiveness and cost, in the past. Current gossip has it that adaptive arithmetic now has an edge over adaptive Huffman encoding, because implementation of arithmetic coding has been improved lately, while tree maintenance techniques for adaptive Huffman encoding remain approximately where Knuth left them. But the pace of technological advance is swift, so it may be that adaptive Huffman encoding may be the less costly of the two, by the time you read this. Of course, many hold the view that arithmetic encoding of any sort has an insurmountable theoretical advantage over the corresponding Huffman encoding, but it is not clear that this theoretical advantage persists in practice.

Exercises 8.3

1. Redo Exercise 6.1.1 adaptively, with all counts initially set at 1.

2. This is about adaptive arithmetic coding using the method of Section 6.4

with an integer interval $[0, M) = \{0, \ldots, M-1\}$. Suppose that $M = 32$. Suppose the source letters are a, b, and EOF, and all source letters start with a count of 1. Suppose that the Gallager fraction by which the letter counts will be occasionally multiplied is $1/2$, with rounding up, as suggested in the text above. Suppose that this fractionalizing of the counts will occur whenever the count sum rises to $11 = (M/4+2)+1$. Give the current counts of a, b, and EOF after each source letter is read, if the source stream is $baabbbbabbaabaaabaaaaEOF$.

8.4 Interval and recency rank encoding

Both encoding methods referred to in the title of this section were introduced by Elias [16], although he gives credit for the independent discovery of recency rank encoding to Bentley, Sleator, Torjan, and Wei.

Both methods are lossless, adaptive methods for encoding text over a source alphabet $S = \{s_1, \ldots, s_m\}$.

8.4.1 Interval encoding

In interval encoding we start with an infinite set $C = \{u_0, u_1, \ldots\}$ of binary words, satisfying the prefix condition and $\text{lgth}(u_0) \leq \text{lgth}(u_1) \leq \cdots$. A letter s_j occurring the source text is encoded by u_i, where i is the number of source letters between this occurrence of s_j and its last occurrence, in previously scanned source text. We get started by imagining the source text to be preceded by the string $s_1 \cdots s_m$.

For example, let us take $C = \{0, 10, 110, \ldots\}$ and consider the source text $(*)$ in Section 8.1.2,

$$s_3 s_3 s_2 s_6 s_1 s_1 s_2 s_6 s_5 s_5 s_1 s_3 s_3 s_5 s_6 s_1 s_2 s_2 s_2 \cdots \qquad (*)$$

The encoder (and the decoder, as well) knows that the source alphabet is $\{s_1, \ldots, s_6\}$. The first s_3 in the source text is encoded by $u_3 = 1110$, because there are three letters, namely s_4, s_5, and s_6, between it and its imaginary first occurrence, in the imagined block $s_1 \cdots s_6$ preceding the real source text. The next s_3 is encoded with $u_0 = 0$. The s_2 following is encoded with $u_6 = 1111110$. (Why?) The code for the first 6 letters of $(*)$, $s_3 s_3 s_2 s_6 s_1 s_1$, will be $11100111110101101^9 00$ (with 1^9 standing for nine ones). Notice that if s_4 ever shows up in this source text, its first occurrence will be represented by a frightfully long code word.

In fact, although interval encoding intuitively seems elegant and efficient, it is rather disappointing in practice, with regard to compression. The source of the infelicity is the choice C of a set of code words, chosen without reference to any particular properties of the source text. You do not need to know any properties of the source text to do adaptive Huffman or arithmetic encoding, but the encoding will reflect statistical properties of the source text discovered or collected as the encoding proceeds. In interval encoding, you are stuck with the set C chosen at the outset. Of course, there are situations involving communications in which it is a good thing for everybody in the conversation to be using the same code word set, so this weakness of interval encoding can also be a strength.

The C we used for the example above, in which $u_k = 1^k0$, $k = 0, 1, 2, \ldots$, is a particularly bad choice. It can be shown (see Exercise 8.4.3) that for any zeroth-order source with alphabet $S = \{s_1, \ldots, s_m\}$, the average length of a code word replacing a source letter in interval encoding of the source text, using the set C of code words above, will be m. This compares very unfavorably with $1 + H(S) \leq 1 + \log_2(m)$ (see Chapters 2 and 5) an upper bound on the average length of a code word replacing a source letter if Huffman encoding were possible, and thus an upper bound on the (long term) average code word length if adaptive Huffman encoding is used.

Elias [16] proposes two infinite code word sets, that we will refer to hereafter as $C_1 = \{u_0, u_1, u_2, \ldots\}$ and $C_2 = \{v_0, v_1, v_2, \ldots\}$, which do a much better job of compression when used in interval encoding than C, above, although they still do not give an average code word length close to $H(S)$, for a zeroth-order source S. The code word u_k is formed by following a string of $\lfloor \log_2(k+1) \rfloor$ zeroes by the binary expansion of $k+1$, which will be of length $1 + \lfloor \log_2(k+1) \rfloor$. Thus $\text{lgth}(u_k) = 1 + 2\lfloor \log_2(k+1) \rfloor$. The first few u_k's are as follows:

$$u_0 = 1 \qquad u_4 = 00101$$
$$u_1 = 010 \qquad u_5 = 00110$$
$$u_2 = 011 \qquad u_6 = 00111$$
$$u_3 = 00100 \qquad u_7 = 0001000$$

The code word v_k is formed by following $u_{g(k)}$, where $g(k) = \lfloor \log_2(k+1) \rfloor$, by the binary expansion of $k+1$. Thus

$$v_0 = 11 \qquad v_4 = 011101$$
$$v_1 = 01010 \qquad v_5 = 011110$$
$$v_2 = 01011 \qquad v_6 = 011111$$
$$v_3 = 011100 \qquad v_7 = 001001000$$

(Verify!) It may appear that the v_k are longer than the u_k, and so it is, in the early going, but

$$\text{lgth}(v_k) = 1 + \lfloor \log_2(k+1) \rfloor + \text{lgth}(u_{g(k)})$$
$$= 2 + \lfloor \log_2(k+1) \rfloor + 2\lfloor \log_2(1 + \lfloor \log_2(k+1) \rfloor) \rfloor,$$

which is asymptotically about half of $\text{lgth}(u_k) = 1 + 2\lfloor \log_2(k+1) \rfloor$, as $k \to \infty$.

It is left to the reader to verify that the sets C_1 and C_2 are prefix-free; i.e., they satisfy the prefix condition.

It can be shown (see Exercise 8.4.4) that in interval encoding using C_1 on text from a zeroth-order source S, the average length of a code word replacing a source letter is no greater than $1 + 2H(S)$; using C_2, that average is no greater than

$$2 + H(S) + 2\sum_{j=1}^{m} f_j \log_2(1 + \log_2(1/f_j)),$$

where the f_j are the relative frequencies of the source letters. If m and $H(S)$ are large, the latter is usually the smaller of the two upper bounds, but for the small source alphabets we use for examples, C_1 is the superior set of code words.

If you do Exercise 8.4.4, you will see that, for m large and the f_j small, the upper bounds on the average code word lengths in the paragraph preceding are not very pessimistic; that is, the true average length of a code word replacing a source letter is not far from the upper bound given. We leave the precise analysis necessary to demonstrate this to the ambitious; the point is that by interval encoding using C_1 or C_2, we cannot hope for the compression achievable with adaptive Huffman or arithmetic coding applied to text from a zeroth-order source.

On the other hand, the upper bounds mentioned above say that interval encoding using C_1 or C_2 results in code text, at worst, no more than twice as long as the code text from Huffman or arithmetic encoding, on the average, when applied to text from a perfect zeroth-order source; and there are offsetting advantages, such as a common, system-wide code word set, and great ease and speed in encoding and decoding.

The decoding procedure should be clear, but just to be sure that we understand it, let's decode

$$011001101001110100100001000100100,$$

the result of applying interval encoding using C_1 to a short passage of source text over $S = \{s_1, s_2, s_3, s_4, s_5, s_6\}$. Recall that it is imagined that the actual source text is preceded by $s_1 \cdots s_6$, to get things started.

Scanning the string above from left to right, we first recognize $u_2 = 011$. (All u_k in the string above are among $u_0, \ldots, u_7$, listed above. We will have a few words to say below about a more systematic way of recognizing the u_k.) Thus the first source letter is s_4. Do you see why? Because s_4 is the choice that puts two letters between it and its previous occurrence in the preliminary block $s_1 \cdots s_6$.

After u_2, we next recognize $u_5 = 00110$. Counting back six places, we come to s_2, and that is the next source letter. The next code word is $u_0 = 1$, so s_2 is repeated. Continuing in this way, you should decode the given string as $s_4 s_2 s_2 s_3 s_2 s_3 s_5 s_5 s_2$.

Given the rules of formation of the u_k, it is quite easy to program so that the u_k can be recognized by reading-left-to-right, without checking any lists. You count the number of zeroes until the first 1 in the word being scanned. If there are t zeroes, then the next $t + 1$ bits give the binary expansion of $k + 1$, which is the number of places you will count back through the source text so far to find the next source letter to be decoded.

8.4.2 Recency rank encoding

Recency rank encoding is quite similar to interval encoding—it applies to the same situations and shares some of the same advantages—but requires only a finite prefix-free set $C = \{u_0, \ldots, u_{m-1}\}$ of $m = |S|$ code words. Given C, an occurrence of a source letter s_j in the source text is encoded by u_k, where k is the number of *distinct* source letters that have appeared in the source text since the last appearance of s_j. As in interval encoding, we pretend that the source text is preceded by $s_1 \cdots s_m$, to get things started.

For example, with $S = \{s_1, \ldots, s_6\}$ and $C = \{0, 10, 110, 1110, 11110, 111110\}$, the source text ($*$) presented earlier in this section will be encoded

$$(s_1 s_2 s_3 s_4 s_5 s_6) \frac{1110}{s_3} \frac{0}{s_3} \frac{11110}{s_2} \frac{110}{s_6} \frac{111110}{s_1} \frac{0}{s_1} \frac{110}{s_2} \frac{110}{s_6} \frac{11110}{s_5} \frac{1110}{s_1} \frac{11110}{s_3} \cdots$$

(We leave the rest of the encoding as an exercise.) Note that s_4, when it first occurs, if ever, will be encoded 111110.

Clearly recency rank encoding shares with interval encoding the advantage of a common code word set to be used by all in a communication environment. Clearly recency rank encoding will compress better than interval encoding, for any reasonable choice of the finite code word set C. (The rigorous analysis of recency rank encoding in this respect remains to be done, but the point is that the number of distinct letters in a given block of source text is certainly no greater than the length of the block.) In fact, the only advantage that interval encoding has over recency rank encoding is in the speed and ease of encoding and decoding: clearly it is a little more trouble to count the number of distinct symbols in a block of symbols, than just to count the length of the block. And in recency rank decoding, having scanned u_k, $0 \leq k \leq m - 1$, you have to go back into the source string decoded so far until you come to the $(k + 1)$st *different* symbol, and clearly this involves some sorting and checking and a good deal more trouble than just counting back $k + 1$ places, as in interval decoding. But when the amount of trouble involved in recency rank encoding and decoding is compared to the corresponding difficulties in adaptive Huffman or arithmetic coding, that trouble does not seem like much.

We are indebted to Greg Hanks, a student, for proposing the following: while encoding either by the interval method or by recency rank, keep counts of the original source letters and of the code words (the u_k) and then, after the source text has been encoded, recode the whole thing by zeroth-order Huffman replacement of S or of C, whichever has the lesser entropy; the entropies will be calculated from the relative frequencies that arise from those counts you kept. Or, you could recode arithmetically using those relative frequencies—again, the recoding is applied either to the original source text or to the code text, regarded as a string of u_k's, depending on which alphabet has lower zeroth-order entropy.

Now, this procedure defaces the speedy online character of the interval and recency rank methods; and, anyway, don't we do adaptive encoding to avoid advance statistical study of the source text? But it is an interesting idea, all the same, and merits some experimentation to see if any significant compression is achievable by such recoding.

Further, such experimentation should provide an interesting test of faith. Recall the discussion at the end of Section 6.2; it is widely believed that no lossless "zeroth-order" method, whatever that may mean, can encode source text over a source alphabet S in fewer than $H(S)$ bits per source letter, on average, with $H(S)$ denoting the zeroth-order entropy of the source. Now, if the source text is encoded by the interval or the recency rank method, each source letter s has been replaced by a code word u belonging to a set C of code words; we can regard C as a source alphabet now, and the encoded text as a new source text. According to the faith about entropy, and by the fact that you can encode text in entropy-plus-epsilon bits per symbol by indisputably zeroth-order lossless methods (see Sections 6.2 and 5.4) either (a) it must be that $H(C) \geq H(S)$; otherwise, if $H(C) < H(S)$, we could encode the text over C, and thus the original source text, in fewer than $H(S)$ bits per symbol; or (b) the coding method is not zeroth-order. There is a good argument for this latter assertion, because in both interval and recency rank encoding, the encoding of each letter s has something to do with the context—either the number of letters since s's last appearance, or the number of different letters since that appearance. Perhaps this objection is sufficient to preserve the faith. In any case, it would be interesting to see if we get $H(C) < H(S)$ in many plausible "real" situations. When we do, Hanks' suggestion provides a way of achieving better compression than could be had by plain zeroth-order Huffman or arithmetic coding of the source text, were the relative source letter frequencies known.

In the silly situation considered at the end of Section 7.4, where $S = \{a, b, c, d\}$ and the source text consists of $abcd$ repeated over and over, all relative frequencies are $1/4$, so the zeroth-order entropy is $H(S) = \log_2 4 = 2$. Encoding either by interval or by recency rank, every source letter gets replaced by $u_3 \in C$; thus the entropy of C, considered as the source alphabet of the resulting text, is $H(C) = 0$. Using Shannon's trick, as described in Section 5.4, and encoding blocks of N of those u_3's by a single bit, we can encode the original text at the rate of $1/N$ bits per original source letter.

True believers (and we are among them, we just wonder what it is we believe) will say that this is evidence either that this is a lousy example, or that interval and recency rank encoding are not zeroth-order methods. We tend to the latter view, without absolutely ruling out the former.

One last comment; should it happen that experiment shows that we frequently have $H(C) < H(S)$ after interval or recency rank encoding, then the idea of Greg Hanks might profitably be put into practice with preservation of the online, no-second-pass character of those coding methods, simply by immediately following the interval or recency rank encoding with adaptive Huffman or arithmetic coding, applied to the symbol set $C = \{u_0, u_1, u_2, \dots\}$ of the first encoding.

Exercises 8.4

1. (a) Encode source sequence (∗), above, by interval encoding using the set
        C_1 of code words defined in the text.
    (b) Complete the encoding of (∗) by recency rank encoding, using $C =$
        $\{0, 10, 110, 1110, 11110, 111110\}$.

2. (a) Supposing $S = \{s_1, \ldots, s_6\}$, decode

        $$0010101001010011000010000101$$

        assuming the encoding was by the interval method, using C_1.
    (b) Supposing $S = \{s_1, \ldots, s_6\}$, decode

        $$11101011100111101101011100011110,$$

        assuming the encoding was by recency rank, using $C = \{0, 10, 110,$
        $1110, 11110, 11110\}$.

3. Before getting to the question, we need some observations.

    (i) Recall the formula for the sum of a geometric series: $\sum_{k=0}^{\infty} \rho^k =$
    $(1 - \rho)^{-1}$, for $|\rho| < 1$.

    (ii) Differentiating both sides of the equation in (i) with respect to ρ, we
    obtain $\sum_{k=1}^{\infty} k\rho^{k-1} = (1 - \rho)^{-2}$.

    (iii) If a symbol s from a perfect zeroth-order source occurs in the source
    text (randomly and independently of all other occurrences) with relative
    frequency f, then, starting from any point in the source text and going
    either forward or backward, assuming the source text extends infinitely in
    both directions, the probability of reading through exactly k letters before
    coming to the first occurrence of s (at the $(k+1)$st place scanned) is $f(1 -$
    $f)^k$. (Thus the average gap between occurrences of s in the source text is
    $\sum_{k=0}^{\infty} kf(1-f)^k = f(1-f)\sum_{k=1}^{\infty} k(1-f)^{k-1} = f(1-f)f^{-2} = \frac{1}{f} - 1$,
    using (ii). To put it another way, s occurs on average once every $1/f$ letters,
    which agrees with intuition, since f is the relative frequency of s.)

    (iv) Suppose that $S = \{s_1, \ldots, s_m\}$ is the alphabet of a perfect zeroth-order
    source, with s_j having relative frequency f_j, $1 \leq j \leq m$. Suppose the
    source text is encoded by the interval method, using some prefix-free set
    $C = \{w_0, w_1, \ldots\}$ of code words. Then the average length of a code word
    replacing s_j will be $\bar{\ell}_j = \sum_{k=0}^{\infty} f_j(1 - f_j)^k \operatorname{lgth}(w_k)$, so the average length
    of a code word replacing a source letter will be $\bar{\ell} = \sum_{j=1}^{m} f_j \bar{\ell}_j =$
    $\sum_{j=1}^{m} f_j \sum_{k=0}^{\infty} (1 - f_j)^k \operatorname{lgth}(w_k)$.

    Finally, the problem. Show that for any zeroth-order source, in interval
    encoding using $C = \{0, 10, 110, \cdots\}$, the average length of a code word
    replacing a source letter will be $m = |S|$.

4. To get at the average length of a code word replacing a source letter in in-
    terval encoding using the code word sets C_1 or C_2, we need to recall some-
    thing about *concave* (some say, concave *down*) functions which played a

Chapter 9

Dictionary Methods

In the previous chapters, lossless compression was obtained by using a probability model to drive a coder. Dictionary methods use a list of phrases (the dictionary), which hopefully includes many of the phrases in the source, to replace source fragments by pointers to the list. Compression occurs if the pointers require less space than the corresponding fragments. (Of course, the method of passing the dictionary must also be considered.)

In many ways, dictionary methods are easier to understand than probabilistic methods. At the simplest level, several (fixed) specialized dictionaries could be made available to both the coder and decoder. For text in English, a few thousand of the most commonly used words could serve as the dictionary; if the source consisted of code in some computer language such as C, then a list of the keywords and standard library functions might serve as a dictionary. Fixed dictionaries may be useful in some situations, but there are at least two serious drawbacks. First, the dictionaries must be known to both the coder and decoder. Changes to the dictionary would have to be propagated to all the sites which use the scheme. Second, fixed dictionary schemes cannot compress "unknown" text. In the case of C code, there would likely be little compression of the variable names created by the programmer.

Our main interest here is methods which adapt to the source; that is, methods which build the dictionary from the source, and which usually do this on-the-fly as the source is scanned. Communication via modem commonly uses such a scheme (V.42*bis*). Fixed dictionaries would be of little use for general-purpose communications, and, in addition, on-the-fly dictionary creation is perhaps essential if the session is interactive.

Adaptive[1] dictionary methods can often be traced to the 1977 and 1978 papers by Ziv and Lempel [85, 86]. The general schemes are known as LZ77 and LZ78, respectively. Applications employing variations on LZ77 include LHarc, PKZIP, GNU zip, Info-ZIP, and Portable Network Graphics (PNG), which is a lossless image compression format designed as a GIF successor.[2] LZ78-type schemes are used in modem communications (V.42*bis*), the Unix *compress* program, and in the GIF graphics format.

---

[1]The use of "adaptive" in the literature has not always been consistent. See Langdon and Rissanen [44] or Williams [82] for some discussion.

[2]See http://www.clione.co.jp/clione/lha, http://www.pkware.com, http://www.gnu.org, http://www.info-zip.org, and http://www.libpng.org, respectively.

The basic difference between LZ77 and LZ78 is in the management of the dictionary. In LZ77, the dictionary consists of fragments from a window (the "sliding window") into recently seen text. LZ78 maintains a dictionary of phrases. In practice, LZ77 and LZ78 schemes may share characteristics. There are distinct advantages of each scheme: roughly speaking, LZ78 uses a more structured approach in managing a slow-growing dictionary (possibly trading compression for speed at the coding stage), and LZ77 has a rapidly changing dictionary (which may offer better matches) and is faster for decoding. In applications, the choice of basic scheme may be complicated by various patent claims (see Appendix C).

If dictionary methods are both simple and popular, the reader may be wondering why they've been presented after the probabilistic methods. Part of the reason is historical, but it should also be noted that, subject to fairly modest restrictions, the compression achieved by a dictionary method can be matched by a statistical method (see Section 9.3). However, dictionary methods continue to be very popular due to their simplicity, speed, relatively good compression, and lower memory requirements compared to the best statistical methods.

A combination of dictionary and probabilistic schemes is possible. An example is provided by the GNU zip program discussed in Section 9.1.2, which uses a statistical method on the output of the dictionary coder.

9.1 LZ77 (sliding window) schemes

In the basic scheme, a two-part window is passed over the source:

```
                      history        lookahead
      ...She s|ells sea shells by t|he seash|ore...
              0987654321098 7654321
```

In the simplest case, the history and lookahead are of fixed length, and the dictionary consists of all phrases (that is, fragments of consecutive characters) which start in the history and which are no longer than the length of the lookahead.[3] With such a dictionary, it is convenient to think of the history as the dictionary; however, we will exhibit schemes where these differ. Typically, the history is much longer than the lookahead.

The idea is to replace an initial segment of the lookahead with a pointer into the dictionary, and then slide the window along. In the example, the initial segment 'he' matches the second two characters of the dictionary phrase 'shell' (at an offset of 10 into the dictionary, counting right to left). The original scheme would output a triple ($offset, length, character$), where the third component is the next character from the source (the "unmatched character"), and then the window is moved:

---

[3] It is understood that the source symbols are also included in the dictionary.

```
                      history         lookahead
        ...She sell|s sea shells by the |seashore|...
                   0 9 8 7 6 5 4 3 2 1 0 9 8 7 6 5 4 3 2 1
```

In this example, the triple is $(10, 2, \Box)$, where '$\Box$' represents the space character. Compression is achieved if this triple requires fewer bits than the three symbols replaced.

Sending the unmatched character in the triple allows the scheme to proceed even in the case of no match in the history. However, it is sometimes wasteful in the case that the character can be part of a match at the next stage. This occurs in the example, with '$\Box$sea' matching the dictionary, and it is common for LZ77 schemes to look for this match. Conceptually, this means that the output consists of two kinds of data, rather than triples: (*offset, length*) pairs, and single (unmatched) characters. The following diagram shows a few steps in the process for the example problem:

```
                    history          lookahead     output
      ...She s|ells sea shells by t|he seash|ore    (10, 2)
      .She sel|ls sea shells by the |seashor|e..     (18, 4)
       sells s|ea shells by the sea|shore...|...     (17, 2)
      ells sea| shells by the seash|ore.....|...      'o'
               0 9 8 7 6 5 4 3 2 1 0 9 8 7 6 5 4 3 2 1
```

The decoder receives the output tokens, from which it can reconstruct the source (by maintaining the same dictionary as the coder). There are several observations which can be made concerning this type of scheme:

- The decoder must be able to distinguish between ordered pairs and characters. This implies that there will be some overhead in transmitting an unmatched character, and hence the scheme can cause expansion (this should not come as a surprise).

- The compression achieved by transmitting an ordered pair depends on the match length and the sizes of the dictionary and lookahead. Too short a match length will cause expansion (in which case it may be desirable to transmit an unmatched character).

- The match can extend into the lookahead. As an example, suppose the window contains the fragment

```
        history lookahead
            ab|aba
            2 1
```

A match for the lookahead is 'aba', beginning at offset 2 into the history and extending into the lookahead.

- At each stage, *greedy* parsing was used: the dictionary was searched for a longest match for the initial segment. There is no guarantee that greedy parsing maximizes compression. It would be preferable if the scheme could search for the "best" combination of ordered pairs and characters, but this is intractable. A very limited form (lazy evaluation) of this is discussed below.

Searching for the longest match in the dictionary could be expensive. A number of LZ77-variants maintain a dictionary (and structures to speed searching) which includes only some of the phrases in the history, thus limiting the amount of searches. This may result in less compression: in the example, instead of the 4-character match at the second stage, we could have matched '□s' with the the first two characters of '□shells' from the dictionary.

- An attractive feature of many of the LZ77-variants (other than their simplicity) is fast decoding: while the coder must do the hard work of finding matches, the decoder need only do simple lookups to rebuild the source.

- The output of the coder could be subject to additional compression. As an example, suppose a fixed number of bits are used to store the *length* component of the ordered pair. If short match lengths are more common, then a probabilistic scheme on the match lengths may be effective.

Among schemes of this form, there is considerable flexibility in choosing the sizes of the history and lookahead, and in management of the dictionary. To understand the process and considerations more clearly, some notes on two specific implementations of LZ77-type schemes are presented. The first of these is a revised version of the LZRW1 scheme proposed by Ross Williams in [83], and illustrates design decisions favoring speed over compression. The second is the well-known GNU zip (*gzip*) utility, which is similar to LZRW1 but uses more advanced techniques in managing a larger dictionary and lookahead.

9.1.1 An LZ77 implementation

Suppose the symbol set consists of the 256 8-bit characters. The history and lookahead are both fixed-length, with offsets represented in 12 bits (giving a history of length $2^{12} = 4096$ bytes) and match lengths in 4 bits. A single "control bit" is used to distinguish (*offset, length*)-pairs from single characters in the output stream.

The cost of transmitting an (*offset, length*)-pair and control bit is then 17 bits, while 2 characters cost 18 bits. For this reason, we transmit a pair only if the match length is at least 2. Accordingly, the 4 bits used for the *length* will represent lengths from 2 to 17. Note that expansion occurs whenever a single character is output, or if a pair is output with match length 2.

As an illustration, the phrase from the previous section is passed through the coder. The portions replaced by (*offset, length*)-pairs are underlined:

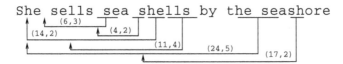

The original string required $36 \cdot 8 = 288$ bits. In the encoded stream, a total of 18 characters are replaced by 6 pairs (leaving 18 unmatched characters to send), giving an encoded stream of length $6 \cdot 17 + 18 \cdot 9 = 264$ bits. It should be noted that the example has been chosen so that matches against the dictionary occur almost immediately—typically, more of the source must make its way into the dictionary before many matches occur.

The parsing has been greedy: at each stage, the dictionary is searched for a longest match to an initial segment of the lookahead. Here, we've assumed that the dictionary includes every phrase of length 2–17 which begins in the history.

The LZ77 variation described is known as LZSS,[4] and an implementation with well-documented source in the C programming language appears in Nelson and Gailly [53]. A tree structure is placed over the history, reducing search times but adding to the complexity and storage requirements. Even so, an exhaustive search for the longest match among every phrase of the history may be prohibitively expensive. Both LZRW1 and GNU zip limit the search for repeated strings in order to improve speed at the coding stage.

LZRW1

The LZRW1 algorithm was presented by Ross Williams in [83]. Its main design goals favored speed over compression, and the result was a very compact and fast method. The sources from a revised version appear in Appendix B.

As above, the history and lookahead are fixed-length, with sizes represented in 12 and 4 bits, respectively. However, the dictionary includes only a subset of the phrases which start in the history, and a hash function is used to speed searches. More specifically, the hash is a function of the first three characters of the lookahead, pointing to the last occurrence of a 3-character string with the same hash. If a match occurs, then this offset is used in the output. This can provide very fast matching, but it significantly reduces the size of the dictionary.[5]

Use of the hash function means that only match lengths of at least 3 will be considered. Accordingly, the 4 bits used for the *length* will represent lengths from 3 to 18.[6] This allows a longer match-length than that used above (18 bytes instead of 17); however, the overhead in representing characters and pairs hasn't changed, so the inability to code a 2-character sequence as a pair may degrade compression (which occurs in the "She sells..." example above).

---

[4]See Storer and Szymanski [75] and Bell [7].

[5]It is essential to identify the dictionary in LZ schemes. Appendix B describes the use of the hash in LZRW1 more completely, and from this the dictionary can be determined.

[6]In LZRW1, match lengths were limited to 16. This was mostly an oversight, and was corrected in the LZRW1-A algorithm.

Table 9.1: *Compression (% remaining) for selected Calgary corpus files.*

File	Kbyte	RW1	(12,4) LZSS	(13,4) LZSS	(13,5) LZSS	(15,4) LZSS	*gzip*	*compress*
bib	109	59.4	48.6	43.0	43.9	37.2	31.5	41.8
book1	751	67.9	56.8	51.8	54.3	46.8	40.8	43.2
geo	100	84.4	83.0	80.3	83.0	81.8	66.9	76.0
obj1	21	61.7	57.3	57.2	57.1	60.0	48.0	65.3
pic	501	25.6	21.3	20.9	16.8	22.0	11.0	12.1
progc	39	54.6	45.4	41.7	42.3	41.6	33.5	48.3
Average		60.0	52.1	49.2	49.6	48.2	38.6	47.8

LZRW1 obtains moderate compression using few resources. Table 9.1 gives compression results on a subset of the "Calgary corpus" [8]. The columns for LZSS are tagged with pairs indicating the number of bits used to represent history and lookahead sizes. For reference, two well-known dictionary-type implementations are included: *gzip* (an LZ77-type scheme) and *compress* (an LZ78 scheme). The speed and simplicity of LZRW1 comes at a price: the compression with LZSS is generally superior (even with the same history and lookahead sizes). The difference is perhaps not as large as we might have expected, given the very minimal dictionary searching used by LZRW1.

In applications, it may be acceptable to sacrifice some speed during coding in order to improve compression. For example, on-line documentation may be viewed frequently—the speed of compression may be less important than amount of compression and the speed of decompression. LZ77 schemes such as LZRW1, LZSS, and that used in *gzip* all have very fast decoding due to the way the history is used as the dictionary. If this feature is to be retained, then the following are perhaps the simplest modifications to improve compression.

Enlarge the history (and dictionary) and/or lookahead. However, more bits will be needed for (*offset, length*)-pairs, and this can increase the breakeven point. In our example, 12 bits were used for the *offset* and 4 bits for the *length* (and 1 bit for control). The breakeven point is just over 2 characters. If, say, a pair requires 15 bits and 8 bits, respectively, then the breakeven is 3 characters; i.e., no compression will be achieved unless the match length is at least 4.

Recall also that increasing either of these can greatly increase search times. The use of a hash function in LZRW1 is fast, but LZSS finds longer matches. A compromise might involve the use of a hash chain to follow (some of) the matches in the history.

Improve the parsing. We've been greedy: there has been no lookahead for the best combination of literals and matches. For our example, greedy parsing was used at the stage

```
She sells sea s|hells by the seashore
```

and the next output items were the pairs corresponding to 'he' and '11s□', respectively, for a total of $17 + 17 = 34$ bits. For this fragment, it would be preferable to use *lazy evaluation*: send 'h' as a literal and then send the pair for 'e11s□' using a total of $9 + 17 = 26$ bits.

Compress the output of the coder. The output of the coder has been described as a mix of literals and (*offset, length*)-pairs. If, say, short match lengths are more common, then a statistical coder may be able to compress the lengths. Of course, this modification may somewhat increase the work for the decoder.

The "*gzip*" column in Table 9.1 is of special interest. It uses an LZ77-type scheme of the same basic form as LZRW1 and LZSS, but typically offers superior compression. To some extent, it implements all three of the modifications mentioned above.

9.1.2 Case study: GNU zip

The *gzip* program is widely used as a general purpose compressor for files (or streams of data), and was designed as a replacement for the *compress* utility (which uses a patented LZ78-type scheme). Sources for *gzip* can be found in the references listed in Appendix C.

As GNU Project software, it was essential to have a patent-free scheme with freely distributable sources. Design goals included portability and acceptable "worst case" performance. The history of the development suggests that compression performance was of less importance (but perhaps decompression speed was essential). Ross Williams' LZRW1 met these conditions, and was to have been used as the basic scheme in *gzip*.[7] To the dismay of Williams, it was discovered that (use of) the algorithm was covered by patent (see Appendix C).

The "deflation" compression method used in the current *gzip* shares many of the features of LZRW1. It cannot match the speed of coding, but it generally gives more compression and faster decompression than that obtained with the *compress* utility (a common reference for dictionary schemes). As noted above, LZRW1 gives somewhat less compression than *compress*.

The algorithm in *gzip* is LZ77-type, with 15 bits reserved for the *offset* (giving a 32K history) and 8 bits for the *length*. An (*offset, length*)-pair then requires $23 + 1$ bits, so the breakeven is 3 characters. Since literals cost 9 bits, only matches of length 3 or more are worth considering. Hence, the 8 bits represent match lengths of 3 to $(2^8 - 1) + 3 = 258$ characters.

A hash of the first three characters from the lookahead is used to speed searches into the history. Unlike LZRW1, a hash chain is maintained in order to permit searching for longer matches. Searching through the chain is expensive, but it is performed only during coding.

---

[7]Confirmed via private email with Jean-loup Gailly, the principal *gzip* developer. Quoted by permission.

Lazy evaluation is used: after finding the longest match at the current symbol, *gzip* looks for a longer match at the next symbol. If a longer match is found, the current symbol is sent as a literal, and the process continues at the next symbol.[8]

To be precise, the lazy evaluation and the search through the hash chain are subject to runtime choices (the '-0' to '-9' compression level options). Several parameters are set, including:

good_length: If the current match has length at least *good_length*, then the maximum depth for a lazy search is reduced (divide *max_chain* by 2).

max_lazy: Do not perform lazy search above this match length.

nice_length: Quit search above this match length.

max_chain: Limit on depth of hash chain search.

Example choices for the parameters appear in Table 9.2. With the exception of the step between compression levels 3 and 4, the parameter values increase with the compression level. It seems reasonable to expect that larger values should correspond to improved compression, but there is no guarantee of this.

Table 9.2: *Parameter values corresponding to* gzip *compression level.*

Parameter	Compression level					
	1	3	4	6[a]	8	9
good_length			4	8	32	32
max_lazy[b]			4	16	128	258
nice_length	8	32	16	128	258	258
max_chain	4	32	16	128	1024	4096

[a]Default value.

[b]No lazy search on levels 0–3. A fifth parameter (*max_insert_length*) limits the updating of the hash table (for speed).

The output of the "dictionary-scheme" portion of *gzip* may be compressible. This is perhaps easiest to see in the match lengths—short matches may be much more common than longer matches. A second "back-end" compressor is used which compresses literals or match lengths with one Huffman tree, and match offsets with another.

Performance results for various choices of the compression level can be seen in Table 9.3 on page 243. These tests were run on a SPARCstation 20, with timings obtained by averaging several blocks of 20 runs. Included are results from the Unix *compress* program, which uses an LZ78 scheme and has become a standard reference for dictionary methods.

---

[8]The notes in 'algorithm.doc' in the *gzip-1.2.4* distribution may be misleading on this point; however, the actual source code explains it well.

Exercises 9.1

1. Williams' paper [83] contains the following bit of poetry:

   ```
   A walrus in Spain is a walrus in vain
   ```

   (a) Find the (*offset*, *length*)-pairs produced by the LZRW1 scheme. Indicate the text fragment corresponding to each pair.
   (b) Calculate the compression (or expansion) in this example.

2. Lazy evaluation can do worse than greedy parsing. Consider an LZSS-type scheme where offsets are represented in 12 bits and lengths are represented in 4 bits with match lengths from 2–17. The use of a control bit means that an (*offset*, *length*)-pair requires 17 bits and a literal requires 9 bits. Suppose the current window contains

history	lookahead
abcbcdedefg	abcdefg

   Show that greedy parsing leads to 34 bits output, while lazy evaluation results in 43 bits.

3. *gzip* searches the hash chains so that the most recent strings are found first, and matches of length 3 which are too distant (more than 4K bytes) are ignored. How does this help compression? (Hint: Consider the back-end Huffman processing.)

4. (Programming exercise) Choose a set of test files and determine if the lazy evaluation of *gzip* is effective (consider both time and compression). It will be necessary to modify the sources of *gzip* so that the parameter values (other than lazy evaluation) remain the same for tests with and without lazy evaluation.

5. The hash function used in LZRW1 can be found in Appendix B. Knuth [39] writes that such functions should be quick to compute and should minimize collisions. Does the choice in LZRW1 satisfy these criteria? The constant 40543 which appears in the definition is prime, but is there any other reason for its choice? (You may wish to consult Knuth's book.)

9.2 The LZ78 approach

The LZ77-schemes discussed in the previous section are attractive for their simplicity, relatively good speed and compression for the resources required, and fast decoding. An exhaustive search through the history during coding can be expensive, but implementations such as LZRW1 and *gzip* illustrate methods to trade compression for speed. Another possible concern with the scheme as presented is that only "recently-seen" strings can be matched. The history (and/or

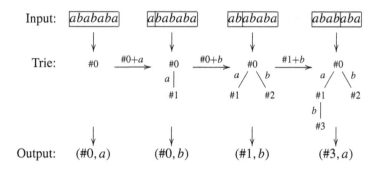

Figure 9.1: *LZ78 coding on 'ababab'.*

lookahead) could be enlarged (and offsets could be represented with variable-length pointers), but this can add complexity and search time to the scheme.

LZ78 takes a different approach in building a slow-growing dictionary. The source is parsed into phrases, and a new phrase enters the dictionary by adding a single character to an existing phrase. In practice, there is a ceiling on the number of phrases, and some action is performed when the dictionary fills. There is a price for this "more structured" approach: the decoder must maintain the dictionary.[9]

In its basic form, LZ78 starts with an empty dictionary (denoted by '#0' in Figure 9.1). At each stage the longest match for the lookahead is sought from the dictionary, and a pointer #n to this phrase, along with the "unmatched" character c, is output as the ordered pair (#n, c). The dictionary is updated, adding c to the phrase represented by #n. The decoder must maintain the same dictionary of phrases.

To see how this works, consider encoding 'ababab'. The top row of Figure 9.1 shows the source, with a vertical line separating the history from the lookahead. The second row illustrates the updating of the dictionary trie.[10] Phrases corresponding to a pointer #n are found by walking up the tree: for example, phrase #3 is 'ab'. At each stage, the dictionary is traversed for the longest match against the lookahead, and then the phrase number and unmatched character are output (the last row in the figure). The notation '#n + c' in the dictionary update means that the character c is to be added to the the phrase represented by #n.

Figure 9.1 shows that four steps are needed to encode the sample source,

---

[9]The LZ77 decoder must maintain the history; however, for the schemes described, this is considerably less than maintaining the LZ78 dictionary.

[10]The name "trie" was suggested by E. Fredkin, according to Knuth [39]. In a footnote to Fredkin's paper (Trie memory, *Communications of the ACM*, 3(9):490-499, September 1960), an editor remarks that "*trie* is apparently derived from re*trie*val." Knuth describes a trie as essentially an n-ary tree, whose nodes are n-vectors. Each node represents the set of keys that begin with a certain sequence of characters; the node specifies an n-way branch, depending on the next character.

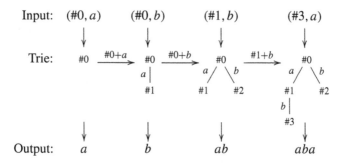

Figure 9.2: *LZ78 decoding on the output of Figure 9.1.*

and the output consists of the ordered pairs (#0, a), (#0, b), (#1, b), and (#3, a). The dictionary at the next stage would consist of the following phrases:

Dictionary trie

Corresponding phrases

Entry	Phrase
#0	null
#1	a
#2	b
#3	ab
#4	aba

Note that phrases always start at the root. For example, the string '*ba*' appears as part of '*aba*' in the trie, but it is not a phrase in the dictionary. Unlike LZ77, the length is not passed as part of the pointer—the length of a phrase is understood from the trie structure. Also, the possibility of $|S|$ children at a given node can make for a more complicated trie structure in the case of a larger symbol set S.

Decoding the output of Figure 9.1 is very simple, and consists of reversing the vertical arrows. Since we wish to compare this carefully with a modified scheme, the decoding is shown in detail in Figure 9.2. The last row gives '*abababa*' as the recovered string, as expected.

In implementations, the dictionary will eventually fill. Indeed, LZ78 variants such as Unix *compress* and the V.42*bis* method commonly used in communication via modem have relatively low ceilings (64K and 2K, respectively) on the total number of phrases. There have been many schemes proposed for handling overflow:

- Freeze the dictionary at this stage. Compression can suffer if the nature of source changes after the dictionary is frozen.

- Flush the dictionary and start over. This can discard much of what was learned about the source.

- Monitor compression, and flush the dictionary only when performance falls below a threshold. This is the approach in Unix *compress*.

- Prune the trie. This could perhaps be a "remove least recently used phrase" scheme. V.42*bis* and the 'shrink' method of PKZIP 1.0 use pruning.

Many variants of the basic LZ78 scheme have been described. One of these, known as LZW [80], drops the explicit transmission of the unmatched character. This variant is the basis for the method used in Unix *compress*, V.42*bis* modem compression, and the GIF graphics format. Unisys currently holds a patent on some portions of the algorithm, and sells the license for use in modems supporting V.42*bis*. In 1995, Unisys announced that it would start pressing its claims in connection with the GIF graphics format. As Greg Roelofs wrote, "GIF became decidedly less popular right around New Year's Day 1995 when Unisys and CompuServe suddenly announced that programs implementing GIF would require royalties, due to Unisys' patent on the LZW compression method used in the GIF format."[11] Since PNG offers technical advantages over GIF, it is likely to receive considerable attention.

9.2.1 The LZW variant

In the basic LZ78 scheme described above, the output of the coder consists of a sequence of ordered pairs $(\#n, c)$, where $\#n$ is a pointer into the dictionary and c is the unmatched character from the search. LZ78 is said to have *pointer guaranteed progress* through the source, since c (known as the *innovation* or *instance*) is part of the output token. Explicit transmission of c may be wasteful in the case that c could be part of a match at the next stage. This same consideration motivated the development of the *deferred innovation* variation (LZSS) of LZ77.

The idea in LZW is to completely drop the transmission of characters c. The dictionary updating process remains essentially unchanged, but the progression through the source will differ from LZ78 (resulting in a different trie). LZW is said to have *dictionary guaranteed progress* through the source. In order for this to work, the dictionary is preloaded with the 1-character phrases from the symbol set.

As an example, the LZW scheme is applied to the string '*abababa*'. The dictionary starts with entries for the 1-character phrases '*a*' and '*b*'. At each stage, the longest match for the lookahead is found, and then the unmatched character c is added to $\#n$, giving a new phrase $\#n + c$. The procedure is illustrated in Figure 9.3.

The process looks quite similar to that in the LZ78 scheme of Figure 9.1. However, we obtain a different collection of phrases, and apparently some compression since only four pointers are output (rather than the four pairs of Figure 9.1). The differences are more dramatic in the decoding process, and this example shows an exceptional case in LZW which must be handled.

---

[11] Quoted from the Portable Network Graphics (PNG) page at http://www.wco.com/˜png/. Used by permission. A short history of PNG appears in [60].

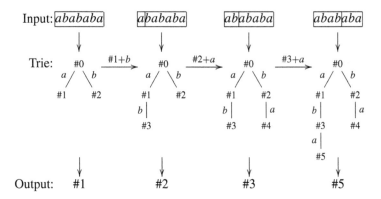

Figure 9.3: LZW coding on 'abababa'.

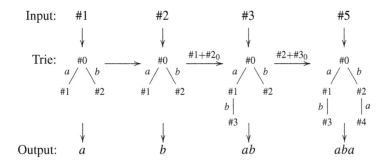

Figure 9.4: LZW decoding on the output of Figure 9.3.

In the LZW decoding of Figure 9.4, the updating of the dictionary is indicated with expressions of the form $\#n + \#m_0$, where the new subscript indicates that the first character corresponding to phrase $\#m$ is to be added to phrase $\#n$. For example, the dictionary update $\#1 + \#2_0$ adds 'b' (the first, and only, character of phrase $\#2$) to 'a' (the phrase corresponding to $\#1$), giving a new phrase 'ab' (phrase $\#3$).

The last column of Figure 9.4 requires some explanation, since phrase $\#5$ is not even listed in the trie (so how did we know the output is 'aba'?). This is the exceptional case in LZW, and can be resolved by noting that the update of the dictionary occurs one step later than in LZ78. The next update of the dictionary would be $\#3 + \#5_0$, so the new phrase $\#5$ satisfies $\#5 = \#3 + \#5_0$. This implies that phrase $\#5$ begins with the first characters of phrase $\#3$; e.g., $\#5_0 = \#3_0$. Hence, $\#5 = \#3 + \#5_0 = \#3 + \#3_0 = $ 'aba'.

Looking back at the coding stage, it can be seen that the exceptional case occurs when the newest node at a given stage is used as the output. This occurs in the last column of Figure 9.3. This case could be avoided (possibly resulting in less compression) if the newest node at a given stage is not considered for output. In the example, the coder could have split the match into #3 and #1.

9.2.2 Case study: Unix *compress*

The *compress* program uses an LZW variant called LZC. The utility compresses single files (or streams of data). Versions exist across many platforms, and it has become a standard reference when comparing compression schemes. It has relatively good performance, given the resource requirements.

The scheme uses pointers of variable size to tag dictionary entries, starting with 9 bits and increasing up to a ceiling of 16 bits (corresponding to a dictionary with $2^{16} = 64K$ entries). An option allows setting a lower ceiling, typically so that files can be uncompressed on small machines.

The dictionary will eventually fill. In this case, *compress* monitors compression, flushing the dictionary when performance drops. This is a simple scheme to implement, and gives *compress* the ability to adjust to changes in the nature of the data after the dictionary fills. Automatically clearing the dictionary on a periodic basis has a similar goal, but may be wasteful if the dictionary is performing well.

Compress can perform rather badly on random data, since the output at each stage consists of a pointer of 9–16 bits (corresponding to an expansion of 9/8 to 2 for a single character match). The parsing of LZ78 with a preloaded dictionary may do better in this special case, since tokens would represent at least 2 characters of the source (corresponding to an expansion of 17/16 to 3/2 for a single character match if *compress* were minimally modified to use an LZ78 approach).

The results are mixed: *compress* gives better compression than LZRW1, but perhaps not as much as expected, given the very minimal resource requirements of LZRW1. Table 9.1 shows a sample from the Calgary corpus. LZRW1 "compresses about 10% absolute worse than [*compress*], but runs four times faster" in the tests run by Williams [83]. Note that LZRW1 actually beat *compress* on 'obj1' (VAX object code).

The GNU zip (*gzip*) utility discussed in Section 9.1.2 was designed as a replacement for *compress*, and Table 9.3 gives performance results on several of the Calgary files. The tests were performed on a SPARCstation 20, with timings obtained by averaging the results from the Unix *time* command on blocks of 20 runs. The decode rates are based on the size of the original file. As expected, *compress* is faster than *gzip* for compression, but considerably slower on decoding.

Although *gzip* can only look at the most recent 32K of history, the dictionary contains many more entries than the 64K phrases maintained by *compress*. For the test files in Table 9.3, *gzip* gives superior compression, even at the lowest (fastest) setting. The increased compression at the higher levels comes at a

Table 9.3: *Performance of* gzip *vs* compress.

File	Kbyte	Encode (10K/s) [compression (% remaining)]				Decode (10K/s)	
		compress	*gzip* (-1)	*gzip* (-3)	*gzip* (-6)	*compress*	*gzip*
bib	109	58 [41.8]	51 [39.4]	39 [35.7]	18 [31.5]	95	158
book1	751	42 [43.2]	42 [47.5]	28 [43.8]	12 [40.8]	102	167
geo	100	45 [76.0]	27 [68.2]	14 [67.9]	6 [66.9]	70	107
obj1	21	44 [65.3]	37 [49.8]	34 [49.2]	23 [48.0]	56	62
pic	501	125 [12.1]	101 [12.8]	84 [12.2]	36 [11.0]	187	313
progc	39	52 [48.3]	46 [39.0]	38 [36.6]	22 [33.5]	76	107
Average		70 [47.8]	51 [42.8]	39 [40.9]	20 [38.6]	98	152

rather steep price, due to the more exhaustive searches (and lazy matching at levels above 3).

Exercises 9.2

1. LZ78 coding produced the pairs: (#0, M), (#0, i), (#0, s), (#3, i), (#3, s), (#2, p), (#0, p), (#0, i). Decode this to obtain 'Mississippi'. Show the final dictionary obtained.

2. Encode 'Mississippi' using an LZW scheme with symbol set {M, i, p, s} and initial dictionary

Entry	Phrase
#1	M
#2	i
#3	p
#4	s

   Show the final trie obtained.

3. The symbol set {a, b} was used for LZW encoding, with initial dictionary

Entry	Phrase
#1	a
#2	b

   Decode the sequence #1, #2, #4. Explain your steps.

4. Some dictionary coders can be converted into a statistical model which gives the same compression. The code space used by a phrase in the dictionary scheme is decomposed into the space used by the individual characters. As an example, consider a greedy parsing scheme with $S = \{a, b\}$ and dictionary $\{a, ba, bb\}$. Suppose the dictionary scheme assigns 1/4 of the code space to 'a', 1/4 to 'ba', and the remaining 1/2 to 'bb'. The

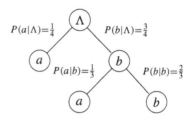

Figure 9.5: *The symbol-wise decomposition in Exercise 4.*

decomposition into a symbol-wise equivalent is shown in Figure 9.5.[12]

The statistical model is determined by the probabilities listed in the figure. Calculate the ideal number of bits assigned to each of the nodes by the symbol-wise equivalent (i.e., determine the number of bits used to encode an '*a*' following '*b*', etc.). Verify that that both the dictionary scheme and the statistical model give the same code length for '*bb*'.

9.3 Notes

Many LZ variants are discussed in *Text Compression* [8], and summarized concisely in [82]. Some schemes possess characteristics from both LZ77 and LZ78: the LZFG algorithm [19] combines the history structure of LZ77 with the phrase structure of LZ78.

In practice, the slow growth of phrases in LZ78 may degrade compression. Horspool [31] proposes modifications to LZW involving more rapid phrase growth and phased in binary numbers, as a way to improve the compression without significantly degrading the speed. Non-greedy parsing schemes (such as the lazy evaluation used in *gzip*) for dictionary schemes are considered in Horspool [32]. These trade time for limited compression improvements, with only minimal (if any) changes needed on the decoding side. The cost of deferred innovation is examined in Cohn [11].

The relationship between statistical and dictionary schemes is sometimes direct: some greedy dictionary methods can be decomposed into statistical methods which give the same compression (see Exercise 9.2.4). Gutmann and Bell [26] present an approach which is the opposite of this decomposition in an attempt to obtain the better compression of statistical methods with the speed of dictionary schemes.

---

[12]This example is adapted directly from Bell and Witten [9]. Their construction of a symbol-wise equivalent for any nonadaptive greedy parsing method is more involved than this simple example might suggest, and the interested reader should consult their paper and also [8, 42] for a much more complete discussion.

Chapter 10

Transform Methods and Image Compression

Images are natural and efficient conveyors of information and have been used throughout history as models for both reality and abstract concepts. Our appetite for visual information seems insatiable, and efficient image management continues to be of pressing concern. An image can contain a large amount of information (more than a thousand words?) and often translates to a data structure whose large size can pose problems to storage and transmission management. The situation only gets worse when several images or animation is involved.

Up to this point we've confined ourselves to compression methods involving zero information loss, i.e., to "lossless" schemes, and any of the schemes from Chapters 5 through 9 could be used on images. However, "images" have certain common features that are better exploited by methods designed specifically for them. Some well-known lossless compression schemes for images include GIF and PNG and each do a respectable compression job. But unless something surprising comes along from the realm of "lossless technology," we are not likely to see anything more than incremental improvements over these two methods. Large gains in compression ratios will come by dropping the requirement that *all* information be retained in the compression process.[1] This quickly brings up the question as to whether or not we can actually remove information from data in a way that allows it to be significantly compressed and yet doesn't thoroughly corrupt it. Fortunately, a moment's reflection is all it takes to convince us that for many applications images have room to give. As an example, consider a black and white photograph in a newspaper. The photograph itself is a (compressed) model of the image it was meant to capture. Close inspection (a magnifying glass will do) reveals simply an array of black and white dots. And yet, this global arrangement of dots presents information in a way that allows us to readily grasp the message it was meant to convey. (sometimes we may need a little help from the caption).

An image can contain more information than necessary to accomplish its purpose. If there was a way of identifying this "unnecessary" detail then we could compress an image by discarding the detail. The message behind an im-

---

[1]The phrase *compression ratio* is used loosely throughout this chapter and generally refers to some way of comparing the size of the source after it has been compressed to its size before or vice versa. At times throughout the chapter we will be more precise.

age can be subjective, and so the "surplus" information. Avoiding issues of artistic representation, there do exist some general principles that lead to acceptable solutions of the compression problem for "everday" images . They're based on observations that the human eye can be rather insensitive to certain fluctuations in an image, and also quite tolerant of a wide range of approximations. Within the environment defined by our subjectivity and the physical characteristics of our visual system, lossy schemes can flourish. The purpose of the chapter is to explore these "tried and true" principles and some of the lossy methods that use them.

Two important such schemes are discussed toward the end of this chapter: the JPEG image compression standard in Section 10.5, and a wavelet technique in Section 10.6. Each one is supported upon a linear algebraic structure called a transform, i.e., a change of basis.. The theme behind compression in the widely-used JPEG standard is based on the observation that local visual information at high (spatial) frequencies is often not as important in our global interpretation of the image as the low frequencies. JPEG tends to suppress this high-frequency information and often eliminates it completely. To gain some understanding of the JPEG process, we'll need to know something about the cosine transform and how it reveals image information in ways that enable us to decide what to keep and what to throw away. Since the cosine transform is an offspring of the Fourier transform, we devote a section to the motivation and development of the classical discrete Fourier transform and how it can be used as a compression device. Along the way we uncover some mathematical structure that is also useful in our discussions of wavelets.

The thrust behind our development and presentation of the Fourier transform is pragmatic rather than theoretical; in brief, we approach it from a classical signal analysis point of view. The function to be transformed will be regarded as a signal in time with the information it contains being composed of several key signals of special frequencies. Later, Fourier transforms are extended to operate on two-dimensional "signals", i.e., images, via a general method that works also with wavelets. The chapter ends with a section devoted to the JPEG image compression method and a section outlining an applied approach to wavelet compression.[2]

Many of the exercises in this chapter are designed to fill in details and to briefly explore what we think to be interesting topics in themselves, e.g., Shannon's Sampling Theorem 10.2.11 and the Fast Fourier Transform 10.2.12. The reader is encouraged to try them.

---

[2]The JPEG 2000 standard is wavelet based. See, for example, [46].

10.1 Transforms

Dividing an image in two and tossing away one of the pieces is a compression method that most of us wouldn't tolerate well. Contrast this with a method that defines a detail structure within an image, tagging details with their level of importance. When information from an image is presented to us in this way, a lossy compression scheme becomes obvious: discard the least important details. Both JPEG and wavelet methods basically do this, even though the way in which they define detail is different.

Images have mathematical representations as rectangular arrays of numbers, typically of integers. Each pixel in the image is assigned an integer whose value, in some way, represents its color. In this chapter we identify arrays with images and images with arrays, often making no distinction between the two. Since arrays of numbers can be scaled (each entry multiplied by the same number) or added together entry by entry without altering their shape, then an image can be thought of as a point in a linear space, that is, as a vector in a vector space. For example, if an image has m rows and n columns of pixels then we can think of it as a member of the vector space of $m \times n$ matrices: a point in a space of dimension mn. It's important to note that the images we usually deal with in practice are not just arbitrary arrays of numbers corresponding to points scattered willy-nilly throughout mn-space. Rather, typical images share certain traits which, when regarded as points in mn-space, translate to a group geometry susceptible to quick approximation by several linear schemes. This observation is at the heart of lossy compression methods based on linear transforms.

Mathematically, image analysis takes place in linear spaces. As such, we have at our disposal all of the processing tools from linear algebra. To use these powerful tools effectively, we'll need to start with a good choice of fundamental or basis images (basis arrays). If chosen properly, these basic images can be effectively used to describe detail levels within a large class of images.

The selection of basis images provides insight into the methods discussed in this chapter. For example, JPEG chooses them purely from a classical frequency content point of view while wavelet techniques attempt to blend "frequency content" together with the location of these frequencies in the image.[3] Once these fundamental detail images have been defined we can then resolve a given image into a linear combination of them and by examining coefficients (i.e., amplitudes) weight the importance of particular detail image to the entire given image.

An image is a special kind of signal, or vector, and the resolution process above is referred to in linear algebra as a change of basis. At the root of any (invertible) *linear* transformation is a change of basis and corresponding to any change of basis is a linear transform. All transformations important to us in this

---

[3] JPEG does this locally, throughout the image.

chapter are linear and so have at their foundation a special set of basis vectors. The Fourier transform is a well-known example and a starting point for us, but before we define it, let's briefly look at some general reasons for transforming information in the first place.

By a signal, we loosely mean some sort of ordered collection of information (ordered data) indexed by what will generically be referred to as time (usually discrete time). When such a signal is presented to us, we normally wish to do something with it; change it somehow, or extract information from it. To do the latter it is sometimes necessary to first do the former. For example, in this chapter our goal is to compress a signal by discarding some of the information it contains. How do we determine what to throw away? We *transform* the signal into a form in which, we hope, important features can be distinguished from unimportant and then keep only the important.

Transform techniques for data analysis have been around for some time and scientists often speak of "transforming" their data during data analysis. Here is a definition of the word "transform" taken from the dictionary:

> *transform*: to change in structure, appearance, or character.[4]

A more mathematical definition could read:

> *transform*: a rule used to exchange one set of objects for another.
> A function from one space to itself or another, i.e., a function.

Neither definition, by itself, does much to explain how someone goes about obtaining a useful transform. In a practical sense a useful transform will be more than just an arbitrary function: it should have some additional properties, e.g., perhaps it should be invertible (information preserving) or "easy" to compute. Changing variables or coordinates is usually done simply because the new coordinates turn out to be more convenient to work with than the old.

As is an example consider the integral $\int_D e^{1-(x^2+y^2)}\,dx\,dy$, where D is the unit disk in the plane $\mathbb{R}^2$. If the exact value of this integral is the goal then the standard polar coordinate transformation $x = r\cos\theta$, $y = r\sin\theta$ turns out to be a good choice:

$$\int_D e^{1-(x^2+y^2)}\,dx\,dy = \int_0^{2\pi}\int_0^1 e^{1-r^2}r\,dr\,d\theta = \pi(e-1).$$

This is not the sort of transform we have in mind for image processing (for one thing, it's not a *linear* transformation) but it does the job, and it's suggested by the circular symmetry of the disk D and by the argument to the exponential integrand: it is suggested by information contained in the problem.

The polar coordinate transformation is often an appropriate choice for a large class of problems exhibiting radial symmetries. An arbitrary change of variable would have been unlikely to produce anything useful. A useful change

---

[4]The "The Merriam-Webster Dictionary," 1974.

of variable not only can simplify form but can characterize it in ways that enable decisions to be made based on the signal's "new look." These decisions can be of a nature that would have been difficult, maybe impossible, to accurately make before the transformation.

For our purposes a transform will be an invertible linear transformation on a single vector space. Mathematically, this is the same as exchanging one set of basis vectors of the space for another. Although this restriction eliminates polar coordinate and other nonlinear transforms, it is not as confining as may first appear.

10.2 Periodic signals and the Fourier transform

Consider a basic sine wave $t \mapsto A \sin \omega t$, $t \in \mathbb{R}$. It carries two pieces of information: the scaling factor A ($|A|$ is known as the *amplitude* of the wave), and its oscillation frequency $\omega/2\pi$ (or, equivalently, its period $2\pi/\omega$). These two bits of information, along with the knowledge that the original signal was a sine function, allows it to be perfectly reconstructed for all time t, i.e., a "sine-wave" is completely characterized by its frequency and amplitude coefficient.[5] Not all signals are so simple. How do we distill down to the essential information that they contain? It might be nice to have our signals all defined on some common domain. The sine wave above is defined for all time t whereas most signals we observe have a finite life. This "defect" can be fixed if we imagine extending what has been observed over a finite time interval to a periodic function defined for all time.

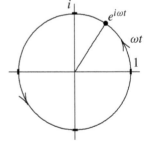

The Unit Circle

In order to analyze periodic signals more complex than a particular sine wave, we start with perhaps the most fundamental of oscillations: $\theta \mapsto e^{i\theta}$, $\theta \in \mathbb{R}$. It maps the interval $0 \le \theta < 2\pi$ (or any interval of length 2π) onto the unit circle and is the basic starting point for classical Fourier analysis. Fix a real number ω and let $\theta = \omega t$ (although not necessary, think of t as representing time). This gives the map $t \mapsto e^{i\omega t}$, an oscillation about the unit circle that completes precisely one revolution (clockwise if $\omega < 0$ and counterclockwise if $\omega > 0$) of 2π radians in $T = 2\pi/|\omega|$ units of time. The constant ω can be thought of as angular velocity with units of radians (unit-less) per unit time. T is the period of the oscillation and the frequency of oscillation is $f = \omega/2\pi = \pm 1/T$ cycles per second. Thus, a basic oscillation with frequency f can be described by the function $t \mapsto e^{2\pi i f t}$, $t \in \mathbb{R}$.

---

[5]We should consider here a phase-shift also, but let's assume that our sine-waves are zero when time is zero.

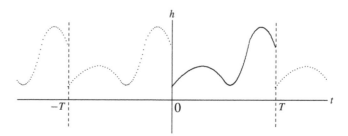

Figure 10.1: *A periodic signal.*

Imagine a signal, that is, a function h of time t. It it's helpful, you can think of h as an audio signal, i.e., a voltage level, fluctuating with time. Or scan from left to right along a horizontal line in a greyscale photograph; in this case t is not a temporal variable but a spatial measurement and $h(t)$ the shade of grey at position t. We can't observe a signal forever (even if we can imagine it lasting that long) so we watch it for awhile, say $T > 0$ units of time. By replaying this piece of signal over and over again we can think of it as defined for all time, i.e., we regard the portion of the signal sampled as just one period of a period T function defined on all of $\mathbb{R}$, c.f., Figure 10.1. It could be possible that this period T signal is built up from a few basic, more elementary period T signals. But before we try to find out exactly what these elementary period T signals are and how they can be used to synthesize h, we'll first try and make precise our notion of an elementary signal.

What could be a simpler example of an elementary period T signal than the oscillation $t \mapsto e^{i\omega t}$ referred to and pictured in the figure of the unit circle above. If, for each integer n, we set $\omega_n = 2\pi n/T = 2\pi f_n$, then the map

$$t \mapsto e^{i\omega_n t} = e^{2\pi i f_n t}, \quad t \in \mathbb{R}$$

completes n trips around the unit circle (counterclockwise if n is positive and clockwise if n is negative) during the time interval $0 \le t \le T$.

Exactly one of these *basic* signals exists for each integer n; that is, for each $n \in \mathbb{Z}$, $t \mapsto e^{2\pi i f_n t}$ is a signal on $\mathbb{R}$ of period T and frequency $f_n = n/T$. The collection of all of these fundamental signals

$$\mathcal{A} = \left\{ t \mapsto e^{2\pi i f_n t} \mid n \in \mathbb{Z} \right\}$$

is a set of raw material we can use to generate other period T signals. The collection $\mathcal{A}$ contains infinitely many *different* signals and they each have a common period T of oscillation. However, even though $\mathcal{A}$ is large, it does not contain enough functions for our purposes, in that it's easy to write down a signal with period T that does not belong to $\mathcal{A}$. We need a bigger set. It doesn't seem prudent to enlarge $\mathcal{A}$ any more than necessary, and since we're trying to generate period T signals there seems to be no reason to add signals other than period T signals. One natural way to add more such signals is to combine those

period T extension

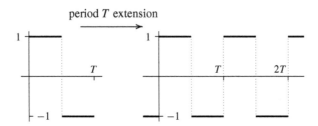

Figure 10.2: *A square wave.*

that are already in $\mathcal{A}$. For each complex number a_n, the map $t \mapsto a_n e^{2\pi i f_n t}$ is still an oscillation with period T (frequency f_n), it's just not on the unit circle any more; instead the motion takes place on a circle of radius $|a_n|$, that is, the amplitude of the oscillation is $|a_n|$.[6] Going a step further, take two amplitudes; $|a_k|$ and $|a_n|$, and two frequencies; f_k and f_n, and sum their corresponding oscillations. This signal

$$t \mapsto a_k e^{2\pi i f_k t} + a_n e^{2\pi i f_n t}, \quad t \in \mathbb{R}$$

has period T and generally differs from anything in $\mathcal{A}$.[7]

To reach other period T signals, combine more than just two oscillations: select k numbers $a_1, \ldots, a_k$ and k frequencies (basic signals) $f_{n_1}, \ldots, f_{n_k}$ and form the function $t \mapsto \sum_{j=1}^{k} a_j e^{2\pi i f_{n_j} t}$. This *linear combination* of $e^{2\pi i f_{n_1} t}, \ldots,$ $e^{2\pi i f_{n_k} t}$ from $\mathcal{A}$ will always have period T (Exercise 10.2.2). Since $\mathcal{A}$ contains an infinite number of signals then we should expect this linear combination process to generate a tremendous number of new period T functions. It does—but the resulting set, let's call it span $\mathcal{A}$, is still not large enough to contain the signals we might be interested in. For example, extend the map, $t \mapsto 1$ on $0 < t \leq T/2$ and $t \mapsto -1$ on $T/2 < t \leq T$ to a period T map on $\mathbb{R}$; see Figure 10.2. This square wave jumps at $0, \pm T/2, \pm T, \ldots$ and does not belong to span $\mathcal{A}$ because span $\mathcal{A}$ contains only continuous functions.

However, we can obtain square waves and other discontinuous functions, if we allow linear combinations of infinite numbers of oscillations from $\mathcal{A}$. There are convergence issues and subtleties associated with infinite series of functions that arise when we do this but these are issues we will avoid. In fact, in a few paragraphs we'll be back to considering only finite sums. For the interested reader though, good resources dealing with convergence questions and basic (Fourier) analysis include [61] and [67].

---

[6]If $|a_n| = 1$ then the resulting oscillation is still on the unit circle but, unless $a_n = 1$, it starts out of phase with the rest of the signals in $\mathcal{A}$.

[7]Multiplying basic signals $t \mapsto e^{2\pi i f_k t}$ and $t \mapsto e^{2\pi i f_n t}$ together results in a basic signal, i.e., is an element of $\mathcal{A}$ (cf. Exercise 10.2.1) and doesn't give us anything new. *Scaling* a basic signal, however, is a way to escape the group $\mathcal{A}$.

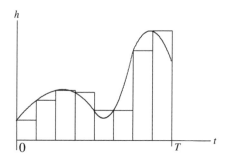

Figure 10.3: *A step function approximation to h.*

Denote by $\mathcal{B}$ the formal collection of all functions generated from $\mathcal{A}$ in this manner. Thus,

$$\mathcal{B} = \Big\{ t \mapsto \sum_{n \in \mathbb{Z}} a_n e^{2\pi i f_n t} \mid a_n \in \mathbb{C} \Big\}.$$

It turns out that $\mathcal{B}$ will include just about any period T signal we are likely to encounter in practice.[8] The sequence $n \mapsto a_n$ of coefficients is sometimes called the Fourier transform of a signal $\sum_{n \in \mathbb{Z}} a_n e^{2\pi i f_n t}$ from $\mathcal{B}$; however, we normally start with a signal h and not its Fourier coefficients a_n. We need a recipe for converting a signal h to its Fourier coefficients, i.e., a way to compute from h a sequence of coefficients $\langle a_n \rangle_{n \in \mathbb{Z}}$ so that

$$h(t) = \sum_n a_n e^{2\pi i f_n t}, \quad 0 \leq t < T. \tag{10.1}$$

Exercise 10.2.4 leads to the simple formula

$$a_n = \frac{1}{T} \int_0^T h(t) e^{-2\pi i f_n t} \, dt \tag{10.2}$$

which allows each a_n to be computed directly from h. The formula holds when h is known to have a convergent expansion (10.1).

Equations (10.1) and (10.2) are interesting formulae and find their way into a variety of engineering applications, but we're after something different. For one thing, in practice we can hardly ever hope to know a signal h for every value of t in some continuous time interval $0 \leq t \leq T$. Rather, a signal is typically measured or sampled at discrete moments t_n, $n = 0, \ldots, N$ with $0 \leq t_n < T$. It's quite common to keep the sampling interval constant and T some multiple of it. The result is a collection of sampling times t_n that are equi-spaced between 0 and T. More precisely, if N samples of the signal are desired, then the sampling

---

[8]As long as the signal h is of finite *power*, i.e., $\int_0^T |h|^2 < \infty$, and provided that we are willing to relax the condition that equality in (10.1) hold for *all* t in the interval $0 < t \leq T$ to that of holding for nearly or *almost all* t in this interval.

interval is T/N and the sample times are $0, T/N, \ldots, (N-1)T/N$. The effect of this discrete sampling procedure is equivalent to replacing the continuous signal h with the step function approximation $\tilde{h}$ defined by[9]

$$\tilde{h}(t) = h(kT/N) \quad \text{for} \quad kT/N \le t < (k+1)T/N \quad \text{and} \quad k = 0, \ldots, N-1.$$

Figure 10.3 contains an illustration of this process and suggests that we should really be using a discontinuous step function $\tilde{h}$ in place of the ideal signal h in equations (10.1) and (10.2).

How will the use of $\tilde{h}$ instead of h change things? If $k/T \le t < (k+1)T/N$, then $\tilde{h}(t) = h(kT/N)$ and (10.1) becomes

$$\tilde{h}(t) = h(kT/N) = \sum_{n \in \mathbb{Z}} a_n e^{2\pi i f_n kT/N} = \sum_{n \in \mathbb{Z}} a_n e^{2\pi i n k/N}. \tag{10.3}$$

Here is how to get a finite sum from this last expression: since $e^{2\pi i j} = 1$ for each integer j then $e^{2\pi i(n+jN)k/N} = e^{2\pi i n k/N}$, so grouping terms and factoring reduces (10.3) to a finite sum

$$h_k = h(kT/N) = \sum_{n=0}^{N-1} b_n e^{2\pi i n k/N}$$

where b_n is defined from the sequence $\langle a_n \rangle$ by the relationship $b_n = \sum_{j \in \mathbb{Z}} a_{n+jN}$. The coefficients $\langle a_n \rangle_{n \in \mathbb{Z}}$ of (10.3) now fade into the background and we no longer worry about them; our problem is now finite dimensional.

Problem Given a *discrete* signal $\mathbf{h} = (h_0, \ldots, h_{N-1})$, find a vector (of amplitude coefficients) $\mathbf{b} = (b_0, \ldots, b_{N-1})$ so that for each $k = 0, \ldots, N-1$

$$h_k = \sum_{n=0}^{N-1} b_n e^{2\pi i n k/N}. \tag{10.4}$$

This is a well-posed mathematical problem and for a given vector (signal) $\mathbf{h}$ we can solve these N equations for the N unknowns in $\mathbf{b}$ by methods learned in any linear algebra course, e.g., Gaussian elimination. However, there is more to system (10.4) than can be seen at first glance because the *orthogonality* relationship[10]

$$\sum_{n=0}^{N-1} e^{2\pi i n k/N} e^{-2\pi i n j/N} = \begin{cases} N, & \text{if } k = j, \\ 0, & \text{if } k \neq j, \end{cases} \tag{10.5}$$

developed in Exercise 10.2.5, permits us to easily describe its solution. From

---

[9]In practice, the values of this step function are also determined by the number of bits used to measure or resolve the value of the signal at the moment it is being sampled—a quantization effect which we shall not be concerned with.

[10]The word *orthogonal* is used to generalize the notion of *perpendicular* and is typically used in dimensions higher than 3 or when the inner product is different than the usual dot product.

(10.4) and (10.5) one can show (cf., Exercise 10.2.6) that

$$b_k = \frac{1}{N} \sum_{n=0}^{N-1} h_n e^{-2\pi i n k/N}, \quad k = 0, \ldots, N-1. \tag{10.6}$$

This equation defines a map $\mathbf{h} \mapsto \mathbf{b}$, from $\mathbb{C}^N$ back into $\mathbb{C}^N$, and the vector $\mathbf{b}$ is often called the (discrete) Fourier transform of $\mathbf{h}$.

In this text, though, we reserve this title for a slightly modified form of (10.6), which has more symmetry. Set $\widehat{\mathbf{h}} = \sqrt{N}\mathbf{b}$ in equations (10.4) and (10.6) to get the form of what, henceforth, will be called the *discrete Fourier transform*

$$\widehat{\mathbf{h}}(v) = \frac{1}{\sqrt{N}} \sum_{k=0}^{N-1} \mathbf{h}(k) e^{-2\pi i k v/N}, \quad v = 0, \ldots, N-1, \tag{10.7}$$

$$\mathbf{h}(k) = \frac{1}{\sqrt{N}} \sum_{v=0}^{N-1} \widehat{\mathbf{h}}(v) e^{2\pi i k v/N}, \quad k = 0, \ldots, N-1. \tag{10.8}$$

The vector $\widehat{\mathbf{h}} = (\widehat{h}_0, \ldots, \widehat{h}_{N-1})$ defined by (10.7) is called the Fourier transform of $\mathbf{h} = (h_0, \ldots, h_{N-1})$.[11] The vector $\mathbf{h}$ defined by (10.8) is called the inverse Fourier transform of $\widehat{\mathbf{h}}$.[12] The two systems (10.7) and (10.8) enable us to compute either $\widehat{\mathbf{h}}$ from $\mathbf{h}$ or $\mathbf{h}$ from $\widehat{\mathbf{h}}$, that is, given one of them, we can compute the other.

Each of (10.7) and (10.8) defines a linear transformation on $\mathbb{C}^N$ (Exercise 10.2.3), and hence have matrix expressions. If we let W be the $N \times N$ matrix whose entry in the jth row and kth column is $W(j,k) = (1/\sqrt{N})e^{2\pi i j k/N}$, then (10.7) and (10.8) assume simple forms:

$$\widehat{\mathbf{h}} = \overline{W}\mathbf{h} \tag{10.7'}$$

$$\mathbf{h} = W\widehat{\mathbf{h}}. \tag{10.8'}$$

$\overline{W}$ is notation for the matrix whose entries are just the complex conjugates of the entries in W. It's worth noting that (10.7') and (10.8') together imply that $\mathbf{h} = W\overline{W}\mathbf{h}$ for all $\mathbf{h} \in \mathbb{C}^N$ or that $\widehat{\mathbf{h}} = \overline{W}W\widehat{\mathbf{h}}$ for each $\widehat{\mathbf{h}} \in \mathbb{C}^N$; either implies that $W^{-1} = \overline{W}$. This observation, together with symmetry of W, allows for an easy proof showing that the Fourier transform is an isometry on $\mathbb{C}^N$, i.e., a map from $\mathbb{C}^N \to \mathbb{C}^N$ that preserves lengths of vectors (Exercises 10.2.7 and 10.2.8 have the details).

There is another way to regard the Fourier transform, one which has the advantage of enabling us to identify vectors as the frequency elements that compose the signal. Let $\mathbf{W}_k$ denote the kth column vector of the matrix W, i.e., for

---

[11] In reference to sequences and vectors, we'll interchangeably use the notation $\mathbf{x}(k)$ and x_k to denote the same kth entry of the sequence or vector $\mathbf{x}$.

[12] Be warned: some sources may refer to (10.8) as the Fourier transform and (10.7) as the inverse transform.

$k = 0, \ldots, N-1$ put

$$\mathbf{W}_k = \frac{1}{\sqrt{N}} \begin{bmatrix} 1 \\ e^{2\pi i k/N} \\ \vdots \\ e^{2\pi i j k/N} \\ \vdots \\ e^{2\pi i (N-1)k/N} \end{bmatrix}. \tag{10.9}$$

Equations (10.7) and (10.8) can now be written as

$$\widehat{\mathbf{h}} = \sum_{k=0}^{N-1} \mathbf{h}(k)\overline{\mathbf{W}}_k \tag{10.7''}$$

$$\mathbf{h} = \sum_{v=0}^{N-1} \widehat{\mathbf{h}}(v)\mathbf{W}_v. \tag{10.8''}$$

The columns of W, then, are the basic signals or elements associated with the Fourier transform. From a linear algebra perspective, the columns of W form a basis for the vector space $\mathbb{C}^N$ and the Fourier transform $\widehat{\mathbf{h}}$ of $\mathbf{h}$ is just the (ordered) collection of coefficients needed to expand $\mathbf{h}$ with this basis.

Note that the jth component of the basis vector $\mathbf{W}_k$ is just

$$\mathbf{W}_k(j) = \frac{1}{\sqrt{N}}(e^{2\pi i k/N})^j$$

and, hence, the entries in the kth column of W are generated by taking successive powers of $e^{2\pi i k/N}$. The column (vector) $\mathbf{W}_k$, like any vector, is just a function of its index j,[13] and since the right-hand side of the above equation makes sense for any integer j then it provides a natural extension of $\mathbf{W}_k$ from $j = 0, 1, \ldots, N-1$ to all of $\mathbb{Z}$. Also, since $(e^{2\pi i k/N})^{j+N} = (e^{2\pi i k/N})^j$ for all integers j, then the extension is a sequence with period N.

When regarded as a function on $\mathbb{Z}$, each "column" $\mathbf{W}_k$ oscillates with period N and, because time is measured discretely, the frequency of oscillation increases as the argument $2\pi k/N$ gets closer to π, that is, when $k \approx N/2$.[14] Consequently, high frequency oscillations correspond to columns at the "middle" of the matrix W, i.e., $\mathbf{W}_k$ with k at or near $N/2$. The columns on the left and right of W oscillate at lower frequencies. Also, the orthogonality re-

---

[13] An N-vector $\mathbf{v}$ is a function of its index: $\mathbf{v} : \{0, 1, \ldots, N-1\} \to \mathbb{C}$ with $\mathbf{v}(k) = v_k$.

[14] For each j, $(e^{2\pi i k/N})^j$ is a point on the unit circle. The map $j \mapsto (e^{2\pi i k/N})^j$ defines a sequence of points that march around the unit circle with time j. To see that these points march "faster" when $k \approx N/2$ consider the following argument. For any real number Θ, there is a unique θ with $-\pi \le \theta < \pi$ and $e^{i\Theta} = e^{i\theta}$ (in fact, if you like, you can write $\theta = \Theta \pmod{2\pi} - \pi$). If $e^{i\Theta}$ is close to 1, then θ must be close to 0 and the point $(e^{i\theta})^j = e^{ij\theta}$ marches slowly around the unit circle as j ranges over the integers. At the other extreme, if $e^{i\Theta}$ is close to -1, then θ is close to either π or $-\pi$ and the argument $j\theta$ changes more dramatically with j; the net effect is that the point $e^{ij\theta}$ jumps rapidly around the unit circle as j clicks from one integer to the next.

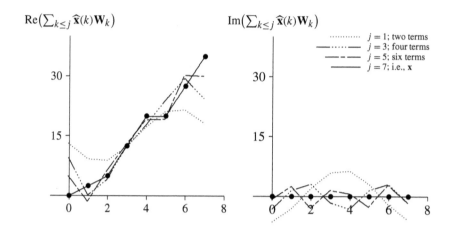

Figure 10.4: *Real and imaginary parts of partial sum approximations to* **x**.

lationships (10.5) imply the orthogonality of the vectors $\mathbf{W}_k$, $k = 0, \ldots, N - 1$. Since each $\mathbf{W}_k$ has unit length, then the collection $\{\mathbf{W}_k \mid k = 0, \ldots, N - 1\}$ is an orthonormal basis for $\mathbb{C}^N$, cf. Exercise 10.2.9.[15]

We end this section with a quick example of a vector **x** and its Fourier transform $\widehat{\mathbf{x}}$:

$$
\mathbf{x} = \begin{bmatrix} 0.0 \\ 2.5 \\ 5.0 \\ 12.5 \\ 20.0 \\ 20.0 \\ 27.5 \\ 35.0 \end{bmatrix}, \quad
\widehat{\mathbf{x}} = \begin{bmatrix} 43.31 \\ -5.82 + 17.96i \\ -4.42 + 8.84i \\ -8.32 + 2.05i \\ -6.19 \\ -8.32 - 2.05i \\ -4.42 - 8.84i \\ -5.82 - 17.96i \end{bmatrix}.
$$

Figure 10.4 contains some of the partial sums from the Fourier expansion (10.8″) of **x** (split into real and imaginary parts). The solid piecewise linear curve is a graph of **x**, the dotted curve is obtained from the first 2 terms from the sum (10.8″), the dash-dots curve from the first 4, and the dashed line from the first 6.

10.2.1 The Fourier transform and compression: an example

The Fourier transform $\widehat{\mathbf{x}}$ of a signal **x** is a vector containing the amplitudes of the fundamental frequencies that make up **x**. Each component of $\widehat{\mathbf{x}}$ indicates the strength of a particular frequency in **x**. In certain classes of signals, e.g.,

---

[15] Unless otherwise stated, we take the usual inner product in $\mathbb{C}^N$: $\langle \mathbf{v}, \mathbf{w} \rangle = \sum_{\nu=0}^{N-1} v_\nu \overline{w}_\nu$ for vectors $\mathbf{v} = (v_0, \ldots, v_{N-1})$ and $\mathbf{w} = (w_0, \ldots, w_{N-1})$ in $\mathbb{C}^N$. A collection of vectors is *orthonormal* if the vectors are mutually orthogonal and have unit length.

audio signals, entire frequency ranges may not be relevant or meaningful to our interpretation of the signal's quality. The Fourier transform gives us direct control over these frequencies: replacing an entry in $\widehat{\mathbf{x}}$ with zero "removes" the corresponding frequency from $\mathbf{x}$.

The decision to remove frequency information may be based on mathematical or physical importance of the frequencies. Coefficients in $\widehat{\mathbf{x}}$ of small magnitude indicate frequencies with weak mathematical presence in $\mathbf{x}$, and discarding them may be done with relative impunity. In applications, there may be physical considerations which allow the suppression of certain frequency information, even if these frequencies have significant mathematical presence. For example, if very high frequencies are suppressed in an audio signal, then the signal changes, but we may not be aware of it.[16] The approximation obtained by zeroing certain coefficients is a special case of *quantizing*, a method which reduces the precision of coefficients, and which will be discussed in connection with JPEG in Section 10.5. Thoughtful quantizing can help suppress both non-meaningful and weak mathematical frequencies simultaneously. A similar story holds for wavelet transforms and the process of selectively eliminating or approximating transform coefficients provides a foundation for the lossy schemes discussed in this book.

Let's examine the action of the Fourier transform on the two signals

$$\mathbf{x} = \begin{bmatrix} 0.0 \\ 2.5 \\ 5.0 \\ 12.5 \\ 20.0 \\ 20.0 \\ 27.5 \\ 35.0 \end{bmatrix} \quad \text{and} \quad \mathbf{y} = \begin{bmatrix} 20.0 \\ 14.2 \\ 0.0 \\ -10.0 \\ -15.0 \\ -10.0 \\ 0.0 \\ 14.2 \end{bmatrix}.$$

Think of $\mathbf{x}$ and $\mathbf{y}$ as single periods of two larger signals whose graphs appear in Figure 10.5. The Fourier transforms $\widehat{\mathbf{x}}$ and $\widehat{\mathbf{y}}$ are approximately

$$\widehat{\mathbf{x}} = \begin{bmatrix} 43.31 \\ -5.82 + 17.96i \\ -4.42 + 8.84i \\ -8.32 + 2.05i \\ -6.19 \\ -8.32 - 2.05i \\ -4.42 - 8.84i \\ -5.82 - 17.96i \end{bmatrix} \quad \text{and} \quad \widehat{\mathbf{y}} = \begin{bmatrix} 4.74 \\ 24.47 \\ 1.77 \\ 0.27 \\ -1.20 \\ 0.27 \\ 1.76 \\ 24.47 \end{bmatrix}.$$

One clear difference in these two vectors is that $\widehat{\mathbf{y}}$ is real and $\widehat{\mathbf{x}}$ isn't. A glance at (10.7) shows that the Fourier transform generally outputs complex vectors even if the input vectors are real. So why is $\widehat{\mathbf{y}}$ real? Is it just by chance or is $\mathbf{y}$ special in some way that makes vectors like it have real Fourier transforms? This question is something we'll return to in the next section, but for now we're content to emphasize that each entry of $\widehat{\mathbf{x}}$ and $\widehat{\mathbf{y}}$ measures the amplitude of a par-

---

[16] A dog might be able to notice the change.

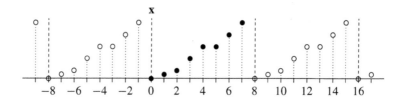

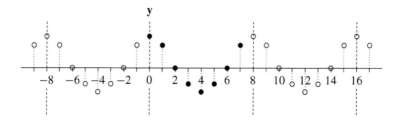

Figure 10.5: *Signals* **x** *and* **y**, *and their periodic extensions.*

ticular frequency component of **x** and **y**, respectively, and that high-frequency components correspond to middle entries of $\widehat{\mathbf{x}}$ and $\widehat{\mathbf{y}}$.

Before we do additional analysis, some notation is required. If $\mathbf{z} \in \mathbb{C}^N$ is any N-vector of complex numbers, then define Abs(**z**) to be the vector of absolute values or magnitudes of **z**. Thus, the kth component Abs(**z**)(k) of Abs(**z**) is just $|\mathbf{z}(k)|$, that is, Abs(**z**)$(k) = |\mathbf{z}(k)|$, $k = 0, 1, \ldots, N-1$. With this notation, consider

$$\text{Abs}(\widehat{\mathbf{x}}) = \begin{bmatrix} 43.31 \\ 18.88 \\ 9.88 \\ 8.57 \\ 6.19 \\ 8.57 \\ 9.88 \\ 18.88 \end{bmatrix} \quad \text{and} \quad \text{Abs}(\widehat{\mathbf{y}}) = \begin{bmatrix} 4.74 \\ 24.47 \\ 1.77 \\ 0.27 \\ 1.20 \\ 0.27 \\ 1.77 \\ 24.47 \end{bmatrix}.$$

From these magnitudes it's clear that high frequencies are more prevalent in **x** than in **y**. We could wonder whether this difference was apparent before their transforms were taken. Look again to the graphs of **x** and **y** in Figure 10.5. High frequencies can be "spotted" by looking for abrupt changes in values over small changes in (in our case, discrete) time, rather than gentle trends. If our attention is fixed to only the part of the graphs over the integers $0, 1, \ldots, 7$, then we might be led to believe that **x** is as smooth, if not smoother, than **y**. Over *just* these eight integers this may be true, but remember, the Fourier transform sees these signals as defined on all of $\mathbb{Z}$, not just $0, 1, \ldots, 7$.

Since each $\mathbf{W}_k$ has the same length (unit length in this case) then the kth Fourier coefficient of a signal is all we need to determine the importance of its frequency component in the $\mathbf{W}_k$ "direction." If the vectors $\mathbf{W}_k$ were of differing lengths then the relative importance of a particular frequency $\mathbf{W}_{k_0}$ could not be reliably determined by examining its Fourier coefficient alone. In this case, throwing out "small" coefficients, thinking the corresponding frequencies are unimportant, could be a mistake.[17] This is a good argument for normalizing basis signals.

Because the high-frequency entries in $\widehat{\mathbf{y}}$ are quite small when compared to other coefficients (particularly when compared to the low-frequency entries) they may not play much of a role and we could ask what would happen to $\mathbf{y}$ if we left them out. More specifically, how does it affect $\mathbf{y}$ to set $\widehat{\mathbf{y}}(2) = \widehat{\mathbf{y}}(3) = \widehat{\mathbf{y}}(4) = \widehat{\mathbf{y}}(5) = \widehat{\mathbf{y}}(6) = 0$? Doing this gives a vector, say $\widehat{\mathbf{z}}$, where

$$\widehat{\mathbf{z}} = \begin{bmatrix} 4.74 \\ 24.47 \\ 0 \\ 0 \\ 0 \\ 0 \\ 0 \\ 24.47 \end{bmatrix}.$$

$\widehat{\mathbf{z}}$ is not the Fourier transform of $\mathbf{y}$, but it's natural to ask which vector $\mathbf{z}$ has $\widehat{\mathbf{z}}$ as its Fourier transform? $\widehat{\mathbf{z}}$ is not too far from $\widehat{\mathbf{y}}$ so we could hope that $\mathbf{z}$ is not far from $\mathbf{y}$. It is easy to compute $\mathbf{z}$ using the inverse Fourier transform, (10.8), on $\widehat{\mathbf{z}}$. Then $\mathbf{z}$ and the entry-by-entry *error* $\mathbf{z} - \mathbf{y}$ are given by

$$\mathbf{z} = \begin{bmatrix} 18.98 \\ 13.91 \\ 1.68 \\ -10.56 \\ -15.63 \\ -10.56 \\ 1.68 \\ 13.91 \end{bmatrix} \quad \text{and} \quad \mathbf{z} - \mathbf{y} = \begin{bmatrix} 1.02 \\ 0.28 \\ -1.68 \\ 0.56 \\ 0.63 \\ -0.56 \\ -1.68 \\ 0.29 \end{bmatrix}.$$

Whether or not this error is acceptable depends on the purpose of the original signal and on how accurately it needs to be known.[18] In any event, consider this: we threw away $5/8$ or 62.5% of the components of $\widehat{\mathbf{y}}$ and we were able to invert and get something that appears to be fairly "close" to the original $\mathbf{y}$.

Now let's repeat the above procedure on $\mathbf{x}$. In fact, let's keep even *more* of $\widehat{\mathbf{x}}$ than we did of $\widehat{\mathbf{y}}$ by setting just the three smallest components $\widehat{\mathbf{x}}(3)$, $\widehat{\mathbf{x}}(4)$,

---

[17]For example, in a compression scheme, tossing out an innocuous looking coefficient could prove dangerous if the associated basis element's magnitude is much larger than some of the others. More precisely, suppose $\mathbf{u}$ and $\mathbf{v}$ have the same length, say $|\mathbf{u}| = |\mathbf{v}| = 1$, and that $\mathbf{w} = a\mathbf{u} + b\mathbf{v}$. If $|a| \ll |b|$ then $a/b \approx 0$ and $\mathbf{w} = b[(a/b)\mathbf{u} + \mathbf{v}] \approx b\mathbf{v}$. On the other hand, if $|\mathbf{u}|$ is much larger than $|\mathbf{v}|$ then, even though $a/b \approx 0$, it could be the case that $|(a/b)\mathbf{u}| \gg 0$. The statement $\mathbf{w} \approx b\mathbf{v}$ could then be extremely misleading.

[18]Often engineers will use the l^2 or Euclidean norm to measure error between two vectors $\mathbf{v}$ and $\mathbf{w}$: $\|\mathbf{v} - \mathbf{w}\| = \sqrt{\sum_k |v_k - w_k|^2}$. Since the Fourier transform is an isometry (in this l^2 sense, see Exercise 10.2.8) then the l^2-error found after reconstruction of $\mathbf{y}$ from a modification of its transform will be exactly the same as the error introduced into its transform $\widehat{\mathbf{y}}$. This practical feature is shared by all orthogonal (and unitary) transforms.

and $\widehat{\mathbf{x}}(5)$ to zero (note that these coefficients are not, relative to the rest of the components, as small as the smallest components of $\widehat{\mathbf{y}}$). Call this modification $\widehat{\mathbf{z}}$ again and invert to obtain

$$
\mathbf{z} = \begin{bmatrix} 8.07 \\ -2.83 \\ 5.74 \\ 15.50 \\ 16.30 \\ 20.95 \\ 31.13 \\ 27.63 \end{bmatrix} \quad \text{and} \quad \mathbf{z} - \mathbf{x} = \begin{bmatrix} 8.07 \\ -5.33 \\ 0.74 \\ 2.30 \\ -3.7 \\ 0.95 \\ 3.63 \\ -7.37 \end{bmatrix}.
$$

We didn't do as well in this case even though we altered less frequency information. Does this mean that $\mathbf{x}$ cannot be compressed effectively? No, it could just mean that the Fourier transform is not the right compression tool to use on $\mathbf{x}$. If signals like $\mathbf{x}$ need to be compressed, then perhaps better results could be obtained by using a different set of basis signals than the Fourier $\{\mathbf{W}_0, \dots, \mathbf{W}_{N-1}\}$.

Just how to construct such a "compression" basis can be a problem that is not easily solved. The above strategy can be thought of as a "projection" scheme in the sense that setting Fourier coefficients to zero *projects* the transform vector (orthogonally) into a subspace of smaller dimension (dimension 3 in the case of $\widehat{\mathbf{y}}$, 5 in the case of $\widehat{\mathbf{x}}$). Projection methods can work if the subset of the data type space from which we will select vectors to compress is "thin" in several directions.[19] In this setting, the job of choosing basis elements amounts to finding these special "directions."

The following example may better illustrate our meaning here. Consider the set

$$
\mathcal{G} = \left\{ \begin{bmatrix} \alpha & a_{12} & a_{13} \\ a_{21} & \alpha & a_{23} \\ a_{31} & a_{32} & \beta \end{bmatrix} : \alpha, \beta \geq 1, \; |a_{ij}| \ll 1 \text{ if } i \neq j \right\}.
$$

$\mathcal{G}$ is a subset of the (9-dimensional) space of 3 by 3 matrices and can be described "well" using only two matrices;

$$
M_1 = \begin{bmatrix} 1 & 0 & 0 \\ 0 & 1 & 0 \\ 0 & 0 & 0 \end{bmatrix}, \quad M_2 = \begin{bmatrix} 0 & 0 & 0 \\ 0 & 0 & 0 \\ 0 & 0 & 1 \end{bmatrix}.
$$

To fill out a basis, select seven more 3 by 3 arrays having directions as different as possible from M_1 and M_2, i.e., choose them orthogonal to the span of M_1 and M_2.[20] The nine arrays will form a basis for the space of 3 by 3 matrices. With respect to this basis, the subset $\mathcal{G}$ is "thin" in any basis direction other than M_1 and M_2. One could ask why it has been suggested that the other basis elements be chosen orthogonal to both M_1 and M_2—after all, to get a basis they

---

[19] An effect of quantizing can be to turn a "thin" set into a "thinner" discrete set.

[20] Think of 3 by 3 matrices as vectors in $\mathbb{R}^9$ and use the inner product there to determine orthogonality. Also, see Exercise 10.2.10 for more on why these directions are as different as possible from each other.

only need to be linearly independent of M_1, M_2 and each other. In answer to this suppose we take as a basis element an array that is not too different from M_2 (and definitely not orthogonal to it), e.g.,

$$M = \begin{bmatrix} 0 & 0 & 0 \\ 0 & 0 & 0 \\ 0.001 & 0 & 1 \end{bmatrix}.$$

M doesn't belong to the span of $\{M_1, M_2\}$ but its close proximity to M_2 means that it plays a significant role in describing any array that also relies on M_2, in particular, in describing arrays in $\mathcal{G}$. In other words, $\mathcal{G}$ isn't "thin" in the direction of M. To see this more precisely, the matrix

$$A = \begin{bmatrix} 2 & 0 & 0 \\ 0 & 2 & 0 \\ 0.001 & 0 & 2 \end{bmatrix}$$

certainly belongs to $\mathcal{G}$, but since $A = 2M_1 + M_2 + M$, its construction from any basis containing M_1, M_2 and M requires the "same amount" of M as M_2.[21]

In summary, we were able to suppress much of the information contained in $\widehat{\mathbf{y}}$ (setting 5 of the 8 coefficients to zero), still having something whose inverse transform looked like the original $\mathbf{y}$. The Fourier transform may not have been the best tool to use in compressing $\mathbf{x}$. We can't dismiss the Fourier transform so easily though, and interestingly enough it will provide us with a better tool to use on vectors like $\mathbf{x}$. We just need to better understand the way in which the Fourier transform looks at signals as single periods of larger signals defined for all (discrete) time. In this context, the Fourier transform will also aid us in understanding why $\widehat{\mathbf{y}}$ is real and $\widehat{\mathbf{x}}$ is not. In the next section we pursue this thread, leading us to the cosine transform.

Exercises 10.2

Throughout these exercises, T is a nonzero real number and, for each $n \in \mathbb{Z}$, $f_n := n/T$.

1. The collection $G = \{e^{2\pi i f_n t} \mid t \in \mathbb{R}, n \in \mathbb{Z}\}$ forms a (commutative) group under pointwise multiplication.[22] To check this, show the following:

    (a) $e^{2\pi i f_n t} e^{2\pi i f_m t} = e^{2\pi i f_{n+m} t}$; multiplication of two elementary signals is an elementary signal,

---

[21] Both M and M_2 are about the same magnitude so their coefficients are comparable.

[22] A *group* is a basic algebraic structure. It consists of a set G together with a *multiplication on* G such that

   (a) $ab \in G$ whenever a and b are in G,

   (b) $a(bc) = (ab)c$ whenever a, b, and c belong to G,

   (c) there is an identity element e in G, i.e., there exists $e \in G$ such that $ae = ea = a$ for every $a \in G$,

   (d) each element in G has an inverse in G, i.e., for each element $a \in G$ there is an element $a^{-1} \in G$ with $aa^{-1} = a^{-1}a = e$.

(b) $e^{2\pi i f_0 t} = 1$; the identity element $t \mapsto 1$, is an elementary signal, and

(c) $e^{2\pi i f_n t} e^{2\pi i f_{-n} t} = 1$; each elementary signal has an inverse.

To reach functions outside of G requires some operation other than just multiplying its elements together.

2. (a) If $a_1, \ldots, a_N$ are any N complex numbers and $n_1, \ldots, n_N$ any N integers, then show that the map $t \mapsto \sum_{k=1}^{N} a_k e^{2\pi i f_{n_k} t}$ has period T.

   (b) If $a_n \in \mathbb{C}$ for each $n \in \mathbb{Z}$ then (formally) show that the map $t \mapsto \sum_{n \in \mathbb{Z}} a_n e^{2\pi i f_n t}$ still has period T.

3. Show that the Fourier transform map $\widehat{}: \mathbb{C}^N \to \mathbb{C}^N$ defined by (10.7) is a linear map; that is, show that $\widehat{\alpha \mathbf{g} + \beta \mathbf{h}} = \alpha \widehat{\mathbf{g}} + \beta \widehat{\mathbf{h}}$ for any $\alpha, \beta \in \mathbb{C}$ and $\mathbf{g}, \mathbf{h} \in \mathbb{C}^N$.

4. (a) Show that the functions $e^{2\pi i f_n t}$ are orthogonal over the interval $0 \le t \le T$; that is, show that

$$\int_0^T e^{2\pi i f_n t} e^{-2\pi i f_m t} \, dt = \begin{cases} T, & \text{if } m = n \\ 0, & \text{if } m \ne n. \end{cases}$$

   (b) Now suppose that $h(t) = \sum_{k \in \mathbb{Z}} a_k e^{2\pi i f_k t}$, for $0 \le t \le T$. Multiply each side by $e^{-2\pi i f_n t}$ and then integrate both sides over the interval $0 \le t \le T$. Use part (a) above to conclude that

$$a_n = \frac{1}{T} \int_0^T h(t) e^{-2\pi i f_n t} \, dt.$$

5. Equation (10.5)

$$\sum_{n=0}^{N-1} e^{2\pi i n k / N} e^{-2\pi i n j / N} = \begin{cases} N, & \text{if } j = k, \\ 0, & \text{if } j \ne k \end{cases}$$

is a discrete version of the orthogonality relationship in Exercise 4(a) above. To see this, fix $N \in \mathbb{Z}$ and, for each $k \in \mathbb{Z}$, put $z_k = e^{2\pi i k / N}$. Now do the following:

   (a) Show that $z = z_k$ is a solution to the equation $z^N - 1 = 0$. (The N distinct complex numbers $z_0, \ldots, z_{N-1}$ are known as the Nth *roots of unity*.)

   (b) Plot z_k for $k = 0, \ldots, N-1$ for several values of N, say $N = 2, 3, 4,$ and 8.

   (c) Argue that, if $z_k \ne 1$, then $\sum_{m=0}^{N-1} (z_k)^m = 0$; i.e., that $\sum_{m=0}^{N-1} e^{2\pi i m k / N} = 0$. *Hint*: The expression $z^N - 1$ can be factored: $z^N - 1 = (z-1)(1 + z + z^2 + \cdots + z^{N-1})$.

   (d) Now prove the orthogonality relationship above.

6. Starting with equation (10.4) and the orthogonality relationship (10.5) show that equation (10.6) follows, i.e., show that

$$b_k = \frac{1}{N} \sum_{n=0}^{N-1} h_n e^{-2\pi i n k / N}, \quad k = 0, \dots, N-1$$

whenever $h_k = \sum_{n=0}^{N-1} b_n e^{2\pi i n k / N}$.

7. Let $W(j,k) = (1/\sqrt{N}) e^{2\pi i j k / N}$ for $0 \le j, k \le N-1$ and show that the pair of equations (10.7) and (10.8) can be written as the pair (10.7′) and (10.8′).

8. If $\mathbf{v} = (v_0, \dots, v_{N-1})$ and $\mathbf{w} = (w_0, \dots, w_{N-1})$ are vectors in $\mathbb{C}^N$ then we define their inner product $\langle \mathbf{v}, \mathbf{w} \rangle$ to be

$$\langle \mathbf{v}, \mathbf{w} \rangle = \sum_{k=0}^{N-1} \mathbf{v}(k) \overline{\mathbf{w}(k)}.$$

The length $|\mathbf{v}|$ of a vector $\mathbf{v}$ is defined in the usual way

$$|\mathbf{v}| = \sqrt{\sum_{k=0}^{N-1} |\mathbf{v}(k)|^2}.$$

(a) If $\mathbf{v} \in \mathbb{C}^N$ then show that $\langle \mathbf{v}, \mathbf{v} \rangle$ is always a nonnegative real number and that $|\mathbf{v}| = \sqrt{\langle \mathbf{v}, \mathbf{v} \rangle}$.

(b) If A is an $N \times N$ matrix of complex numbers then show that $\langle A\mathbf{v}, \mathbf{w} \rangle = \langle \mathbf{v}, A^*\mathbf{w} \rangle$ where the matrix A^* is the conjugate transpose of A.

(c) Using equations (10.7′), (10.8′) and parts (a) and (b) above, argue that $|\widehat{\mathbf{h}}| = |\mathbf{h}|$ for any $\mathbf{h} \in \mathbb{C}^N$. This shows that the Fourier transform is an isometric automorphism on $\mathbb{C}^N$.

9. Use the orthogonality relationship (10.5) or the relationship $W^{-1} = \overline{W}$ from page 254 to show that the columns $\mathbf{W}_k, k = 0, \dots, N-1$, of the matrix W exhibit the following property

$$\langle \mathbf{W}_j, \mathbf{W}_k \rangle = \begin{cases} 1, & \text{if } j = k, \\ 0, & \text{if } j \ne k. \end{cases}$$

Thus, the columns of W form an orthonormal set of vectors in $\mathbb{C}^N$. What about the rows of W?

10. Let $\mathbf{u}$ and $\mathbf{v}$ be orthogonal unit vectors in $\mathbb{C}^N$. Suppose that $\mathbf{w} \in \mathbb{C}^N$ is another unit vector not orthogonal to $\mathbf{u}$. Then show that there is a complex number α, with $|\alpha| = 1$, such that $\|\alpha \mathbf{u} - \mathbf{w}\| < \|\mathbf{u} - \mathbf{v}\|$.

Remark: The exercise shows that, direction-wise, orthogonal vectors are further from each other than nonorthogonal vectors.

11. *Shannon's sampling theorem.* Suppose that h is a continuous signal defined on the interval $[-T/2, T/2]$ of length T. Then, from equations (10.1) and (10.2)

$$h(t) = \sum_{n\in\mathbb{Z}} a_n e^{2\pi i f_n t}, \qquad (10.10)$$

where $f_n = n/T$ and

$$a_n = \frac{1}{T} \int_{-T/2}^{T/2} h(t) e^{-2\pi i f_n t}\, dt. \qquad (10.11)$$

The sequence of coefficients, $\langle a_n \rangle_{n\in\mathbb{Z}}$, is called the Fourier transform $\widehat{\mathbf{h}}$ of h, i.e., $\widehat{\mathbf{h}}(n) = a_n$. If, instead of being defined on some finite interval, h is defined on all of $\mathbb{R}$ then it may still have a Fourier transform.[23] Generally, to synthesize such a signal we need to use basic signals of *all* frequencies f:

$$h(t) = \int_{\mathbb{R}} \widehat{h}(f) e^{2\pi i f t}\, df, \qquad (10.12)$$

$$\widehat{h}(f) = \int_{\mathbb{R}} h(t) e^{-2\pi i f t}\, dt. \qquad (10.13)$$

A signal $h : \mathbb{R} \to \mathbb{R}$ is called "band-limited" if for some frequency f_c, its Fourier transform $\widehat{h}(f) = 0$ whenever $|f| > f_c$, that is, if h is composed of only a finite range of frequencies.

(a) If $\widehat{h}(f) = 0$ for $|f| > f_c$, then use (10.12) to show that

$$h(t) = \int_{-f_c}^{f_c} \widehat{h}(f) e^{2\pi i f t}\, df. \qquad (10.14)$$

Use (10.10) and (10.11), with $\widehat{h}$ in place of h, to show that

$$\widehat{h}(f) = \sum_{n\in\mathbb{Z}} c_n e^{2\pi i \frac{n}{2f_c} f}, \qquad (10.15)$$

where

$$c_n = \frac{1}{2f_c} \int_{-f_c}^{f_c} \widehat{h}(f) e^{-2\pi i \frac{n}{2f_c} f}\, df. \qquad (10.16)$$

---

[23] It turns out that $h \in L^2[-T/2, T/2]$ (i.e., $\int_{-T/2}^{T/2} |h|^2 < \infty$) if and only if there is a sequence $\langle a_n \rangle_{n\in\mathbb{Z}} \in \ell^2$ (i.e., $\sum_{n\in\mathbb{Z}} |a_n|^2 < \infty$) and (10.10) holds. In this case the sequence $\langle a_n \rangle_{n\in\mathbb{Z}}$ (the Fourier transform of h) is given by (10.11) and (Parseval's theorem) $\sum_{n\in\mathbb{Z}} |a_n|^2 = \int_{-T/2}^{T/2} |h|^2$, i.e., $\|\langle a_n \rangle_{n\in\mathbb{Z}}\|_{\ell^2} = \|h\|_{L^2[-T/2, T/2]}$. Similarly, $h \in L^2(\mathbb{R})$ (i.e, $\int_{\mathbb{R}} |h|^2 < \infty$) if and only if there exists a function $\widehat{h}$ also in $L^2(\mathbb{R})$ such that (10.12) holds. In this case the function $\widehat{h}$ (the Fourier transform of h) is given by (10.13) and (the Plancherel theorem) $\int_{\mathbb{R}} |h|^2 = \int_{\mathbb{R}} |\widehat{h}|^2$, i.e., $\|h\|_{L^2(\mathbb{R})} = \|\widehat{h}\|_{L^2(\mathbb{R})}$.

(b) Now argue from (10.14) and (10.16) that

$$h\left(-\frac{n}{2f_c}\right) = 2f_c c_n$$

and, hence, from (10.15)

$$\widehat{h}(f) = \sum_{n\in\mathbb{Z}} \frac{1}{2f_c} h\left(-\frac{n}{2f_c}\right) e^{2\pi i \frac{n}{2f_c} f} = \sum_{n\in\mathbb{Z}} \frac{1}{2f_c} h\left(\frac{n}{2f_c}\right) e^{-2\pi i \frac{n}{2f_c} f}.$$

(c) Use this last equation in (10.14) and integrate term by term to get

$$h(t) = \frac{1}{2f_c} \sum_{n\in\mathbb{Z}} h\left(\frac{n}{2f_c}\right) \frac{\sin 2\pi f_c(t - \frac{n}{2f_c})}{\pi(t - \frac{n}{2f_c})}.$$

(d) Finally, letting $\Delta = 1/2f_c$ in the above sum, show that

$$h(t) = \Delta \sum_{n\in\mathbb{Z}} h(n\Delta) \frac{\sin 2\pi f_c(t - n\Delta)}{\pi(t - n\Delta)}.$$

This last equation is known as Shannon's sampling theorem. It allows the reconstruction of a continuous signal h everywhere if its values are sampled at a rate at least twice as frequently as the critical value f_c. The cut-off frequency f_c is called a Nyquist critical frequency and Δ the corresponding sampling interval.

12. *A fast Fourier transform.* To compute the Fourier transform $\widehat{\mathbf{h}}$ directly from its definition

$$\widehat{\mathbf{h}}(v) = \frac{1}{\sqrt{N}} \sum_{k=0}^{N-1} \mathbf{h}(k) e^{-2\pi i k v/N}, \quad v = 0, \ldots, N-1 \qquad (10.17)$$

requires basically N^2 add-multiply operations (you should count them!). Fast Fourier transforms (FFTs) are attempts to speed up this computation by more efficiently handling arithmetic.

In parts (a), (b), and (c) of this problem, we'll assume that N can be factored $N = p_1 p_2$.[24]

(a) Convince yourself that each of the indices k and v can be expressed in the following forms:

$$k = k_1 p_1 + k_0; \quad \text{for } k_0 = 0, \ldots, p_1 - 1 \text{ and } k_1 = 0, \ldots, p_2 - 1$$
$$v = v_1 p_2 + v_0; \quad \text{for } v_0 = 0, \ldots, p_2 - 1 \text{ and } v_1 = 0, \ldots, p_1 - 1.$$

---

[24] By the end of the exercise we hope to see that even if the signal length N is prime it could be advantageous to pad its length, to say a power of 2, with zeros and use an FFT.

Consequently, the transform equation (10.17) can be written as

$$\widehat{\mathbf{h}}(v) = \frac{1}{\sqrt{N}} \sum_{k_1=0}^{p_2-1} \sum_{k_0=0}^{p_1-1} \mathbf{h}(k_1 p_1 + k_0) e^{-2\pi i \frac{(k_1 p_1 + k_0)v}{N}}. \tag{10.18}$$

(b) Now argue that

$$e^{-2\pi i \frac{(k_1 p_1 + k_0)v}{N}} = e^{-2\pi i \frac{v k_0}{N}} e^{-2\pi i \frac{v_0 k_1 p_1}{N}}$$

and hence, from (10.18)

$$\widehat{\mathbf{h}}(v) = \frac{1}{\sqrt{p_1}} \sum_{k_0=0}^{p_1-1} e^{-2\pi i \frac{v k_0}{N}} \widetilde{\mathbf{h}}(k_0, v_0)$$

where

$$\widetilde{\mathbf{h}}(x, y) = \frac{1}{\sqrt{p_2}} \sum_{k_1=0}^{p_2-1} \mathbf{h}(k_1 p_1 + x) e^{-2\pi i \frac{y k_1 p_1}{N}}$$

for $x = 0, 1, \ldots, p_1 - 1$ and $y = 0, 1, \ldots, p_2 - 1$.

(c) Argue that there are exactly N different $\widetilde{\mathbf{h}}(x, y)$ and each one takes p_2 add-multiplies to compute. To compute them all requires Np_2 add-multiplies. After they are computed we can go about computing the $\widehat{\mathbf{h}}(v)$. Convince yourself that now p_1 add-multiplies are needed to compute each $\widehat{\mathbf{h}}(v)$. Thus, the total number of add-multiply operations required to compute $\widehat{\mathbf{h}}(v)$, $v = 0, 1, \ldots, N - 1$, in this manner is $Np_1 + Np_2 = N(p_1 + p_2)$. Part (d) compares this number with the N^2 operations required of (10.17) when it is used directly.

(d) This process can be repeated: if $N = p_1 p_2 \cdots p_j$ then the total add-multiplies will be $N(p_1 + p_2 + \cdots + p_j)$. Now take the special case that $p_1 = p_2 = \cdots = p_j = p$ so that $N = p^j$. Show that the total number of add-multiplies is $Njp = pN \log_p N$.

The computational savings using an FFT can be considerable. To get some idea of how much faster the FFT can be, take the case $p = 2$ and look at the ratio of the number of add-multiplies using the FFT ($2N \log_2 N$) to the number N^2: $2N \log_2 N / N^2 = 2 \log_2 N / N$. In a signal with $N = 2^{10} = 1024$ (a relatively short signal) this ratio is

$$\frac{2 \log_2 2^{10}}{2^{10}} = \frac{20}{2^{10}} \approx \frac{1}{50}.$$

The FFT transforms this signal with about 50 times fewer operations than the direct use of (10.17). Approximate the speed increase for a signal with $N = 2^{20}$.

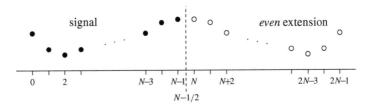

Figure 10.6: *Even extension of a signal about $k = N - 1/2$.*

10.3 The cosine and sine transforms

The Fourier transform is a map between N-periodic sequences of complex numbers. Figure 10.5 illustrates these extensions and so can help explain the high frequencies prevalent in **x** but not in **y**. The sequence **x**, as seen by the Fourier transform, makes a considerable *jump* each time k goes from -9 to -8, -1 to 0, 7 to 8, 15 to 16, etc. The extension contains an artificially introduced "high-frequency" blip; after transforming **x** we see this behavior reflected in significant high-frequency components. On the other hand, **y** was rigged so that it didn't exhibit such large endpoint differences, i.e., compare $\mathbf{x}(7) - \mathbf{x}(0) = 35$ to $\mathbf{y}(7) - \mathbf{y}(0) = -5.8$. Figure 10.6 suggests the possibility of extending a signal in a way that avoids introducing a high-frequency blip.

First start with a signal **x**, defined at times $k = 0, \ldots, N - 1$ and then extend to $k = N, N + 1, \ldots, 2N - 1$ by reflecting its graph across the vertical line that passes through the horizontal axis at the point $k = N - 1/2$ (see Figure 10.6). Mathematically, this amounts to the definition

$$\mathbf{x}(N + k) := \mathbf{x}(N - (k + 1)), \quad k = 0, \ldots, N - 1. \qquad (10.19)$$

The resulting signal, still call it **x**, is defined for $k = 0, \ldots, 2N - 1$ and extends the original **x** symmetrically. It has the property that $\mathbf{x}(0) = \mathbf{x}(2N - 1)$, i.e., the endpoint values now match. This type of extension is usually called an "even" extension or, more precisely, an even extension centered at $k = N - 1/2$, and the Fourier transform sees it as now having period $2N$ instead of N.

Apply the Fourier transform to this new (period $2N$) signal and then use Euler's identity $2\cos\theta = e^{i\theta} + e^{-i\theta}$ repeatedly. The result is eventually a linear combination of cosine functions in place of exponential functions; ergo, the name "cosine transform." Details of this process are outlined in Exercise 10.3.3 and result in the following pair of equations:

$$\widehat{\mathbf{x}}(v) = \sum_{k=0}^{N-1} \mathbf{x}(k) C(v) \cos \frac{(2k+1)v\pi}{2N}, \quad v = 0, \ldots, N-1, \qquad (10.20)$$

$$\mathbf{x}(k) = \sum_{v=0}^{N-1} \widehat{\mathbf{x}}(v) C(v) \cos \frac{(2k+1)v\pi}{2N}, \quad k = 0, \ldots, N-1, \qquad (10.21)$$

where $C(0) = \sqrt{1/N}$ and $C(k) = \sqrt{2/N}$ if $k \neq 0$. System (10.20) is often called the forward cosine transform and system (10.21) the backward or inverse cosine transform. One difference between the cosine transform and the Fourier transform is that the cosine transform is *real*, in the sense that if $\mathbf{x}$ is real then so is its cosine transform $\widehat{\mathbf{x}}$.[25]

Both of the above systems have matrix representations. Define the $N \times N$ matrix A whose vth column $\mathbf{A}_v$ is given by

$$\mathbf{A}_v = C(v) \begin{bmatrix} \cos \frac{v\pi}{2N} \\ \cos \frac{3v\pi}{2N} \\ \vdots \\ \cos \frac{(2N-1)v\pi}{2N} \end{bmatrix}. \qquad (10.22)$$

Then (10.20) and (10.21) can be written simply as

$$\widehat{\mathbf{x}} = A^t \mathbf{x} \qquad (10.20')$$

$$\mathbf{x} = A\widehat{\mathbf{x}} \qquad (10.21')$$

Combining (10.20') and (10.21') shows that $\mathbf{x} = AA^t\mathbf{x}$ for each $\mathbf{x} \in \mathbb{R}^N$. Thus, $AA^t = I_{N \times N}$, and hence $A^{-1} = A^t$. A matrix of real numbers with this property is known as an orthogonal matrix. Their columns (and rows) form a set of mutually orthogonal unit vectors.

From a linear algebra perspective, the Fourier transform process is equivalent to a change of basis—the new basis vectors are just the columns of the transform's matrix W (cf., Section 10.2). In similar fashion, the cosine transform is a change in basis, the new basis vectors being the columns of its matrix A. When the vth column $\mathbf{A}_v$ of A is extended to a function on $\mathbb{Z}$, i.e., $\mathbf{A}_v(k) = C(v) \cos[(2k+1)v\pi/2N]$ for $k \in \mathbb{Z}$, then $\mathbf{A}_v$ is periodic with period $2N$ and its frequency, $v/2N$, increases with the (column) index v. This orders $\widehat{\mathbf{x}}$ compatibly with the frequencies of $\mathbf{x}$, i.e., $\widehat{\mathbf{x}}(0)$ is the amplitude of $\mathbf{A}_0$, the lowest frequency component of $\mathbf{x}$, $\widehat{\mathbf{x}}(1)$ is the amplitude of $\mathbf{A}_1$, the next-to-lowest frequency component, and so on with $\widehat{\mathbf{x}}(N-1)$ the amplitude of $\mathbf{A}_{N-1}$, the highest frequency in $\mathbf{x}$. This ordering is in contrast with the Fourier transform of $\mathbf{x}$ where the middle entries of $\widehat{\mathbf{x}}$ give information about its high-frequency components, cf., footnote 14 on page 255.

---

[25] Also, if $\mathbf{x}$ has nonzero imaginary part and extended in this manner, then $\widehat{\mathbf{x}}$ will also have nonzero imaginary part. This gives a partial answer to the question posed in Section 10.2.1 concerning the form a real vector must have to be transformed back to a real vector.

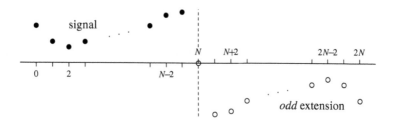

Figure 10.7: *The odd extension of a signal to all of* $k = 0, \ldots, 2N$.

Sine transforms In a similar fashion, sine transforms can be developed by applying the Fourier transform to an appropriate extension of $\mathbf{x}$. For example, a sine transform can be obtained by setting $\mathbf{x}(N) = 0$ and defining

$$\mathbf{x}(N+k) := -\mathbf{x}(N-k) \tag{10.23}$$

for $k = 1, \ldots, N$. Figure 10.7 contains a picture of this odd extension of $\mathbf{x}$. The equations for the corresponding sine transform are

$$\widehat{\mathbf{x}}(v) = \sqrt{\frac{2}{N+1}} \sum_{k=0}^{N-1} \mathbf{x}(k) \sin \frac{\pi(k+1)(v+1)}{N+1}, \quad v = 0, \ldots, N-1 \tag{10.24}$$

$$\mathbf{x}(k) = \sqrt{\frac{2}{N+1}} \sum_{v=0}^{N-1} \widehat{\mathbf{x}}(v) \sin \frac{\pi(k+1)(v+1)}{N+1}, \quad k = 0, \ldots, N-1. \tag{10.25}$$

The $N \times N$ sine transform matrix B has its vth column given by the vector

$$\mathbf{B}_v = \sqrt{\frac{2}{N+1}} \begin{bmatrix} \sin \frac{\pi(v+1)}{N+1} \\ \sin \frac{2\pi(v+1)}{N+1} \\ \vdots \\ \sin \frac{N\pi(v+1)}{N+1} \end{bmatrix}. \tag{10.26}$$

B is real and symmetric, so the transform equations (10.24) and (10.25) take the form

$$\widehat{\mathbf{x}} = B\mathbf{x} \tag{10.24'}$$

$$\mathbf{x} = B\widehat{\mathbf{x}}. \tag{10.25'}$$

These two equations imply that $B^2 = I_{N \times N}$ and, hence, that B is its own inverse.[26] Exercise 4 has more on the sine transform.

The cosine transform controls high frequencies resulting from endpoint differences, but Figure 10.7 suggests that a sine transform could possibly exacerbate the problem, introducing high-frequencies not only at endpoints but across

---

[26]If B is the matrix $\begin{bmatrix} 1 & 0 \\ 2 & -1 \end{bmatrix}$ then $B^2 = I_{2 \times 2}$. Thus, a matrix B with $B^2 = I_{N \times N}$ can be quite different from the identity.

the line $k = N$ as well.

As an example we compute cosine and sine transforms of the sample vector $\mathbf{x}$ from last section and compare them with each other and its Fourier transform $\widehat{\mathbf{x}}$. For convenience, $\mathbf{x}$ and the magnitudes (amplitudes) of its Fourier transform are reproduced here:

$$\mathbf{x} = \begin{bmatrix} 0.0 \\ 2.5 \\ 5.0 \\ 12.5 \\ 20.0 \\ 20.0 \\ 27.5 \\ 35.0 \end{bmatrix} \qquad \text{Abs}(\widehat{\mathbf{x}}) = \begin{bmatrix} 43.31 \\ 18.88 \\ 9.88 \\ 8.57 \\ 6.18 \\ 8.57 \\ 9.88 \\ 18.88 \end{bmatrix}.$$

Using $\mathcal{C}\mathbf{x}$ and $\mathcal{S}\mathbf{x}$ to denote the cosine and sine transform of $\mathbf{x}$, respectively, with $\widehat{\mathbf{x}}$ reserved for its Fourier transform, then (10.20) and (10.24) imply

$$\mathcal{C}\mathbf{x} = \begin{bmatrix} 43.31 \\ -32.46 \\ 2.11 \\ -2.67 \\ 4.42 \\ -2.04 \\ -1.83 \\ 0.97 \end{bmatrix} \quad \text{and} \quad \mathcal{S}\mathbf{x} = \begin{bmatrix} 40.03 \\ -29.54 \\ 13.27 \\ -11.88 \\ 11.05 \\ -7.14 \\ 1.64 \\ -0.71 \end{bmatrix}.$$

Entries in $\mathcal{S}\mathbf{x}$, like the cosine transform, are order compatible with increasing frequencies of $\mathbf{x}$. Thus, in this one case anyway, of the three transforms, the cosine transform seems to be the winner if the race is to represent $\mathbf{x}$ with small high-frequency terms.

10.3.1 A general orthogonal transform

The Fourier, cosine, and sine transforms are all examples of orthogonal transformations. Each could have been developed starting with an appropriate orthonormal basis for $\mathbb{C}^N$, i.e., a set of vectors $\{\mathbf{e}_0, \mathbf{e}_1, \ldots, \mathbf{e}_{N-1}\} \subset \mathbb{C}^N$ with

$$\langle \mathbf{e}_u, \mathbf{e}_v \rangle = \begin{cases} 1, & \text{if } u = v, \\ 0, & \text{if } u \neq v. \end{cases}$$

Here $\langle \mathbf{e}_u, \mathbf{e}_v \rangle$ denotes the usual inner product of $\mathbf{e}_u$ and $\mathbf{e}_v$ in $\mathbb{C}^N$,

$$\langle \mathbf{e}_u, \mathbf{e}_v \rangle = \sum_{k=0}^{N-1} \mathbf{e}_u(k)\overline{\mathbf{e}_v(k)}. \tag{10.27}$$

For example, the Fourier transform on $\mathbb{C}^N$ has basis vectors $\{\mathbf{W}_0, \ldots, \mathbf{W}_{N-1}\}$ from Section 10.2.

If $\{\mathbf{e}_0, \ldots, \mathbf{e}_{N-1}\}$ is a basis for $\mathbb{C}^N$, then for each vector $\mathbf{v} \in \mathbb{C}^N$ there is a unique vector $\widehat{\mathbf{v}} \in \mathbb{C}^N$ such that

$$\mathbf{v} = \sum_{j=0}^{N-1} \widehat{\mathbf{v}}(j)\mathbf{e}_j. \tag{10.28}$$

We call $\widehat{\mathbf{v}}$ the transform of $\mathbf{v}$ with respect to the basis $\{\mathbf{e}_0, \mathbf{e}_1, \dots, \mathbf{e}_{N-1}\}$ and to compute $\widehat{\mathbf{v}}$ requires solving the above linear system.

Define an $N \times N$ matrix E by letting its kth column be the (column) vector $\mathbf{e}_k$. Thus, $E = [\mathbf{e}_0 \ \mathbf{e}_1 \ \cdots \ \mathbf{e}_{N-1}]$ and $\mathbf{v} = E\widehat{\mathbf{v}}$. Suppose now, in addition to being a basis, the set $\{\mathbf{e}_0, \dots, \mathbf{e}_{N-1}\}$ is also orthonormal. Then the columns of E are orthonormal, i.e., $\overline{E}^t E = I_{N \times N}$, hence, $E^{-1} = \overline{E}^t$ and the relationship

$$\widehat{\mathbf{v}} = \overline{E}^t \mathbf{v} \tag{10.29}$$
$$\mathbf{v} = E\widehat{\mathbf{v}} \tag{10.30}$$

holds for all $\mathbf{v} \in \mathbb{C}^N$.[27] To obtain the Fourier, cosine, and sine transforms from these general transform equations, take the basis vectors $\mathbf{e}_k$ to be $\mathbf{W}_k$, $\mathbf{A}_k$, and $\mathbf{B}_k$ respectively, i.e., see 10.9, 10.22, and 10.26.

10.3.2 Summary

At this point in the chapter our signals $\mathbf{x} \in \mathbb{R}^N$ have been one-dimensional; the kth component $\mathbf{x}(k)$ of $\mathbf{x}$ the value of whatever it is we're recording at time k (a good example might be a simple audio signal sampled N times). But $\mathbf{x}$ can be just about any ordered list of data, for instance, it could represent a sequence of N daily observations of the snow depth at the Alta ski area in the Wasatch mountains near Salt Lake City, Utah. The transforms that have been discussed are not concerned with the physical nature of the vector $\mathbf{x}$.

The Fourier, cosine, and sine transforms each take as input a vector $\mathbf{x}$ and output a vector $\widehat{\mathbf{x}}$ whose components $\widehat{\mathbf{x}}(k)$ contain information about the fundamental frequency make-up of $\mathbf{x}$. The use of any of these classical transforms can be thought of as a decomposition process, i.e., a process of breaking up a signal into fundamental frequency "pieces," and each of them holds a steady place in the general study, description, and further analysis of signals.

We've also indicated how the information in $\widehat{\mathbf{x}}$ can help to define a "compression" of $\mathbf{x}$, provided we are willing to allow some error into its reconstruction. In the next section our goal is to extend these one-dimensional tools to two-dimensional signals; i.e., to *images*.

Exercises 10.3

1. Compute cosine and sine transforms, $\mathcal{C}\mathbf{y}$ and $\mathcal{S}\mathbf{y}$, of the sample vector $\mathbf{y}$ of Section 10.2.1 and compare results with each other and $\widehat{\mathbf{y}}$.

---

[27] A matrix E satisfying $\overline{E}^t = E^{-1}$ is called a unitary matrix. If E is also real then it is called an orthogonal matrix.

2. If E is an $N \times N$ matrix whose columns are orthonormal, then its rows are, too. *Hint:* $\overline{E}^t E = I_{N \times N}$ implies that $\overline{E}^t = E^{-1}$.

3. This exercise develops the cosine transform from the Fourier transform.

   If $\mathbf{x} = (x_0, \ldots, x_{N-1}) \in \mathbb{R}^N$ then we've seen that $\mathbf{x}$ and its Fourier transform $\widehat{\mathbf{x}}$ are related by

   $$\widehat{\mathbf{x}}(v) = \frac{1}{\sqrt{N}} \sum_{k=0}^{N-1} \mathbf{x}(k) e^{-2\pi i k v / N} \qquad (10.31)$$

   $$\mathbf{x}(k) = \frac{1}{\sqrt{N}} \sum_{v=0}^{N-1} \widehat{\mathbf{x}}(v) e^{2\pi i k v / N}. \qquad (10.32)$$

   Extend $\mathbf{x}$ to a larger interval $k = 0, \ldots, 2N - 1$ by defining

   $$\mathbf{x}(N + k) = \mathbf{x}(N - (k + 1)) \qquad (10.33)$$

   and think of $\mathbf{x}$ now as a vector in $\mathbb{R}^{2N}$. Figure 10.6 is helpful here.

   (a) Transform $\mathbf{x}$ (remember, it's now in $\mathbb{R}^{2N}$) and use (10.33) to show that

   $$\widehat{\mathbf{x}}(v) = \sqrt{\frac{2}{N}} e^{\pi i v / 2N} \sum_{k=0}^{N-1} \mathbf{x}(k) \cos \frac{(2k+1)v\pi}{2N}.$$

   Hint: Apply (10.31) to obtain $\widehat{\mathbf{x}}(v) = (1/\sqrt{2N}) \sum_{k=0}^{2N-1} \mathbf{x}(k) e^{-\pi i k v / N}$. Split the sum $\sum_{k=0}^{2N-1}$ into $\sum_{k=0}^{N-1} + \sum_{k=N}^{2N-1}$ and, in the second sum, use the extension definition $\mathbf{x}(N + v) = \mathbf{x}(N - (v+1))$ together with the fact that $\cos \theta = (e^{i\theta} + e^{-i\theta})/2$.

   Remark: The right-hand side of this equation is defined for any integer v and has period $2N$, providing an extension of $\widehat{\mathbf{x}}$ from $v = 0, \ldots, 2N - 1$ to all of $\mathbb{Z}$ with the property $\widehat{\mathbf{x}}(v + 2N) = \widehat{\mathbf{x}}(v)$.

   (b) Define $\mathbf{y}(v) = e^{-\pi i v / 2N} \widehat{\mathbf{x}}(v)$, for $v \in \mathbb{Z}$ so that the result in part (a) can be written as

   $$\mathbf{y}(v) = \sqrt{\frac{2}{N}} \sum_{k=0}^{N-1} \mathbf{x}(k) \cos \frac{(2k+1)v\pi}{2N}, \qquad v = 0, \ldots, N - 1, \quad (10.34)$$

   Now show that

   (i) when $\mathbf{x}$ is real, so is $\mathbf{y}$,
   (ii) $\mathbf{y}(N) = 0$,
   (iii) and, for all $v \in \mathbb{Z}$, $\mathbf{y}(-v) = \mathbf{y}(v)$ and $\mathbf{y}(v + 2N) = -\mathbf{y}(v)$.

   (c) Argue that $\mathbf{x}$ can be recovered from $\mathbf{y}$ from the relation

   $$\mathbf{x}(k) = \frac{1}{\sqrt{2N}} \mathbf{y}(0) + \sum_{v=1}^{N-1} \sqrt{\frac{2}{N}} \mathbf{y}(v) \cos \frac{(2k+1)v\pi}{2N}. \qquad (10.35)$$

Hint: From (10.31),

$$\mathbf{x}(k) = \frac{1}{\sqrt{2N}} \sum_{v=0}^{2N-1} \widehat{\mathbf{x}}(v)e^{2\pi ikv/2N}$$

$$= \frac{1}{\sqrt{2N}} \sum_{v=0}^{2N-1} e^{\pi iv/2N}\mathbf{y}(k)e^{2\pi ikv/2N}$$

so proceed with an argument similar in spirit with that of part (a).

(d) By defining $\widetilde{\mathbf{y}}(0) = \mathbf{y}(0)/\sqrt{2}$, $C(0) = \sqrt{1/N}$ and $\widetilde{\mathbf{y}}(v) = \mathbf{y}(v)$, $C(v) = \sqrt{2/N}$ for $v = 1, \ldots, N-1$, rewrite (10.34) and (10.35) as

$$\widetilde{\mathbf{y}}(v) = \sum_{k=0}^{N-1} \mathbf{x}(k)C(v)\cos\frac{(2k+1)v\pi}{2N}$$

$$\mathbf{x}(k) = \sum_{v=0}^{N-1} \widetilde{\mathbf{y}}(v)C(v)\cos\frac{(2k+1)v\pi}{2N}.$$

Relabeling $\widetilde{\mathbf{y}}$ with $\widehat{\mathbf{x}}$ gives the symmetric discrete cosine transform (10.20), (10.21).

4. Consider the sine transform on C^N defined by (10.24) and (10.25).

   (a) The right-hand sides of these two equations are defined for all integers $\mathbb{Z}$. Show that these *extensions* of $\widehat{\mathbf{x}}$ and $\mathbf{x}$ have period $2N+2$.
   (b) Show that $\mathbf{x}(N) = 0$ and that $\mathbf{x}(2N+1) = 0$.
   (c) Show that $\mathbf{x}(N+k) = -\mathbf{x}(N-k)$, for $k = 0, \ldots, N$ (see Figure 10.7).

5. Show the general orthogonal transform defined in Section 10.3.1 is an isometry on C^N, i.e., if $\widehat{\mathbf{v}}$ is the (orthogonal) transform of $\mathbf{v}$ then $\|\widehat{\mathbf{v}}\| = \|\mathbf{v}\|$. This shows, at one stroke, that the Fourier, cosine, and sine transforms are all isometries on either C^N or $\mathbb{R}^N$.

10.4 Two-dimensional transforms

The JPEG image compression scheme employs a 2D cosine transform as part of its specification. In this section we develop the 2D cosine transform as a special case of a more general type of 2D transform. Section 10.5 discusses the role this cosine transform plays in the JPEG compression scheme.

The Fourier, cosine, and sine transforms are, at this stage, one-dimensional orthogonal transforms. However, each can easily be extended to a two dimensional transform. A separate 2D extension process could be argued for each one, but it's more efficient (and insightful) to do the argument only once, obtaining a general two-dimensional orthogonal transform. The 2D Fourier, cosine, and sine transforms will then follow as special cases.

There are three ways we've looked at transforms, all of them equivalent:

1. as a list of equations giving explicit instructions on how to compute each component of the transformed vector, e.g., (10.7) and (10.8),

2. as an operator or matrix expression, e.g., (10.7′) and (10.8′), or

3. as a change of basis, e.g., (10.7″) and (10.8″).

The last approach emphasizes a basis choice and defines the path we'll follow here.

Basis elements for 2D orthogonal transforms can be constructed from the basis vectors of orthogonal one-dimensional transforms. The procedure starts with a basis of mutually orthonormal N-vectors, $\{e_0, e_1, \ldots, e_{N-1}\} \subset \mathbb{C}^N$. Thus,

$$\langle e_u, e_v \rangle = \begin{cases} 1, & \text{if } u = v, \\ 0, & \text{if } u \neq v, \end{cases}$$

where the inner product $\langle \cdot, \cdot \rangle$ is defined in Subsection 10.3.1. For each pair of indices u and v from $\{0, \ldots, N-1\}$ we can define an $N \times N$ array f_{uv}, whose entry in the jth row and kth column is given by

$$f_{uv}(j, k) = e_u(j)\overline{e_v(k)}, \quad 0 \leq j, k \leq N - 1. \tag{10.36}$$

There are N^2 of these matrices and they turn out to be mutually orthogonal when regarded as members of $\mathbb{C}^{N^2}$, i.e.,

$$\langle f_{uv}, f_{u'v'} \rangle = \sum_{j,k=0}^{N-1} f_{uv}(j,k)\overline{f_{u'v'}(j,k)}$$

$$= \sum_{j,k=0}^{N-1} e_u(j)\overline{e_v(k)}\overline{e_{u'}(j)\overline{e_{v'}(k)}}$$

$$= \sum_{j=0}^{N-1} e_u(j)\overline{e_{u'}(j)} \sum_{k=0}^{N-1} \overline{e_v(k)}e_{v'}(k)$$

$$= \langle e_u, e_{u'} \rangle \overline{\langle e_v, e_{v'} \rangle} = \begin{cases} 1, & \text{if } u = u' \text{ and } v = v', \\ 0, & \text{otherwise.} \end{cases}$$

This computation also shows that each f_{uv} has unit length, i.e., that $\|f_{uv}\| = 1$. The collection $\{f_{uv} \mid 0 \leq u, v \leq N - 1\}$ of $N \times N$ arrays forms an orthonormal subset of $\mathbb{C}^{N^2}$ and we'll use it as a basis for the space of all $N \times N$ arrays of complex numbers.

Now let f be an $N \times N$ matrix (think of f as a 2D signal—an image). The transform of f, corresponding to the N^2 basis matrices $\{f_{uv} \mid 0 \leq u, v \leq N - 1\}$, is another $N \times N$ matrix $\widehat{f}$, with entries $\widehat{f}(u, v)$ defined by the equation

$$f = \sum_{u,v=0}^{N-1} \widehat{f}(u, v) f_{uv}. \tag{10.37}$$

We've seen this kind of equation before (cf., (10.28)). It uniquely defines $\widehat{f}$ but it doesn't explicitly tell us how to compute the entry in its uth row and vth column, i.e., the entry $\widehat{f}(u, v)$. This is again a linear algebra problem (with N^2 unknowns $\widehat{f}(u, v)$, $0 \leq u, v \leq N - 1$). To solve it we'll again exploit orthonormality of the construction material, i.e., of the basis matrices $\{f_{uv}\}$.

Pick a row index u_0 and a column index v_0. To compute the entry $\widehat{f}(u_0, v_0)$ in the array $\widehat{f}$, proceed in the usual way by taking the ($\mathbb{C}^{N^2}$) inner product of both sides of (10.37) with $f_{u_0 v_0}$:

$$\langle f, f_{u_0 v_0} \rangle = \left\langle \sum_{u,v=0}^{N-1} \widehat{f}(u, v) f_{uv}, f_{u_0 v_0} \right\rangle$$

$$= \sum_{u,v=0}^{N-1} \widehat{f}(u, v) \langle f_{uv}, f_{u_0 v_0} \rangle = \widehat{f}(u_0, v_0).$$

Thus $\widehat{f}(u_0, v_0) = \langle f, f_{u_0 v_0} \rangle$, i.e., the transform coefficient $\widehat{f}(u_0, v_0)$ is just the inner product of f with the basis array $f_{u_0 v_0}$. Expanding this inner product results in the transform formula

$$\widehat{f}(u_0, v_0) = \sum_{j,k=0}^{N-1} f(j, k) \overline{f_{u_0 v_0}(j, k)}.$$

Since u_0 and v_0 are arbitrary row and column indices, then, together with the system (10.37), we have a general transform pair

$$\widehat{f}(u, v) = \sum_{j,k=0}^{N-1} f(j, k) \overline{f_{uv}(j, k)}, \quad 0 \leq u, v \leq N - 1, \tag{10.38}$$

$$f(j, k) = \sum_{u,v=0}^{N-1} \widehat{f}(u, v) f_{uv}(j, k), \quad 0 \leq j, k \leq N - 1. \tag{10.39}$$

10.4.1 The 2D Fourier, cosine, and sine transforms

In this section, we apply (10.38) and (10.39) to the development of two-dimensional versions of the Fourier, cosine, and sine transforms.

The 2D Fourier transform Recall, from the latter part of Section 10.2, that the columns from the Fourier transform transform matrix W form the basis vectors corresponding to the one-dimensional Fourier transform. The kth column $\mathbf{W}_k$

is

$$\mathbf{W}_k = \frac{1}{\sqrt{N}} \begin{bmatrix} 1 \\ e^{2\pi i k/N} \\ e^{2\pi i 2k/N} \\ \vdots \\ e^{2\pi i (N-1)k/N} \end{bmatrix}.$$

Using (10.36), with $\mathbf{W}_k$ in place of $\mathbf{e}_k$, we can construct the basis matrices f_{uv} for the 2D Fourier transform: the (j,k) entry in f_{uv} is given by

$$f_{uv}(j,k) = \mathbf{e}_u(j)\overline{\mathbf{e}_v(k)} = \mathbf{W}_u(j)\overline{\mathbf{W}_v(k)} = \frac{1}{N} e^{2\pi i j u/N} e^{-2\pi i k v/N}.$$

It follows that $f_{uv}(j,k) = (1/N)e^{2\pi i(ju-kv)/N}$ and from (10.38) and (10.39), the two-dimensional Fourier transform takes the form

$$\widehat{f}(u,v) = \frac{1}{N} \sum_{j,k=0}^{N-1} f(j,k)e^{-2\pi i(ju-kv)/N} \tag{10.40}$$

$$f(j,k) = \frac{1}{N} \sum_{u,v=0}^{N-1} \widehat{f}(u,v)e^{2\pi i(ju-kv)/N}. \tag{10.41}$$

We've emphasized before how the one-dimensional Fourier transform is really a map between periodic sequences. The exponential functions in the right-hand sides of equations (10.40) and (10.41) give us a 2D analog: they are defined for any pairs of integers u and v or j and k, and, as such, extend definitions of both $\widehat{f}$ and f to all of $\mathbb{Z} \times \mathbb{Z} = \mathbb{Z}^2$. Moreover, the extension is periodic with period N in both directions.

In the same manner then that the one-dimensional Fourier transform sees signals of length N as single periods of periodic signals of period N, the 2D Fourier transform sees both $\widehat{f}$ and f as maps from $\mathbb{Z}^2$ to $\mathbb{C}$ with the property that $\widehat{f}(u+N,v) = \widehat{f}(u,v+N) = \widehat{f}(u,v)$ and $f(j+N,k) = f(j,k+N) = f(j,k)$ for any (u,v) or (j,k) in $\mathbb{Z}^2$.

The 2D cosine transform From Section 10.3, the vth basis vector $\mathbf{e}_v$ corresponding to the 2D cosine transform is the vth column of the cosine transform matrix A. Thus in (10.36), $\mathbf{e}_k = \mathbf{A}_k$ where

$$\mathbf{A}_k = C(k) \begin{bmatrix} \cos \frac{k\pi}{2N} \\ \cos \frac{3k\pi}{2N} \\ \vdots \\ \cos \frac{(2N-1)k\pi}{2N} \end{bmatrix}.$$

for $0 \leq k \leq N-1$ with $C(0) = 1/\sqrt{N}$ and $C(k) = \sqrt{2/N}$ if $k \neq 0$. The (j,k)

entry in the basis matrix for the 2D cosine transform then looks like

$$f_{uv}(j,k) = \mathbf{A}_u(j)\overline{\mathbf{A}_v(k)} = C(u)\cos\frac{(2j+1)u\pi}{2N}C(v)\cos\frac{(2k+1)v\pi}{2N}.$$

Equations (10.38) and (10.39) then give us the two-dimensional cosine transform pair

$$\widehat{f}(u,v) = \sum_{j,k=0}^{N-1} f(j,k)C(u)\cos\frac{(2j+1)u\pi}{2N}C(v)\cos\frac{(2k+1)v\pi}{2N} \qquad (10.42)$$

$$f(j,k) = \sum_{u,v=0}^{N-1} \widehat{f}(u,v)C(u)\cos\frac{(2j+1)u\pi}{2N}C(v)\cos\frac{(2k+1)v\pi}{2N}. \qquad (10.43)$$

Once again we emphasize that these equations extend $\widehat{f}$ and f to two-dimensional periodic signals (now with period $2N$) defined on $\mathbb{Z}^2$. This is the global signal the cosine transform "sees" when given either $\widehat{f}$ or f.

The extension provided by (10.43) is also *even*, in the sense that it extends f to an even function in both horizontal and vertical directions. This extended f is generally smoother than that provided by the 2D Fourier transform or sine transform. To visualize this extension, picture f as an array and then reflect it across each of its four boundaries. This creates four more arrays, each having the same dimensions. Now reflect each one of these "new" arrays across their boundaries and so on. Continuing this tiling process throughout $\mathbb{Z}^2$ yields the signal that the 2D cosine transform regards as f.

To gain further insight into the 2D-cosine transform, rewrite (10.43) as

$$f = \sum_{u,v=0}^{N-1} \widehat{f}(u,v)B_{uv} \qquad (10.44)$$

where B_{uv} is the $N \times N$ basis element whose entry in the jth row and kth column is

$$B_{uv}(j,k) = C(u)\cos\frac{(2j+1)u\pi}{2N}C(v)\cos\frac{(2k+1)v\pi}{2N}.$$

The collection $\{B_{uv} \mid 0 \le u, v \le N-1\}$ contains the building blocks used to construct f. Each B_{uv} can be regarded as a basic image element.

A greyscale image can be represented as a table of grey-levels, i.e., an array of integers. Conversely, an array of integers can be thought of as a table of grey-levels. Figure 10.8(a) shows the 16 basic image elements B_{uv} arranged in an array indexed by u and v, corresponding to a 4×4 transform. Any given 4×4 image can be written uniquely as a linear combination of these basic images. The (transform) coefficient $\widehat{f}(u,v)$ is a measure of the "presence" of B_{uv} in the overall image. Figure 10.8(b) is an illustration of equation (10.44), showing, in this case, a randomly generated 4×4 image being built up in stages by linear combinations of basic images.[28] The partial sums of (10.44) are formed by

---

[28] Figure A.1 in Appendix A shows a similar example with an 8×8 image.

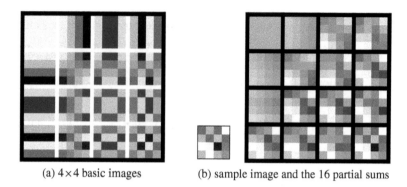

(a) 4×4 basic images   (b) sample image and the 16 partial sums

Figure 10.8: *Image elements for the 2D cosine transform (N = 4), sample image, and the 16 partial sums.*

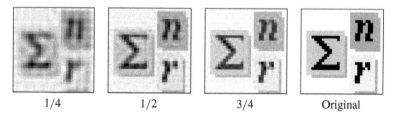

1/4   1/2   3/4   Original

Figure 10.9: *Partial sums build up to the original image.*

following a zigzag sequence through the basis images; Figure 10.12 indicates this type ordering for the 8×8 case.

In another example, Figure 10.9 shows how the cosine transform might be applied to an image with 32×32 pixels. In this case there are $32 \times 32 = 1024$ basic image elements (the B_{uv}s) so it's impractical to display them all in an array setting like Figure 10.8(a). We can still "plot" some of the partial sums though. The original image is on the right and is a representation of the complete sum in (10.44). The three images to its left each correspond to a certain fraction of this sum; that is, they are partial sums. The leftmost image contains the first 1/4, or 256, terms of the sum, the next image contains an additional 256 (the halfway stage), and the next adds 256 more. Scanning from left to right in the figure illustrates how the image sharpens as more and more basic image terms get added to this "running" sum.

The 2D sine transform A two-dimensional sine transform is left as Exercise 10.4.3.

10.4.2 Matrix expressions for 2D transforms

Not surprisingly, the general 2D transform equations (10.38), (10.39) have matrix forms:

$$\widehat{f} = \overline{E}^t f E \tag{10.38'}$$

$$f = E \widehat{f} \overline{E}^t. \tag{10.39'}$$

E has for columns the orthonormal vectors $\mathbf{e}_k$, $k = 0, 1, \ldots, N-1$. Any such matrix E, whose columns form an orthonormal set of vectors, is called a unitary matrix (orthogonal, if E is real) and has the property that $\overline{E}^t E = E \overline{E}^t = I_{N \times N}$, or equivalently, $E^{-1} = \overline{E}^t$.

The two-dimensional versions of the Fourier, cosine, and sine transforms have matrix representations of this kind and are quickly obtained from (10.38') and (10.39'). Before we do this, we need to check that (10.38') and (10.39') are correct (we haven't done this yet!). This turns out to be, for the most part, bookkeeping, but we feel it's instructive. To proceed we'll start with the basic transform (10.38)

$$\widehat{f}(u, v) = \sum_{j,k=0}^{N-1} f(j, k) \overline{f_{u,v}(j, k)}$$

$$= \sum_{j,k=0}^{N-1} f(j, k) \overline{\mathbf{e}_u(j)} \overline{\mathbf{e}_v(k)} = \sum_{j=0}^{N-1} \overline{\mathbf{e}_u(j)} \sum_{k=0}^{N-1} f(j, k) \mathbf{e}_v(k). \tag{*}$$

The inner sum of the above expression can be interpreted as the entry in a matrix product:

$$\sum_{k=0}^{N-1} f(j, k) \mathbf{e}_v(k) = \left[j\text{th row of } f \right] \begin{bmatrix} v\text{th} \\ \text{column} \\ \text{of } E \end{bmatrix} = \left[f E \right] (j, v)$$

where $\left[f E \right] (j, v)$ denotes the (j, v) entry of the (matrix) product $f E$. Inserting this expression back into (*) gives

$$\widehat{f}(u, v) = \sum_{j=0}^{N-1} \overline{\mathbf{e}_u(j)} \left[f E \right] (j, v)$$

$$= \left[u\text{th column of } \overline{E} \right]^t \begin{bmatrix} v\text{th} \\ \text{column} \\ \text{of } f E \end{bmatrix}$$

$$= \left[u\text{th row of } \overline{E}^t \right] \begin{bmatrix} v\text{th} \\ \text{column} \\ \text{of } f E \end{bmatrix} = \left[\overline{E}^t f E \right] (u, v).$$

Thus, $\widehat{f} = \overline{E}^t f E$, for any $N \times N$ matrix f. This is, of course, equation (10.38′) and, since E is unitary, it easily inverts to equation (10.39′).

The matrix equations (10.38′), (10.39′) provide clean descriptions of the 2D Fourier, cosine, and sine transforms and are found below. However, implementation of these transforms are usually coded in a more efficient manner (cf., the FFT in Exercise 10.2.12). Nevertheless, the matrix forms above do make it easy to experiment with images, especially if using a matrix-oriented mathematical software package such as MATLAB or Octave.

2D Fourier transform The one-dimensional Fourier transform matrix W in (10.8′) is symmetric, i.e., has the property that $W^t = W$. From (10.38′) and (10.39′), with $E = W$, we see that the 2D Fourier transform can be written as

$$\widehat{f} = \overline{W} f W \tag{10.40′}$$

$$f = W \widehat{f} \overline{W}. \tag{10.41′}$$

2D cosine transform The cosine transform matrix A is real (see Section 10.3), hence $\overline{A} = A$. The matrix form of the 2D cosine transform is then

$$\widehat{f} = A^t f A \tag{10.42′}$$

$$f = A \widehat{f} A^t. \tag{10.43′}$$

2D sine transform The sine transform is even simpler. The matrix B corresponding to the sine transform (Section 10.3 again) is real and symmetric, i.e., $\overline{B} = B$ and $B^t = B$. Thus, the 2D sine transform is

$$\widehat{f} = B f B \tag{10.45}$$

$$f = B \widehat{f} B. \tag{10.46}$$

Exercises 10.4

1. Show that the process of cosine transforming an $N \times N$ image f is equivalent to first taking the 1D cosine transform of each column followed by the 1D cosine transform of the resulting rows. *Hint:* $\widehat{f} = A^t f A = (A^t (A^t f)^t)^t$.

   Does a similar process describe the 2D Fourier transform? The general 2D transform (10.38′)?

2. In Section 10.4.1 the even 2D signal f seen by the cosine transform was described in a visual or geometric fashion. Describe the extension of f provided by the 2D Fourier transform.

3. Show that the two-dimensional sine transform equations have the form

$$\widehat{f}(u,v) = \frac{2}{N+1} \sum_{j,k=0}^{N-1} f(j,k) \sin \frac{\pi(j+1)(u+1)}{N+1} \sin \frac{\pi(k+1)(v+1)}{N+1}$$

$$f(j,k) = \frac{2}{N+1} \sum_{u,v=0}^{N-1} \widehat{f}(u,v) \sin \frac{\pi(j+1)(u+1)}{N+1} \sin \frac{\pi(k+1)(v+1)}{N+1}$$

10.5 An application: JPEG image compression

The compression methods discussed in Chapters 5–9 can be used on image data. In fact, the popular GIF format uses an LZW scheme to compress 256-color images. Portable Network Graphics (PNG) is more sophisticated and capable, using a predictor (or filter) to prepare the data for a *gzip*-style compressor. However, applications using high resolution images with thousands of colors may require more compression than can be achieved with these lossless methods.

Lossy schemes discard some of the data in order to obtain better compression. The problem, of course, is deciding just what information is to be compromised. Loss of information in compressing text is typically unacceptable, although simple schemes such as elimination of every vowel from English text may find application somewhere. The situation is different with images and sound: some loss of data may be quite acceptable, even imperceptible.

In the 1980s, the Joint Photographic Experts Group (JPEG) was formed to develop standards for still-image compression. The specification includes both lossless and lossy modes, although the latter is perhaps of the most interest (and is usually what is meant by "JPEG compression"). This section will consider only the ideas of the lossy mode, applied to greyscale images.[29]

The method in lossy JPEG depends for its compression on an important mathematical and physical theme: local approximation. Both mathematical and physical objects are often easier to understand and examine when analyzed locally. The JPEG group took this idea and fine-tuned it with results gained from studies on the human visual system. The resulting scheme enjoys wide use, in part because it is an open standard, but mostly because it does well on a large class of images, with fairly modest resource requirements.

The ideas can be illustrated with a greyscale image; that is, a matrix of integer values representing levels of grey. The range of values isn't important in understanding the mathematical ideas, although it is common to restrict values to the interval $[0, 255]$, giving a total of 256 levels of grey. The 'bird' at left in Figure 10.10 shows an image containing 256×256 pixels with 145 shades of grey represented.

Portions of this image appear to contain relatively constant levels of grey. Working *locally*, we could collapse these almost-constant regions to their average shade of grey. Aesthetic questions aside for now, suppose we do this, that is, partition the 256×256 'bird' image into 1024 8×8 blocks and replace each of the 8×8 pixel blocks with its average shade of grey. The resulting image appears on the right in Figure 10.10. The original 256×256 array of numbers has been reduced to a 32×32 array, or to $1/64$ of its original size ($64 = 256^2/32^2$).

On certain (mostly uninteresting) portions of the image this simple method works quite well but, of course, considerable detail has been lost in several key

---

[29]See Section 10.7 for some remarks on color.

Figure 10.10: *Block-averaging applied to 'bird'.*

areas. This idea of working locally, though, does seem to have merit. However, as Figure 10.10 so clearly shows, at the very least it needs considerable refinement before it can be thought of as a viable method. We could refine the block size, i.e., go to a smaller than 8×8 block, but doing so could sacrifice compression; in fact, it would perhaps be better to use the largest block size we could get away with.[30] It's tempting to imagine some sort of adaptive method that uses large blocks when possible but goes to smaller blocks in image areas of high detail,[31] perhaps carving the picture up into odd shapes (like a jigsaw puzzle) as the method progresses through the image. However, unless done elegantly, this complication could add considerable baggage to the information required for image reconstruction and thus prove self-defeating. Instead of block size modification, JPEG simply chooses to preserve more detail in an 8×8 block whenever it determines detail is too important to throw away.

The "detail detector" built into JPEG is the 2D cosine transform of Section 10.4. The cosine transform (or, for that matter, any Fourier transform) exchanges raw image (spatial) information directly for information about frequency content. An 8×8 block is built up with basic 8×8 cosine block images of increasing detail. There are 64 of these image elements, each of which is displayed in Figure A.1 of Appendix A.

Figure 10.9 illustrates approximation by sums of the basic cosine block images ($N = 32$) described in that section. In the sum, terms have been ordered so that the "tail" contains the high-frequency information. Roughly speaking, each successive term in the sum adds a little more detail. Stopping the sum at a certain point amounts to truncating subsequent (high) frequencies from the original block, and is equivalent to replacing the appropriate entries in the trans-

---

[30]Of course, in the extreme case of reducing block size down to 1×1, the image will not change nor will there be any compression.

[31]In a sense, wavelet techniques attempt this.

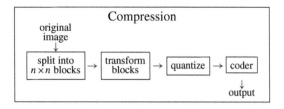

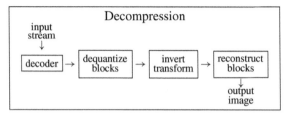

Figure 10.11: *A schematic of the JPEG process.*

formed matrix with zeros. Discarding these trailing zeros and retaining only the nonzero coefficients corresponds to a compression method—and can even be considered a special case of JPEG.[32]

JPEG exploits the idea of local approximation for its compression: 8×8 portions of the complete image are transformed using the cosine transform and then the information retained in each block is *quantized* by a method which tends to suppress higher-frequency elements. Figure 10.11 is a schematic of JPEG and JPEG-like compression schemes. Below is a quick summary of the ideas behind JPEG.

1. *Work locally.* Carve the image into smaller $k \times k$ blocks. In the case of JPEG, an $m \times n$ image is split up into 8 pixel by 8 pixel blocks, i.e., $k = 8$. These blocks are usually very small pieces of the entire image, e.g., in a 256×256 pixel image, an 8×8 block occupies only $100(8^2/256^2) = .098\%$ of the picture area.

2. *Transform.* Each block is transformed to expose spatial frequencies (detail) within. JPEG uses the cosine transform and expresses the original image in terms of 64 basic "cosine" images of fixed horizontal and vertical spatial frequencies; see Figure A.1.

3. *Quantize.* A "rounding" procedure is performed which reduces magnitudes of the transformed coefficients. Typically more aggressive reduction is performed on coefficients corresponding to high-frequency components. The coefficients which quantize to zero correspond to frequen-

---

[32]The amount of compression is complicated by the fact that the entries in the original and transformed matrices are not in the same range, but the main idea is correct. Also, to be precise, the approximation is a special case of JPEG only if the image is 8×8 and the entries in the quantizer can be chosen sufficiently large.

cies omitted during block reconstruction. This is the "lossy" step of JPEG.

4. *Encode*. The output of step 3 is compressed with a lossless scheme. Huffman and arithmetic coding are specified in the JPEG standard.

We've already said something about the benefits of working on small blocks within an image. But there is an inherent weakness in any local approach that does *not* take into account the rest of the image: if small blocks are processed one at a time, removing information in a way that ignores the rest of the image, then we shouldn't be surprised to find discontinuities between neighboring blocks after the image has been, block by block, reassembled. The question is whether or not they are noticeable—they certainly can be. For example, in Figure 10.10 these *blocking artifacts* can be seen just about everywhere (of course, the simplistic block-averaging scheme used there wiped clean all detail from every block).

In searching for a tool that would selectively allow more detail to remain in a block the JPEG group found the cosine transform to have some desirable properties. It's relatively easy to compute, depending only on the dimension of the block to be transformed, and is computed in the same way throughout the image.[33] The coefficients of a cosine transformed array are also arranged in a "natural" order from the low to high frequencies. However, it is perhaps the cosine transform's "smoothing" effect, as much as anything, that helps us to see that the JPEG group made a good choice. Each of the Fourier transforms, including the cosine transform, views an image (signal) as defined everywhere on $\mathbb{Z}^2$, but the cosine transform does not generally introduce sharp transients the others may, cf., Section 10.4.1. This property allows for the design of reliable quantizers and their stable implementation: if the cosine transform sees a block as containing high frequencies, then the high frequencies are likely to be genuine, that is, they probably haven't been artificially introduced by the transform process itself. Typical images are largely continuous and locally smooth so this "extended vision" of the cosine transform often does a decent job at predicting or guessing a few pixels into a block's immediate surrounding.[34] Even so, the very act of quantizing transformed block by transformed block results in abrupt changes to these blocks, producing discontinuities that ultimately will be passed back onto the image. In practice, artifacts between blocks from a JPEG processed image are not very noticeable unless aggressive quantizing has been used. In this case further smoothing may be desirable.[35]

In the JPEG procedure, information is lost at the quantizing stage; the other steps are invertible.[36] After the transformed coefficients of a block have been

---

[33]The KLH transform depends on the data being transformed and requires on-the-fly adjustment from one block to the next. [2]

[34]The "smoothing" effect of the cosine transform is discussed and compared with the Fourier and sine transforms in Section 10.3.

[35]A smoothing strategy is discussed in Appendix A.

[36]In applications, there are also roundoff errors when transforming, but these are usually minor compared with the information loss from quantizing.

quantized, the original block cannot generally be recovered. This trade-off allows JPEG to obtain typical compression ratios of 20:1 or better with little noticeable image degradation. Compare this to ratios of, say, 2 or 3 to 1 for the lossless GIF or PNG methods.

Generically, the word *quantize* refers to the process of slicing up or partitioning continuous objects (intervals of real numbers for us) into sub-pieces (subintervals) and matching each sub-piece with some member of a discrete set. We do this all the time when working with numbers. Here are some examples: rounding to the nearest integer, flooring, truncating a real number after its third decimal place, or replacing a positive real number with the integer part of its logarithm. In each of these cases, either the real numbers or some subinterval of real numbers has been replaced by a discrete set together with a map containing instructions on which member of the discrete set we should assign to a given real number. If we were to sketch the graph of the map associated with any of the above examples we would see a series of "steps," i.e., a *step-function*.

JPEG's scheme quantizes individual ranges of each coefficient in the cosine transform with high frequencies more aggressively quantized than low frequencies. The scheme can be described as follows: each 8×8 transformed block Tx is associated with an 8×8 array q of positive integers—an array of "quantizers" referred to as a quantizing matrix. In the simplest case, the matrix q is fixed for each block in the image. Each entry in Tx is then divided by its corresponding integer entry in q and the result rounded to the nearest integer. Provided the quantizer entries are large enough, the effect of this process is, quite frequently, a very sparse matrix.

One quantizer that is frequently used with JPEG is the luminance matrix

$$q = \begin{bmatrix} 16 & 11 & 10 & 16 & 24 & 40 & 51 & 61 \\ 12 & 12 & 14 & 19 & 26 & 58 & 60 & 55 \\ 14 & 13 & 16 & 24 & 40 & 57 & 69 & 56 \\ 14 & 17 & 22 & 29 & 51 & 87 & 80 & 62 \\ 18 & 22 & 37 & 56 & 68 & 109 & 103 & 77 \\ 24 & 35 & 55 & 64 & 81 & 104 & 113 & 92 \\ 49 & 64 & 78 & 87 & 103 & 121 & 120 & 101 \\ 72 & 92 & 95 & 98 & 112 & 100 & 103 & 99 \end{bmatrix}.$$

Each entry in this array is based on a visual threshold of its corresponding basis element, see Figure A.1 and [57]. The smaller entries of q are generally found in its upper left-hand corner and the larger entries in the lower right. In any (cosine) transformed block Tx, the "low frequency" coefficients are located towards its upper left-hand corner and the "high frequency" coefficients towards its lower right-hand corner. Thus, the effect of quantizing Tx with q is to suppress the higher frequency signals in the original block x. The design of the luminance table q is typical of other "JPEG quantizers."

After the quantizing step is finished, the entries in the output array are ordered, from low to high frequency, trailing zeros are truncated, and the resulting string encoded. Figure 10.12 indicates the low to high frequency ordering of a transformed (and transformed-quantized) block. In this ordering, an entry cor-

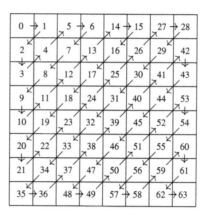

Figure 10.12: *Ordering of the block entries.*

responds to the amplitude of a frequency that is at least as high as the frequency of its predecessors.

Quantizing considerably reduces the number of distinct values that quantized coefficients can assume. Repetition of values among quantized coefficients is likely to be found. In practice, zero is the most commonly repeated value and it's usually the case that all but a small handful of the 64 coefficients in the block get quantized to zero. In such cases the block will have a string representation consisting of a few nonzero quantized coefficients delimiting "long" strings of zeros and, since high frequencies are targeted aggressively, trailed by zeros to the end of the block. There is no need to encode the trailing zeros, only to mark where they begin in the block. The typical result is a transformed-quantized block with a string representation containing far fewer than 64 coefficients. Moreover, a JPEG encoder, designed to exploit the form of this string, is waiting to compress it even further.

The process (omitting the encoder) on a given matrix x follows the diagram

$$x \xrightarrow{transform} Tx \xrightarrow{quantize} QTx \xrightarrow{dequantize} T\tilde{x} \xrightarrow{invert} \tilde{x}$$

where T is the cosine transform, defined by $Tx = A^t x A$ with A, the 8×8 cosine transform matrix, given (to a few places of accuracy) by:

$$A = \begin{bmatrix} .35 & .35 & .35 & .35 & .35 & .35 & .35 & .35 \\ .49 & .42 & .28 & .10 & -.10 & -.28 & -.42 & -.49 \\ .46 & .19 & -.19 & -.46 & -.46 & -.19 & .19 & .46 \\ .42 & -.10 & -.49 & -.28 & .28 & .49 & .10 & -.42 \\ .35 & -.35 & -.35 & .35 & .35 & -.35 & -.35 & .35 \\ .28 & -.49 & .10 & .42 & -.42 & -.10 & .49 & -.28 \\ .19 & -.46 & .46 & -.19 & -.19 & .46 & -.46 & .19 \\ .10 & -.28 & .42 & -.49 & .49 & -.42 & .28 & -.10 \end{bmatrix} .$$

Let's follow the process on a particular 8×8 matrix x, taken from part of the smooth background in the 'Lena' image in Figure A.3. The background in

'Lena' has little shade variation (translation: little or no high-frequency presence), so we expect x to compress well. x is at top left of the following diagram:[37]

$$
\begin{bmatrix}
102 & 104 & 105 & 110 & 111 & 110 & 118 & 115 \\
104 & 106 & 107 & 103 & 107 & 106 & 109 & 108 \\
104 & 107 & 107 & 111 & 108 & 107 & 109 & 116 \\
104 & 104 & 110 & 107 & 113 & 113 & 112 & 111 \\
105 & 104 & 108 & 106 & 111 & 111 & 108 & 108 \\
104 & 107 & 108 & 109 & 109 & 107 & 107 & 108 \\
105 & 103 & 108 & 109 & 108 & 108 & 111 & 111 \\
104 & 107 & 106 & 109 & 107 & 112 & 105 & 111
\end{bmatrix}
\qquad
\begin{bmatrix}
104 & 105 & 106 & 107 & 109 & 110 & 111 & 112 \\
104 & 105 & 106 & 107 & 109 & 110 & 111 & 112 \\
104 & 105 & 106 & 107 & 109 & 110 & 111 & 112 \\
104 & 105 & 106 & 107 & 109 & 110 & 111 & 112 \\
104 & 105 & 106 & 107 & 109 & 110 & 111 & 112 \\
104 & 105 & 106 & 107 & 109 & 110 & 111 & 112 \\
104 & 105 & 106 & 107 & 109 & 110 & 111 & 112 \\
104 & 105 & 106 & 107 & 109 & 110 & 111 & 112
\end{bmatrix}
$$

$T\downarrow$ $\uparrow (QT)^{-1}$

$$
\begin{bmatrix}
864.0 & -17.0 & -3.8 & -3.4 & 0.5 & -1.1 & 0.7 & 1.2 \\
1.9 & -5.2 & 2.4 & -0.8 & -0.6 & 0.9 & -3.8 & 3.0 \\
-0.9 & -2.6 & 2.7 & -1.4 & 0.3 & 1.5 & -2.5 & -3.0 \\
-0.5 & -0.9 & -1.5 & 1.6 & -0.8 & 2.4 & -2.5 & 2.9 \\
3.8 & -4.5 & -2.6 & 4.1 & -1.2 & -0.6 & 1.6 & -0.1 \\
5.9 & -6.1 & -0.6 & -1.5 & 1.4 & 3.5 & -1.3 & 1.1 \\
2.5 & -0.3 & -0.3 & -3.3 & 2.6 & -1.3 & -1.9 & -4.5 \\
1.0 & -1.9 & 1.3 & -1.4 & 2.6 & 1.3 & -0.2 & -1.4
\end{bmatrix}
\xrightarrow{Q}
\begin{bmatrix}
54 & -2 & 0 & 0 & 0 & 0 & 0 & 0 \\
0 & 0 & 0 & 0 & 0 & 0 & 0 & 0 \\
0 & 0 & 0 & 0 & 0 & 0 & 0 & 0 \\
0 & 0 & 0 & 0 & 0 & 0 & 0 & 0 \\
0 & 0 & 0 & 0 & 0 & 0 & 0 & 0 \\
0 & 0 & 0 & 0 & 0 & 0 & 0 & 0 \\
0 & 0 & 0 & 0 & 0 & 0 & 0 & 0 \\
0 & 0 & 0 & 0 & 0 & 0 & 0 & 0
\end{bmatrix}
$$

The encoder receives its information from the sparse matrix QTx. The entries in the block QTx are ordered (cf., Figure 10.12) and the tokens: 54, -2, EOB are sent on to the encoder for further compression (EOB denotes an end of block marker). The information contained in this string permits the array QTx to be reconstructed exactly. At the receiving end, this is the only information there is about the original x. From this information, the decoder can produce the block $\widetilde{x}$ at the top right of the diagram. We remark that because 8×8 blocks make up such a relatively small piece of most images, the sparseness of QTx may be typical of a large fraction of the total number of blocks in an image.

It's interesting to note that approximation $\widetilde{x}$ has only shade variation in the horizontal direction and not the vertical—this observation is easy to spot from the transformed-quantized matrix QTx and the cosine basis image shown in the first row and second column of Figure A.1. The entry-by-entry difference between x and its JPEG replacement $\widetilde{x}$ is

$$
x - \widetilde{x} =
\begin{bmatrix}
-2 & -1 & -1 & 3 & 2 & 0 & 7 & 3 \\
0 & 1 & 1 & -4 & -2 & -4 & -2 & -4 \\
0 & 2 & 1 & 4 & -1 & -3 & -2 & 4 \\
0 & -1 & 4 & 0 & 4 & 3 & 1 & -1 \\
1 & -1 & 2 & -1 & 2 & 1 & -3 & -4 \\
0 & 2 & 2 & 2 & 0 & -3 & -4 & -4 \\
1 & -2 & 2 & 2 & -1 & -2 & 0 & -1 \\
0 & 2 & 0 & 2 & -2 & 2 & -6 & -1
\end{bmatrix}.
$$

The eye, of course, is the best device to measure this error.

---

[37]Note that $(QT)^{-1}$ is used here to indicate dequantizing followed by the inverse transform; however, the quantizing step $Tx \mapsto QTx$ is not invertible, so this is a slight abuse of notation. Also, the result of the inverse transform has been rounded.

Here is another example. The 8×8 block z in the top left of the following diagram is taken from the same 'Lena' image but from a region where detail is prevalent:

$$\begin{bmatrix} 150 & 151 & 155 & 169 & 164 & 149 & 156 & 171 \\ 156 & 158 & 161 & 162 & 156 & 157 & 172 & 174 \\ 161 & 145 & 150 & 160 & 164 & 175 & 168 & 155 \\ 140 & 133 & 154 & 163 & 163 & 154 & 159 & 164 \\ 140 & 151 & 163 & 156 & 145 & 156 & 168 & 172 \\ 158 & 154 & 142 & 141 & 155 & 165 & 165 & 146 \\ 154 & 142 & 142 & 154 & 161 & 152 & 149 & 157 \\ 147 & 145 & 151 & 156 & 144 & 141 & 158 & 168 \end{bmatrix} \quad \begin{bmatrix} 146 & 158 & 170 & 168 & 158 & 153 & 159 & 168 \\ 161 & 153 & 148 & 156 & 169 & 173 & 164 & 153 \\ 158 & 150 & 145 & 153 & 167 & 173 & 165 & 155 \\ 139 & 151 & 162 & 161 & 154 & 152 & 162 & 173 \\ 139 & 150 & 160 & 158 & 151 & 150 & 161 & 173 \\ 157 & 147 & 139 & 145 & 159 & 167 & 162 & 154 \\ 159 & 148 & 139 & 143 & 156 & 164 & 159 & 151 \\ 144 & 153 & 159 & 154 & 144 & 142 & 154 & 166 \end{bmatrix}$$

$$\Big\downarrow T \qquad\qquad\qquad\qquad \Big\uparrow (QT)^{-1}$$

$$\begin{bmatrix} 1245.9 & -37.3 & 0.5 & -6.4 & 10.6 & 10.4 & -1.1 & 1.8 \\ 26.4 & -3.7 & -7.7 & -5.0 & 2.6 & 1.4 & 0.0 & -0.5 \\ -0.8 & 9.3 & 6.7 & -9.4 & 13.9 & -3.5 & -5.7 & -1.9 \\ -1.8 & 5.6 & 4.1 & 1.4 & -5.5 & -8 & -4.5 & -0.3 \\ -6.9 & -8.2 & -3.4 & -31.7 & 6.4 & 1.3 & 0.7 & 0.5 \\ -10.6 & 1.0 & -17.7 & 12.9 & 18.4 & -3.6 & 2.7 & 2.0 \\ -0.9 & 4.1 & -0.2 & 19.3 & -6.7 & 1.8 & -0.9 & -1.7 \\ 4.8 & -0.2 & -0.2 & -1.8 & -0.6 & 0.3 & 3.7 & -4.1 \end{bmatrix} \xrightarrow{Q} \begin{bmatrix} 78 & -3 & 0 & 0 & 0 & 0 & 0 & 0 \\ 2 & 0 & -1 & 0 & 0 & 0 & 0 & 0 \\ 0 & 1 & 0 & 0 & 0 & 0 & 0 & 0 \\ 0 & 0 & 0 & 0 & 0 & 0 & 0 & 0 \\ 0 & 0 & 0 & -1 & 0 & 0 & 0 & 0 \\ 0 & 0 & 0 & 0 & 0 & 0 & 0 & 0 \\ 0 & 0 & 0 & 0 & 0 & 0 & 0 & 0 \\ 0 & 0 & 0 & 0 & 0 & 0 & 0 & 0 \end{bmatrix}$$

QTz is still sparse but a much longer string

$$78, -3, 2, 0, 0, 0, 0, -1, 1, \underbrace{0, \dots, 0}_{23 \text{ zeros}}, -1, EOB$$

is sent to the encoder. However, the JPEG standard requires the encoder to run-length encode the two stretches of zeros, so, in the end, the block will still compress well. Detail is more important in z than in x and, correspondingly, the JPEG process keeps more of it.

Software from the Independent JPEG Group was used to compress 'bird' at several "quality" levels, and the results are displayed in Figure 10.13. The sizes are given in bits per pixel (bpp); i.e., the number of bits, on average, required to store each of the numbers in the matrix representation of the image. The sizes for the GIF and PNG versions are included for reference.[38]

Exercises 10.5

1. Define a quantizing matrix

$$q = \begin{bmatrix} 3 & 5 & 7 & 9 \\ 5 & 7 & 9 & 11 \\ 7 & 9 & 11 & 13 \\ 9 & 11 & 13 & 15 \end{bmatrix}.$$

For each x below, compute the transformed matrix Tx and then the quantized matrix $QTx = \mathrm{round}(Tx./q)$, where we have borrowed the following

----

[38] 'bird' is part of a proposed collection of standard images at the Waterloo BragZone, and has been modified for this textbook.

(a) original test image
4.9 bpp GIF, 4.3 bpp PNG

(b) 1.3 bpp JPEG

(c) .74 bpp JPEG

(d) .34 bpp JPEG

(e) .16 bpp JPEG

(f) difference between (a) and (e)

Figure 10.13: *GIF, PNG, and JPEG compression on 'bird'.*

MATLAB notation.

- If A is a matrix, then round(A) is the matrix obtained from A by round-ing each of its entries to the nearest integer.
- If B is a matrix of the same dimensions as A, then $A./B$ is the matrix obtained by dividing each entry of A by the corresponding entry of B.

(a)

$$x = \begin{bmatrix} 160 & 160 & 160 & 160 \\ 160 & 160 & 160 & 160 \\ 160 & 160 & 160 & 160 \\ 160 & 160 & 160 & 160 \end{bmatrix}$$

(b)

$$x = \begin{bmatrix} 160 & 160 & 160 & 161 \\ 160 & 160 & 161 & 162 \\ 160 & 161 & 162 & 163 \\ 161 & 162 & 163 & 164 \end{bmatrix}$$

(c)

$$x = \begin{bmatrix} 160 & 0 & 0 & 0 \\ 160 & 0 & 0 & 0 \\ 160 & 160 & 0 & 0 \\ 160 & 160 & 160 & 0 \end{bmatrix}$$

It has been said that JPEG doesn't do well on cartoons. Why would someone say this and what do they really mean? Is it a "true" state-ment? Does this block and how it transforms shed any light on the matter? The block in part (a) "transforms" well and you could easily find many like it in a cartoon image. Dequantize QTx to get $T\tilde{x}$, and compare with Tx. Now do an inverse cosine transform on $T\tilde{x}$. Does $\tilde{x}$ "look" anything like the matrix x you started with?

(d)

$$x = \begin{bmatrix} 54 & 70 & 182 & 81 \\ 183 & 1 & 240 & 227 \\ 33 & 106 & 61 & 167 \\ 23 & 7 & 46 & 38 \end{bmatrix}$$

From a compression viewpoint QTx doesn't look very promising. What do you think happened? Hint: MATLAB's rand() command was used to generate x. Is this result expected?

(e) Exchange the quantizing matrix q with a more aggressive quantizer of your own design. Using it, repeat items (a)–(d). For example, you could simply scale up q (multiply q by a number larger than 1), use a piece (corner) of the JPEG's luminance matrix, selectively grab thresh-olds from the luminance quantizer by matching frequencies with the 4×4 basis elements, or even make up one of your own.

2. (*Project*) The 8×8 transform size in JPEG was chosen for several reasons, including hardware considerations and the desire to take advantage of local behavior. Larger transform sizes may offer the possibility of better compression at a given "quality" level, especially in high-resolution images. Section A.1 contains a simple example of using other transform sizes on the 'Lena' image.

Can JPEG benefit from a larger transform size (ignoring hardware costs)? This is a more difficult question than can be answered here, but is well suited for experiment. Choose one or two test images, and compress with a "typical" JPEG 8×8 scheme. Then attempt to match the image quality, but obtain superior compression, with larger transform sizes. You will have to find suitable quantizing matrices for the larger transforms. In addition, you will need to determine a way to measure compression. This could be a simple counting of trailing zeros, a compression with some off-the-shelf lossless compressor on the output of your scheme, or a modification of the JPEG entropy coder (the first two of these are easy but not ideal; the third could involve some time). Include some samples, and write a short summary on the experiments.

3. (*Project*) One troublesome aspect of JPEG-like schemes is the appearance of blocking artifacts. Section A.1 discusses a smoothing procedure proposed in [57]. In brief, the scheme on a specific 8×8 block looks at nearest-neighbor block averages in order to adjust some of the low-frequency AC coefficients (subject to a certain clamping). Implement such a scheme (or adapt the supplied MATLAB scripts). There are several areas for experimentation: the number of coefficients considered for smoothing, the clamping condition, and the polynomial approximation itself. Attempt to do better than the example given in Figure A.5.

10.6 A brief introduction to wavelets

The issue at heart in this chapter is really one of signal representation. From a compression point of view, we would like to represent a signal efficiently, that is, as a linear combination of basis elements using as few as possible. Since our signals are discrete and finite, then the signal representation problem is naturally modeled from a linear algebra approach with the problem's mathematical setting being a linear space (of "signals") together with a suitable choice of basis elements. The linear space needs to be large enough to include all signals we expect to encounter. For us, this has amounted to selecting $\mathbb{R}^N$ (or $\mathbb{C}^N$) for some large enough integer N.[39] Thus, the real artwork in the subject comes down to choosing basis vectors with which to describe the signals of the space.

---

[39] An $N \times M$ array (image) can be identified with a vector in $\mathbb{R}^{MN}$.

As alluded to in the preface to this chapter, there are many choices.

Features of a signal that we especially wish to examine can guide us in our quest for the "right" basis vectors. For example, our development of the Fourier transform basis vectors in Section 10.2 was, in a sense, a consequence of our search for basic "frequencies" with which we could resolve periodic signals. Before starting off on this search we needed to query just what it is we should regard as "pure" tones or signals of fundamental frequencies. Although there were details left to stumble over, once this question was answered, the hard work was really finished because it was at this point where the "linear algebra" machine took control and eventually led us to the (discrete) Fourier basis elements $\mathbf{W}_k$, $k = 0, 1, \ldots, N - 1$ described at the end of the section. Interestingly enough, and not obvious in the process of their development, the vectors $\{\mathbf{W}_k\}$ ended up being orthogonal. We've since seen that much of the mathematical convenience of the Fourier transform stems from the linear algebra of orthogonal expansions. However, requiring the basis to be orthogonal was not something we imposed and did nothing to guide us to these special vectors: the matter at issue was the synthesis of arbitrary signals with linear combinations of a small, fundamental set of signals with "known" frequencies; orthogonality was a bonus.

The Fourier transform is a very important tool, indispensable in the realm of signal analysis. When used as a compression device, though, we might sometimes wish it had the additional capacity of being able to highlight local frequency information—generally, it doesn't. The coefficient of $\mathbf{W}_k$ in the Fourier expansion of a signal may yield information about the overall strength of the frequency (vector) $\mathbf{W}_k$ in the signal, but this information is global: even if a coefficient is substantial, it doesn't normally give us any clue as to what time interval(s) over which the corresponding frequency is significant.

As an example, consider a signal that is flat for a time, then rises to oscillate rapidly over a short period of time, and then again becomes flat.[40] Omit a coefficient from the Fourier transform of such a signal and you may have trouble reconstructing it well. In such a situation, it could be advantageous to have, at our disposal, basis elements that reflect this sort of localization property. Perhaps then we would need fewer of them to describe such a signal—certainly a desirable situation from a compression standpoint.

The Fourier transform is a general signal analysis tool and as such it is not too difficult to find special cases where it may not be optimal to use. There have been attempts to adapt the Fourier transform to better handle local information. They stretch from JPEG's approach at cutting signals into small pieces, processing them one at a time, to generating new basis elements from the Fourier elements by taking their product with smooth cutoff functions ("windowed" Fourier transforms, cf. [13]) to the study of wavelets.

The interest in and use of wavelet transforms has grown appreciably in the recent years since Ingrid Daubechies [12] demonstrated the existence of

---

[40]Scanning left to right along a horizontal line in the bird image could yield such a signal.

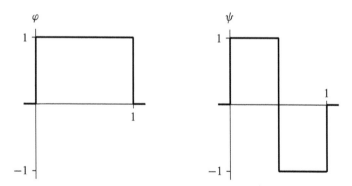

Figure 10.14: *The Haar scaling function φ and mother wavelet ψ.*

continuous (and smoother) wavelets with compact support.[41] They have found homes as theoretical devices in mathematics and physics and as practical tools applied to a myriad of areas including the analysis of surfaces, image editing and querying, and, of course, image compression.

Our goal in this section is to introduce a very simple wavelet family, the Haar wavelet, and apply it to an image compression problem. We'll use the 'bird' image from the last section as our test image and keep the presentation in line with the "linear algebra of orthogonal expansions" theme used in the rest of this chapter. The Haar example can be presented, somewhat superficially, without the theoretical structure necessary to understand wavelets more fully; for this same reason it is also an incomplete introduction to the method. We use it here mainly because of its accessibility and also because it really does work as an image compression device. However, the Haar wavelet is not nearly the whole story on wavelets and we refer the interested reader to several excellent sources on theory and further applications of wavelets, in particular see [13, 14, 45, 52, 54, 58, 68, 78, 79].

Perhaps the simplest example of wavelets are the Haar wavelets. Start with the Haar *scaling* function

$$\varphi(x) = \begin{cases} 1, & \text{if } 0 \leq x < 1, \\ 0, & \text{otherwise.} \end{cases} \tag{10.47}$$

The mother Haar wavelet, ψ, is defined by

$$\psi(x) = \begin{cases} 1, & \text{if } 0 \leq x < 1/2, \\ -1, & \text{if } 1/2 \leq x < 1, \\ 0, & \text{otherwise.} \end{cases}$$

The adjective "mother" should become clearer presently. Figure 10.14 contains the graphs of these two functions.

Note that both of these functions are "finitely" supported and orthogonal on

---

[41] The support of a function is defined as the closure of the set of points over which it is nonzero.

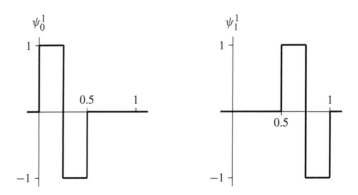

Figure 10.15: *Wavelets ψ_0^1 and ψ_1^1.*

$\mathbb{R}$. In fact, their common support *is* the interval $[0, 1]$ and they are orthogonal there as well. The unit interval is the place where all activity occurs and from now on we confine ourselves to it. The wavelet ψ takes on the two values 1 and -1 on $[0, 1]$ so it seems natural to identify ψ with the ordered pair $(1, -1)$. Over $[0, 1]$, φ is constant $(= 1)$ and so could be regarded, by itself, as a basis for the real numbers $\mathbb{R}$, or, if φ is identified with the vector $(1, 1)$, the pair

$$\{\varphi, \psi\} = \left\{ \begin{bmatrix} 1 \\ 1 \end{bmatrix}, \begin{bmatrix} 1 \\ -1 \end{bmatrix} \right\}$$

is an orthogonal basis for $\mathbb{R}^2$.

Further subdivision of the unit interval $[0, 1]$ allows us to generate an orthogonal basis for $\mathbb{R}^4$ by defining two additional wavelets

$$\psi_0^1(x) = \begin{cases} 1, & \text{if } 0 \le x < 1/4, \\ -1, & \text{if } 1/4 \le x < 1/2, \\ 0, & \text{otherwise.} \end{cases} \qquad \psi_1^1(x) = \begin{cases} 1, & \text{if } 1/2 \le x < 3/4, \\ -1, & \text{if } 3/4 \le x < 1, \\ 0, & \text{otherwise.} \end{cases}$$

Graphs of these wavelets appear in Figure 10.15. Note the size of their supports is half that of the mother wavelet ψ. In fact, these new wavelets are just offsprings of ψ in the sense that $\psi_0^1(x) = \psi(2x)$ and $\psi_1^1(x) = \psi(2x - 1)$. The interval $[0, 1]$ is divided into fourths by the two wavelets ψ_0^1 and ψ_1^1 and we may think of φ, ψ, ψ_0^1, and ψ_1^1 as the 4-tuples $(1, 1, 1, 1)$, $(1, 1, -1, -1)$, $(1, -1, 0, 0)$, and $(0, 0, 1, -1)$, respectively. In this case the collection

$$\{\varphi, \psi, \psi_0^1, \psi_1^1\} = \left\{ \begin{bmatrix} 1 \\ 1 \\ 1 \\ 1 \end{bmatrix}, \begin{bmatrix} 1 \\ 1 \\ -1 \\ -1 \end{bmatrix}, \begin{bmatrix} 1 \\ -1 \\ 0 \\ 0 \end{bmatrix}, \begin{bmatrix} 0 \\ 0 \\ 1 \\ -1 \end{bmatrix} \right\}$$

is a orthogonal basis for $\mathbb{R}^4$.

We can continue to add wavelets in this fashion. For example, add four more wavelets ψ_0^2, ψ_1^2, ψ_2^2, and ψ_3^2 to the set $\{\varphi, \psi, \psi_0^1, \psi_1^1\}$ by dividing the unit interval into eighths and defining, for $k = 0, 1, 2, 3$,

$$\psi_k^2(x) = \begin{cases} 1, & \text{if } 2k/8 \le x < (2k+1)/8, \\ -1, & \text{if } (2k+1)/8 \le x < (2k+2)/8, \\ 0, & \text{otherwise.} \end{cases}$$

Observe that $\psi_k^2(x) = \psi(4x - k)$ for $k = 0, 1, 2, 3$. From inspection we can see that the set $\{\varphi, \psi, \psi_0^1, \psi_1^1, \psi_0^2, \psi_1^2, \psi_2^2, \psi_3^2\}$ forms an orthogonal basis for $\mathbb{R}^8$ when each is identified with an 8-vector in the above way

$$(\varphi, \psi, \psi_0^1, \psi_1^1, \psi_0^2, \psi_1^2, \psi_2^2, \psi_3^2)$$

$$= \left(\begin{bmatrix} 1 \\ 1 \\ 1 \\ 1 \\ 1 \\ 1 \\ 1 \\ 1 \end{bmatrix}, \begin{bmatrix} 1 \\ 1 \\ 1 \\ 1 \\ -1 \\ -1 \\ -1 \\ -1 \end{bmatrix}, \begin{bmatrix} 1 \\ 1 \\ -1 \\ -1 \\ 0 \\ 0 \\ 0 \\ 0 \end{bmatrix}, \begin{bmatrix} 0 \\ 0 \\ 0 \\ 0 \\ 1 \\ 1 \\ -1 \\ -1 \end{bmatrix}, \begin{bmatrix} 1 \\ -1 \\ 0 \\ 0 \\ 0 \\ 0 \\ 0 \\ 0 \end{bmatrix}, \begin{bmatrix} 0 \\ 0 \\ 1 \\ -1 \\ 0 \\ 0 \\ 0 \\ 0 \end{bmatrix}, \begin{bmatrix} 0 \\ 0 \\ 0 \\ 0 \\ 1 \\ -1 \\ 0 \\ 0 \end{bmatrix}, \begin{bmatrix} 0 \\ 0 \\ 0 \\ 0 \\ 0 \\ 0 \\ 1 \\ -1 \end{bmatrix} \right).$$

The supports of these new wavelets ψ_k^2 are half that of those just one "resolution" level lower, that is, half that of the ψ_k^1. If this set is used as a basis for $\mathbb{R}^8$ then coefficients of the ψ_k^2 yield information about local detail in a signal.

Normalizing these eight vectors produces an orthonormal basis for $\mathbb{R}^8$

$$\{\varphi/\sqrt{8}, \psi/\sqrt{8}, \psi_0^1/2, \psi_1^1/2, \psi_0^2/\sqrt{2}, \psi_1^2/\sqrt{2}, \psi_2^2/\sqrt{2}, \psi_3^2/\sqrt{2}\}.$$

The corresponding (wavelet) transform matrix (cf. Section 10.3.1) has the form

$$H_3 = \begin{bmatrix} 1/\sqrt{8} & 1/\sqrt{8} & 1/2 & 0 & 1/\sqrt{2} & 0 & 0 & 0 \\ 1/\sqrt{8} & 1/\sqrt{8} & 1/2 & 0 & -1/\sqrt{2} & 0 & 0 & 0 \\ 1/\sqrt{8} & 1/\sqrt{8} & -1/2 & 0 & 0 & 1/\sqrt{2} & 0 & 0 \\ 1/\sqrt{8} & 1/\sqrt{8} & -1/2 & 0 & 0 & -1/\sqrt{2} & 0 & 0 \\ 1/\sqrt{8} & -1/\sqrt{8} & 0 & 1/2 & 0 & 0 & 1/\sqrt{2} & 0 \\ 1/\sqrt{8} & -1/\sqrt{8} & 0 & 1/2 & 0 & 0 & -1/\sqrt{2} & 0 \\ 1/\sqrt{8} & -1/\sqrt{8} & 0 & -1/2 & 0 & 0 & 0 & 1/\sqrt{2} \\ 1/\sqrt{8} & -1/\sqrt{8} & 0 & -1/2 & 0 & 0 & 0 & -1/\sqrt{2} \end{bmatrix}$$

In general, Haar wavelets of arbitrarily fine resolution can be generated from the mother wavelet ψ through dyadic shifts and scales of its argument. More precisely, for a non-negative integer k (resolution level) we can define a wavelet

$$\psi_j^k(x) = \psi(2^k x - j), \quad \text{for } j = 0, 1, \ldots, 2^k - 1.$$

With this notation, $\psi_0^0 = \psi$ and the 2^{k+1} vectors

$$\{\varphi\} \cup \{\psi_j^{k'} \mid k' = 0, 1, \ldots, k; \, j = 0, 1, \ldots, 2^{k'} - 1\}$$

form an orthogonal basis for $R^{2^{k+1}}$. Identifying them with (column) vectors in $\mathbb{R}^{2^{k+1}}$ then normalizing and using them as columns in a matrix H_{k+1} gives a

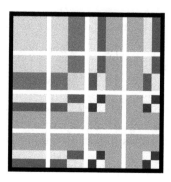

Figure 10.16: *The 4×4 Haar basis elements.*

wavelet transform (on $\mathbb{R}^{2^{k+1}}$)

$$\widehat{\mathbf{v}} = H_{k+1}^t \mathbf{v} \tag{10.48}$$

$$\mathbf{v} = H_{k+1} \widehat{\mathbf{v}}. \tag{10.49}$$

10.6.1 2D Haar wavelets

Because the Haar wavelets are orthogonal, the machinery for producing a 2D Haar wavelet transform from the 1D transform is already in place (cf., Section 10.4). If f is an array (an image) of size $\mathbb{R}^{2^{2(k+1)}}$ and H_{k+1} is the 1D Haar wavelet transform matrix of (10.48)–(10.49), then the 2D Haar transform $\widehat{f}$ of f is given by

$$\widehat{f} = H_{k+1}^t f H_{k+1}.$$

The original image f is recovered from its transform $\widehat{f}$ by

$$f = H_{k+1} \widehat{f} H_{k+1}^t.$$

The basis arrays for the 2D Haar wavelet transform are given by the $2^{2(k+1)}$ matrices

$$\varphi(j)\varphi(j'), \quad \varphi(j)\psi_u^v(j'), \quad \psi_u^v(j)\varphi(j'), \quad \psi_u^v(j)\psi_{u'}^{v'}(j'),$$

where $0 \le v, v' \le k, 0 \le u \le 2^v - 1, 0 \le u' \le 2^{v'} - 1$, and $0 \le j, j' \le 2^{k+1} - 1$ are the row and column indices.

The 16 basis images at resolution level $k = 1$ are shown in Figure 10.16. They form a 2D Haar basis for the set of 4×4 matrices. Compare these with the cosine transform elements in Figure 10.8. One can begin to see the formation of elements with localized supports even at this "coarse" resolution level.

Image compression with wavelets

When an image is expanded with Haar wavelets, the coefficient of the "scaling" array $\varphi(j)\varphi(j')$, $0 \le j, j' \le 2^k - 1$, is just a scaled average of all values within the image. Coefficients of the other 2D wavelet arrays in the expansion are usually called detail coefficients.[42] The simple (lossy) compression scheme that we'll describe here is not as elaborate as the quantizing scheme used in JPEG. Basically, we throw away any detail coefficient meeting a very simple criterion:

(1) start with an image f and a tolerance, say $\epsilon > 0$,

(2) wavelet transform f to $\widehat{f}$ and replace with zero any coefficient in $\widehat{f}$ whose magnitude is less than ϵ.

More elaborate criteria can also be used.[43]   In Figure 10.17 we have used this simple scheme on 'bird', at several tolerance settings. Compare with Figure 10.13, where JPEG has been used at different settings to compress this same image. Setting a coefficient to zero in the transformed image is equivalent to eliminating the corresponding basis array in the expansion of the image—it's another way of saying that that particular basis element is not thought important enough to keep in the expansion of the overall image.

Unlike JPEG, wavelets have been presented as a method that transforms the entire image at once, not a "block" at a time. Pedagogically this makes for a clean description of the process even though it may not always be the best way to think about it. Also, this approach can involve fairly large matrices. However, wavelet matrices are generally quite sparse and not too taxing for machines of even modest performance. Even so, for applications where speed is important, e.g., motion, fast wavelet transform algorithms exist, cf. [54].

Figure 10.17 illustrates a certain kind of simple-minded partial sum (projection) approach to compression. Examples of more sophisticated wavelet schemes appear in Figures 10.18 and 10.19. These were generated using Geoff Davis' Wavelet Image Compression Construction Kit (see Appendix C). Davis cautions, "The coder is not the most sophisticated—it's a simple transform coder—but each individual piece of the transform coder has been chosen for high performance." Figure 10.18 uses a Haar wavelet scheme, and Figure 10.19 uses a wavelet family from [3], which is the default in the Kit and different from the Haar wavelets.

Exercises 10.6

1. Apply the Haar wavelet transform to each of the matrices in Exercise 10.5.1. Choose some threshold of your own and discard the appropriate coefficients. Now invert and compare your results with the JPEG results, in particular, on the array in part (c).

---

[42] This terminology is analogous to the DC and AC coefficients of the cosine transform mentioned in Appendix A.

[43] l^1 and l^2 schemes are discussed in [15] and [68]. Coefficient quantizing schemes are also used.

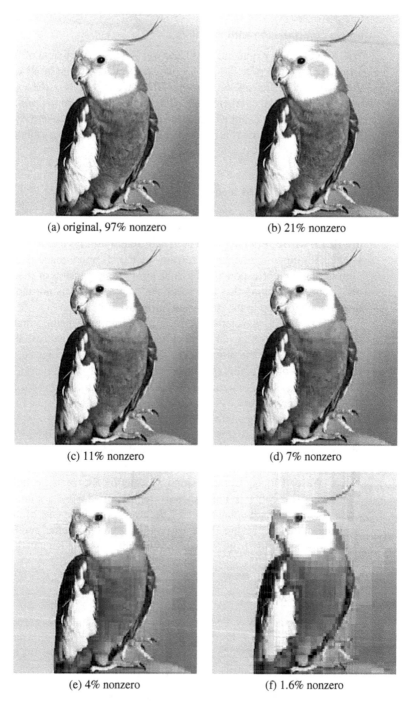

(a) original, 97% nonzero

(b) 21% nonzero

(c) 11% nonzero

(d) 7% nonzero

(e) 4% nonzero

(f) 1.6% nonzero

Figure 10.17: *'bird' (256×256) using Haar wavelet transform with simple thresholding. In (b)–(f), the percentage indicates the number of nonzero coefficients in the transformed array after a threshold condition has been applied.*

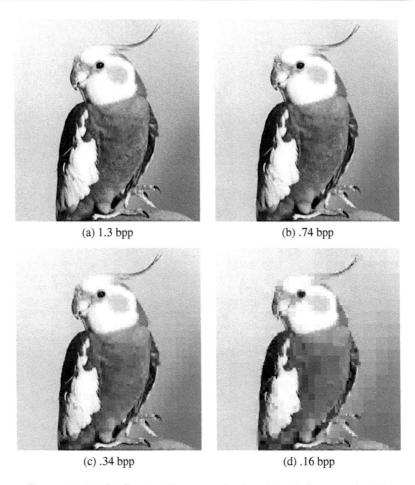

(a) 1.3 bpp   (b) .74 bpp

(c) .34 bpp   (d) .16 bpp

Figure 10.18: *'bird' using Haar wavelet from Davis' Construction Kit.*

2. Find constants c_n so that the Haar scaling function φ in (10.47) satisfies the *scaling equation*

$$\phi(x) = \sum_{n \in \mathbb{Z}} c_n \phi(2x - n).$$

Solutions to this equation are known as scaling functions and used in the construction of wavelet families. *Hint*: Think geometrically, i.e., look at Figure 10.14.

(a) 1.3 bpp   (b) .74 bpp

(c) .34 bpp   (d) .16 bpp

Figure 10.19: *'bird' using Davis' Construction Kit with a wavelet from [3].*

10.7 Notes

On color   The discussion of JPEG and wavelets has centered on greyscale images. Color images may identify a red, green, and blue triple (R, G, B) for each of the pixels, although other choices are possible. Color specified in terms of brightness, hue, and saturation, known as luminance-chrominance representations, may be desirable from a compression viewpoint, since the human visual system is more sensitive to errors in the luminance component than in chrominance [57]. Given a color representation, JPEG and wavelet schemes can be applied to each of the three planes.

On JPEG, fractals, and wavelets JPEG enjoys an open, freely usable standard (other than the arithmetic coding option). This is a significant advantage in comparison with fractal and wavelet methods. To displace JPEG will require a scheme with real gains in speed, compression, and/or quality.

Fractal methods were advertised as one such scheme, and very good compression has been achieved on some test images. There are other interesting features (such as a certain resolution independence), although the scheme can be somewhat intractable in practice, and the early claims for fractal methods were probably exaggerated. Yuval Fisher [20] offers a more cautionary assessment, noting two deficiencies: the encoding is computationally expensive and the encoded image gets very large as perfect reconstruction is approached.

Wavelet and fractal methods are often said to be superior to JPEG at low bit rates, although this generalization needs to be qualified. Tom Lane, organizer of the Independent JPEG Group (IJG), writes:

> ...the limitations of JPEG the standard ought not be confused with the limitations of a particular implementation of JPEG. There are hardly any JPEG codecs available that are optimized for very low quality settings. Certainly the IJG code is not (though I hope to do something about that in the next release). You may see a lot of blockiness in the current IJG encoder's output at low [quality] settings, but you should not conclude that JPEG is incapable of doing better than that.[44]

The problem, in part, is that the quantization is generally done by simple scaling of the suggested JPEG quantizing matrices. In an earlier post, Lane remarked:

> ...the usual technique involves scaling the sample tables mentioned in the JPEG standard up or down by some given ratio. This works [reasonably well] for scale ratios around 0.5 to 1.0 (that's Q 50 to Q 75 in the IJG software, for example) but loses badly at much higher or lower settings. The spec's sample tables are only samples anyway—much more is known about quantization-table design now than was true when the spec was drafted. Not a lot of that research has propagated into shipping products, though.[45]

In addition, compression improvements could be obtained on the encoding side without breaking existing decoders:

> ...just because the decoder will reconstruct DCT coefficients [with simple multiplication] doesn't mean the encoder must form the encoded values by simple division. Adobe's (formerly Storm's) encoder uses this idea a little bit, but it could be taken much further. In particular, you can do "poor man's variable quantization" this way, *without* breaking compatibility with existing decoders, just by zeroing out coefficients that shouldn't be zero according to a strict encoder.

---

[44] comp.compression newsgroup post, 30 Apr 1997.

[45] comp.compression newsgroup post, 2 Sep 1996. Quoted by permission.

This kind of adaptive quantization is discussed [57]. The problem of block-ing artifacts could be addressed on the decoding side, possibly along the lines outlined in [57] and Appendix A.

One can imagine having three open standards, and, perhaps with some human intervention, choosing the best scheme for a given image and given "quality" criteria. However, the climate has changed somewhat since the JPEG standard was developed, now that companies have discovered that the US Patent Office will (in essence) grant patents on algorithms.

Appendix A

JPEGtool User's Guide

This appendix describes the "JPEGtool" package of scripts used to study aspects of JPEG (or JPEG-like) image compression. Scripts for Matlab[1] and Octave[2] (and an optional Maple[3] script) are provided which perform, for example, an $N \times N$ discrete cosine transform of a matrix (image) and quantization. Section A.1 illustrates the use of the package, and Section A.2 provides a synopsis.

At the simplest level, a standard JPEG transform and quantization scheme can be requested with a command of the form

```
jpeg('bird.pgm')
```

The result can easily be displayed on-screen with standard Octave or Matlab commands. More interesting use of the package includes display of partial sums (as matrices or images), experiments with the transform size N and the quantization matrices, and "smoothing" filters to reduce blocking artifacts.

Requirements

Matlab or Octave is required. Matlab is available for Macintosh computers, OS/2, Microsoft Windows, and many Unix-like platforms including Linux. Student editions for some platforms are also available, although these versions may place low ceilings on the size of matrices which can be manipulated. Section A.3 contains information on obtaining Octave.

Maple was used to illustrate the calculation of coefficients in the smoothing program (which attempts to remove blocking artifacts), but it is not required. It should be routine to convert this for use with one of the other mathematical packages (MuPAD[4] is a possibility).

Installation

The scripts, along with various test images in the proper form, can be obtained by anonymous ftp from www.dms.auburn.edu in pub/compression. Users of

---

[1] The Math Works, Inc. On-line information available through http://www.mathworks.com.

[2] Roughly speaking, Octave is a Matlab-like tool running on many Unix-like platforms and OS/2, and is freely-distributable under the GNU Public License (see Section A.3).

[3] Maple is a general-purpose symbolic algebra system from Waterloo Maple Software. On-line information is available through http://www.maplesoft.com.

[4] See http://www.mupad.de or *MuPAD User's Manual*, John Wiley & Sons: 0-471-96716-5, 1996.

web browsers can retrieve these using the location

http://www.dms.auburn.edu/compression

The files are "packaged" in various formats, in order to simplify installation. The files also appear individually in the jpegtool/src subdirectory.

The files should be unpackaged if necessary and placed in a separate directory or folder on your machine. Matlab or Octave must then be made aware of the location of these scripts (on some platforms, it suffices to set the location as the "working directory"). The precise methods for doing these tasks depends on the platform, and will not be discussed in this appendix.

Some sample images are delivered with the scripts, including several which have been widely used as test images for various articles on compression. Other images can be used, but the scripts currently require that these be in *portable graymap* format. Many utilities can provide conversions between various graphics formats and graymaps. Under Unix-like platforms, Poskanzer's Pbmplus toolkit[5] and *xv* are commonly used.

A.1 Using the tools

This section describes some ways that the image tools can be used with Octave or Matlab to study JPEG-like image compression. Strictly speaking, a graphical display is not required, although most users will want to experiment with actual images rather than just looking at the matrices.

An "image" in this context is simply a matrix of integers ranging from 0 to 255 (representing levels of gray). There are many ways to generate such images in Matlab or Octave, but typically the starting point is a "real" image or picture which has been saved in portable graymap format.

Approximation by partial sums

We begin with a simple example of the use of these scripts. As discussed in Chapter 10, the cosine transform exchanges spatial information for frequency information. If the transform is 8×8, then a given 8×8 portion of an image can be written as a linear combination of the 64 basis matrices which appear in Figure A.1(a). The transform provides the coefficients in the linear combination, allowing approximations or adjustments to the original image based on frequency content. Partial sum approximations are a special case. Often, the higher-frequency information in an image is of less importance to the eye, so if the terms are ordered roughly according to increasing frequency, then partial sum approximations may do well even with relatively few terms.

---

[5]Extended Portable Bitmap Toolkit. Netpbm is based on the Pbmplus distribution of 10 Dec 91, and includes improvements and additions.

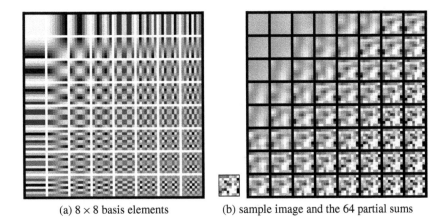

| (a) 8 × 8 basis elements | (b) sample image and the 64 partial sums |

Figure A.1: *8 × 8 basis elements, sample image, and the 64 partial sums.*

Let's take a specific 8×8 example. (Users of the student edition of Matlab may need to use a 5×5 matrix.) We'll use '>' to denote the prompt printed by Matlab or Octave, but this will vary by platform. Define the test image:

```
> x = round(rand(8)*255)   % 8x8 random matrix, integer entries in [0,255]
```

This will display some (random) matrix, perhaps

$$x = \begin{bmatrix} 64 & 80 & 76 & 59 & 157 & 123 & 90 & 237 \\ 252 & 109 & 214 & 220 & 83 & 194 & 181 & 3 \\ 130 & 176 & 10 & 91 & 154 & 148 & 112 & 95 \\ 153 & 124 & 149 & 26 & 29 & 199 & 60 & 228 \\ 92 & 166 & 107 & 166 & 108 & 233 & 234 & 111 \\ 91 & 32 & 10 & 190 & 248 & 231 & 160 & 4 \\ 25 & 128 & 255 & 16 & 198 & 209 & 235 & 1 \\ 89 & 217 & 195 & 107 & 213 & 119 & 103 & 183 \end{bmatrix}$$

and we can view this "image" with

```
> imagesc(x)      % Matlab users
> imagesc(x, 8)   % Octave users
```

Something similar to the smaller image at the lower left in Figure A.1(b) will be displayed.[6] Now ask for the matrix of partial sums (the larger image in Figure A.1(b)):

```
> imagesc(psumgrid(x)); % Display the 64 partial sums
```

The partial sums are built up from the basis elements in the order shown in the zigzag sequence of Figure A.2. This path through A.1 is approximately according to increasing frequency of the basis elements.

Roughly speaking, the image in Figure A.1(b) is the *worst* kind as far as JPEG compression is concerned. Since it is random, it will likely have significant high-frequency terms. We can see these by transforming:

---

[6]Octave users may see a somewhat blurred image; if so, and if *xv* is your viewer, try the "r" command while the image window is active.

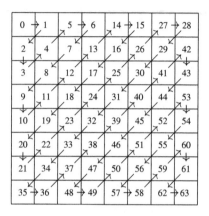

Figure A.2: *The zigzag sequence.*

```
> Tx = dct(x)    % discrete cosine transform of x
```
For the example above, this gives the matrix

$$Tx = \begin{bmatrix}
1062.8 & -69.2 & -68.1 & 117.2 & -107.0 & -33.3 & 22.5 & 5.2 \\
-44.7 & 26.7 & 117.0 & -3.4 & 96.9 & 49.7 & 46.5 & -82.4 \\
17.9 & 25.2 & -39.1 & -81.1 & 18.4 & -54.4 & -5.4 & 112.0 \\
-23.4 & -115.5 & -112.2 & 68.9 & 9.3 & 73.0 & -25.7 & 8.5 \\
11.5 & -65.7 & 146.3 & -149.9 & 43.7 & -126.2 & 58.5 & -41.2 \\
-104.5 & -82.1 & 61.4 & -27.9 & -36.9 & -128.5 & 67.1 & 74.1 \\
-77.0 & -71.7 & -16.9 & 50.6 & 170.4 & -115.3 & -90.4 & -54.3 \\
5.5 & -42.3 & -4.4 & -5.7 & -77.5 & -40.3 & -102.2 & 21.9
\end{bmatrix}$$

of coefficients used to build the partial sums in Figure A.1 from the basis elements. The top left entry gets special recognition as the *DC coefficient*; the
others are the *AC coefficients*, $AC_{0,1}$ through $AC_{7,7}$.

The terms in the lower right of Tx correspond to the high-frequency portion
of the image. Notice that even in this "worst" case, Figure A.1 suggests that a
fairly good image can be obtained with somewhat less than all of the 64 terms.
A similar example with 4×4 basis elements appears in Figure 10.8 on page 278.

It would be a good exercise at this stage to repeat the above steps with
some nicer matrix. A constant matrix might be "too nice," but something "less
random" than the example may be appropriate.

The process of approximation by partial sums is applied to a "real" image in
Figure 10.9 on page 278, where $1/4$, $1/2$, and $3/4$ of the 1024 terms for a 32×32
image are displayed. Our approximations retain all of the frequency information
corresponding to terms from the zigzag sequence below some selected threshold
value; the remaining higher-frequency information is discarded. Although this
can be considered a special case of a JPEG-like scheme, JPEG allows more
sophisticated use of the frequency information.

On to JPEG

The 'Lena' image of Figure A.3 has been widely used in publications on compression. However, it may be too large for some configurations; if so, the 32×32 'math4.pgm' image can be substituted in the examples. Perform the transform and quantization schemes of JPEG, and view the result:

```
> setdefaults;          % Set default quantizer, etc.
> y = jpeg('lena.pgm'); % Do JPEG transform and quantization scheme
> imagesc(y);           % Display the resulting image
```

At this point, the image in Figure A.3(b) should be visible. This isn't a very exciting use of the tools, but it does illustrate the mechanics.[7]

Let's examine the process more carefully. Recall that JPEG compression works by transforming an image so that the frequency information is directly available, and then quantizing in a way that tends to suppress some of the high-frequency information and also so that most of the terms can be represented in fewer bits. To "recover" the image, there is a dequantizing step followed by an inverse transform. (We've ignored the portion of JPEG which does lossless compression on the output of the quantizer, but this doesn't affect the image quality.)

In the above example, the default quantizer stdQ is used. If we do the individual steps which correspond to the above fragment, we might write

```
> setdefaults;            % Set the default quantizer, etc.
> x = getpgm('lena.pgm'); % Get a graymap image
> Tx = dct(x);            % Do the 8x8 cosine transform
> QTx = quant(Tx);        % Quantize, using standard 8x8 quantizer
> Ty = dequant(QTx);      % Dequantize
> y = invdct(Ty);         % Recover the image
> imagesc(y);             % Display the image
```

It should be emphasized that we cannot recover the image completely—there has been loss of information at the quantizing stage. It is illustrative to compare the matrices x and y. The difference image $x - y$ for this kind of experiment appears in Figure 10.13 on page 289. There is considerable interest in measuring the "loss of image quality" using some function of these matrices. This is a difficult problem, given the complexity of the human visual system.

This still isn't very exciting. In JPEG-like schemes, there are (at least) two obvious places to experiment: choice of quantizer, and size of the block used in the transform (which also affects the choice of quantizer).

Adjusting the quantizer

The choice of quantizer can, of course, greatly affect the results, both in terms of compression and quality. The default quantizer in these tools is the standard

---

[7]To be precise, a rounding procedure should be done on the matrix y. In addition, we have ignored the zero-shift specified in the standard, which affects the quantized DC coefficients.

JPEG 8×8 luminance matrix

$$
stdQ =
\begin{bmatrix}
16 & 11 & 10 & 16 & 24 & 40 & 51 & 61 \\
12 & 12 & 14 & 19 & 26 & 58 & 60 & 55 \\
14 & 13 & 16 & 24 & 40 & 57 & 69 & 56 \\
14 & 17 & 22 & 29 & 51 & 87 & 80 & 62 \\
18 & 22 & 37 & 56 & 68 & 109 & 103 & 77 \\
24 & 35 & 55 & 64 & 81 & 104 & 113 & 92 \\
49 & 64 & 78 & 87 & 103 & 121 & 120 & 101 \\
72 & 92 & 95 & 98 & 112 & 100 & 103 & 99
\end{bmatrix}.
$$

We can see the effects of more (or less) quantizing by scaling this matrix (or choosing an entirely different matrix). In addition, we can ask jpeg or quant to report the amount of compression. For example, using the standard quantizer,

```
> x = getpgm('lena.pgm');   % Store the original image in x
> [y, r] = jpeg(x);         % jpeg, result in y and ratio in r
> r                         % Print the compression ratio
r = 73.0
```

where the last line represents the compression achieved at the quantizing stage.[8] Now we'll use a more aggressive quantizer (namely, stdQ*2):

```
> [z, r] = jpeg(x, stdQ*2);  % jpeg, more aggressive quantizer
> r                          % Print the compression ratio
r = 82.8
```

The compression is better, but at the cost of some degradation in the image quality. The images are in x, y, and z, respectively, and can be displayed with the image commands shown earlier. It may also be illustrative to examine the errors $x - y$ and $x - z$.

Many of the tools can be called with different numbers of arguments, with different types of arguments in a given position, and with different return values. This is typical of routines in Octave or Matlab, and can be very convenient. It is important to understand how the tools choose the quantizer and blocksize if these are not explicitly specified in the call. For example, if quant is called without a second argument (i.e., there is no quantizer specified), then a global quantizer QMAT is used (where QMAT defaults to stdQ). The quantizer used by quant becomes the new value for QMAT, and is used by default in other functions such as dequant. Hence, the following sequence of calls would give the expected results:

```
> q = stdQ*2; Tx = dct(x, length(q));
> QTx = quant(Tx, q); dequant(QTx);
```

This "works" because the call to quant will set q as the new value for the global quantizer QMAT. Since dequant is called without specifying a quantizer as the second argument, the routine will use QMAT as the quantizer. (In this example, it would be preferable to set QMAT directly, and then omit passing the blocksize to dct and the quantizer to quant.)

Of course, it would also be "legal" to replace the second line with the sequence

---

[8]See the reference section A.2 for the interpretation of the compression percent.

```
QTx = quant(Tx, stdQ*2); dequant(QTx, stdQ);  % wrong
```

This may lead to interesting results, but it is probably not what was intended, since quant and dequant are using different quantizers. Similarly, the fragment

```
Tx = dct(x, 16); quant(Tx, stdQ);  % wrong
```

is "legal," but probably undesirable since the transform is on 16×16 blocks but stdQ is 8×8. As a final example, note that a call such as jpeg(x,stdQ*2) will set the global quantizer. If this call is followed by jpeg(x), then the result will be the same as the first call.

Adjusting the blocksize

The second obvious area for experimentation is the blocksize used in the transform. There are a number of reasons given for the 8×8 blocksize, including hardware constraints. One heuristic consideration mentioned is that larger blocks are more likely to include portions of the image which are very different, in contrast to the "roughly constant" blocks on which JPEG does best. This is, of course, dependent on the resolution, but it is perhaps reasonable by today's standards.

On the other hand, better compression may be achieved on some images if a larger blocksize is chosen. Changing the blocksize, however, also changes the quantizer. As an experiment, we examine the 256×256 'Lena' image under the standard JPEG 8×8 scheme and with modified schemes using transforms of sizes 16×16 and 32×32. For the 8×8 quantizer, the suggested JPEG luminance quantizer was used. For the other quantizers, matrices were chosen with the typical properties of a quantizer; e.g., entries increase from the top left to the bottom right. Matrices with this rough property can be obtained from the Hilbert matrix, available in Matlab or Octave with the 'hilb' function. The fragment for the 16×16 experiment appears below:

```
> x = getpgm('lena.pgm');  % Get the Lena image
> q = 6 ./ hilb(16) + 26   % Possible 16x16 quantizer
> y = jpeg(x, q);          % Do 16x16 transform
> imagesc(y);              % Matlab users
> imagesc(y, 2);           % Octave users
```

Figure A.3 shows the results. An analysis of the compression and the image quality needs to be done before making any definitive statements. In addition, there are some serious questions about our choice of quantizers for the larger blocksizes. However, it is perhaps safe to say that the 32×32 transform is too large for this particular example. It's not hard to see why there is serious degradation: a 32×32 block covers a relatively large portion of the image, and much of the "local" property on which JPEG relies has been lost.

A JPEG enhancement

As a final example of the use of these scripts, we consider an enhancement to JPEG described in Pennebaker and Mitchell [57]. One troublesome aspect of

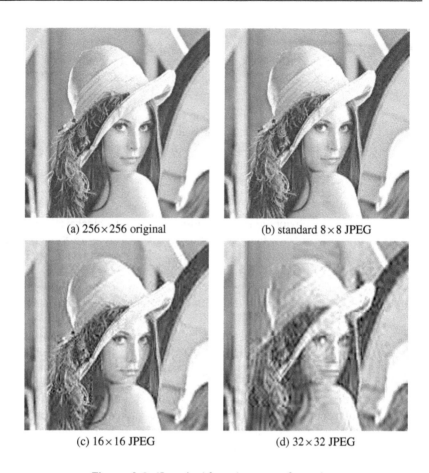

(a) 256×256 original   (b) standard 8×8 JPEG

(c) 16×16 JPEG   (d) 32×32 JPEG

Figure A.3: *'Lena' with various transform sizes.*

JPEG-like schemes is the appearance of "blocking artifacts," the telltale discontinuities between blocks which often follow aggressive quantizing. The image on the left in Figure A.5 was produced using stdQ*4 as the quantizer. Clearly visible blocks can be seen, especially in the "smoother" areas of the image.

Since the DC coefficients represent the (scaled) average value over the block, it might be reasonable to use the nearest-neighbor coefficients to smooth a given block by predicting the low-frequency AC coefficients. Any low-frequency AC coefficients which are zero will be replaced by the predicted values. However, the replacement values should be "clamped" so that (in magnitude) they do not exceed one-half of the corresponding value in the

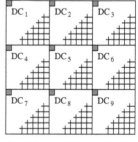

quantizer (values larger than this would not have quantized to zero).

Think of the original image X as a surface with height at (y,x) given by $X(y,x)$. For a given $N \times N$ block (the block corresponding to DC_5 in the grid), the 3×3 superblock consisting of its nearest neighbors contains $3^2 N^2$ total entries. Fit a polynomial

$$p(y,x) = a_1 x^2 y^2 + a_2 x^2 y + a_3 xy^2 + a_4 x^2 + a_5 xy + a_6 y^2 + a_7 x + a_8 y + a_9$$

by requiring that the average value over the kth submatrix equal DC_k (this gives nine equations for the unknowns $a_1, ..., a_9$). The polynomial defines a surface over the center block, which approximates the corresponding portion of the original surface.[9] Figure A.4 shows a surface X in (a) and the polynomial approximations in (b).

The polynomial approximation is fed through the cosine transform, giving AC "predictor" coefficients (in terms of the DC coefficients) for the corresponding portion of the original surface. The first two such predictors (which can be obtained from from the 'deblockc' Maple script) are given by

$$AC_{0,1} = \alpha(DC_4 - DC_6) \quad \text{and} \quad AC_{1,0} = \alpha(DC_2 - DC_8),$$

where

$$\alpha = \tfrac{\sqrt{2}}{128}\left(\cos\tfrac{7\pi}{16} + 3\cos\tfrac{5\pi}{16} + 5\cos\tfrac{3\pi}{16} + 7\cos\tfrac{\pi}{16}\right) \approx 0.14235657.$$

The decoder, which only has the quantized information from the original surface, uses these predictors to "guess" suitable values for the low-frequency AC coefficients (subject to the clamping described above). Figure A.4 illustrates the process, where the lowest five AC coefficients were considered for smoothing. The procedure applied to an aggressively quantized 'bird' appears in Figure A.5.

As an elementary example of this smoothing process, we can consider a single 3×3 superblock. Since the smoothing process is done on the matrix which results from

$$x \xrightarrow{transform} Tx \xrightarrow{quantize} QTx \xrightarrow{dequantize} Ty,$$

we may as well define Ty directly. Let's take Ty to be zero, except for the 9 DC coefficients (which we take to be 1–9, respectively):

```
> Ty = zeros(8*3);          % Initialize 24x24 superblock
> for k=1:9                 % 9 DC coefficients to define
> r = floor((k-1)/3)*8+1;   % Determine proper (row,col) of DC_k
> c = rem(k-1,3)*8+1;
> Ty(r, c) = k;             % Define the DC_k entry
> end
> Ty                        % Display the matrix
```

Since all the AC coefficients are zero, the result of the inverse transform (on one of the 9 blocks) will be a matrix with $k/8$ in every entry, where k is the DC coefficient. Display this with:

```
> imagesc(invdct(Ty,8));    % Display the result
```

---

[9]There is a scaling involved at this stage.

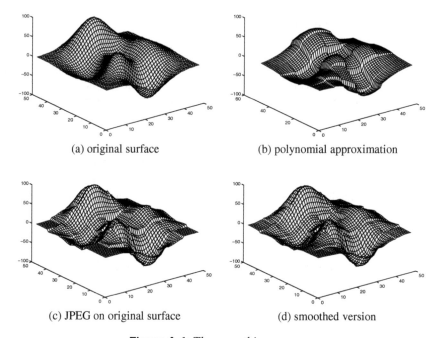

(a) original surface   (b) polynomial approximation

(c) JPEG on original surface   (d) smoothed version

Figure A.4: *The smoothing process.*

Figure A.5: *'bird' with aggressive quantizing, then smoothed.*

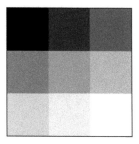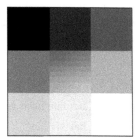

Figure A.6: *Original and the "smoothed" superblock.*

Finally, we can see the results of the smoothing process with

```
> Tz = deblock(Ty, stdQ);  % Smooth
> imagesc(invdct(Tz,8));   % Display the result
```

The two images appear in Figure A.6.

This simple example shows what the smoothing process would do with such a superblock, but it is not very clear if this is a viable process. Looking back at Figure A.5, we might even ask why the process didn't do better. As an experiment, the smoothing procedure can be directed to consider more (or less) than the five lowest-frequency AC coefficients.

A better question might be "How much could be expected?" Recall that the procedure uses only the DC coefficients of nearest neighbors. This scheme is attractive, in part because of its simplicity and the fact that it can be used as a back-end procedure to JPEG (regardless if the original file was compressed with this in mind). However, JPEG achieves its rather impressive compression by discarding information. The smoothing procedure sometimes makes good guesses about the missing data, but it cannot recover the original information.

A.2 Reference

basis

Purpose Calculate basis matrices.

Synopsis basis(v, u)

basis(v, u, N)

Description basis finds the (v, u) basis matrix of size $N \times N$. N defaults to 8.

See also basisgrid, psum, psumgrid

basisgrid

Purpose Create matrix of $N \times N$ basis elements.

Synopsis basisgrid

basisgrid(N)

Description The $N \times N$ basis matrices in a cosine expansion are returned in a matrix (with $(N + 1)N$ rows, due to the space between submatrices). N defaults to 8.

See also basis, psumgrid

dct

Purpose Perform discrete cosine transform on an image.

Synopsis dct

dct(X)

dct(X, N)

Description dct cuts an $m \times n$ image X into $N \times N$ subimages and transforms these images. If X is not given, then $X = $ ans. If N is not given, then N is the size of the global quantizing matrix QMAT, or, if QMAT is undefined or of size 0, then $N = 8$.

See also invdct

dctmat

Purpose Build the $N \times N$ matrix for the cosine transform.

Synopsis dctmat

dctmat(N)

Description The $N \times N$ matrix for the cosine transform is returned. N defaults to 8.

Example If X is $N \times N$, then

$$c = \text{dctmat}(N); \ c * X * c'$$

would give the cosine transform of X.

deblock

Purpose Smooth blocking artifacts.

Synopsis deblock(X)
deblock(X, Q)
deblock(X, k)
deblock(X, Q, k)

Description deblock performs a smoothing procedure in an effort to reduce blocking artifacts (the telltale discontinuities between blocks which often follow aggressive quantizing). For a given block, a polynomial is fitted using the DC coefficients from nearest neighbors. The k (default 5) lowest-frequency zero AC coefficients are replaced by values from the polynomial. Replaced values are clamped so that (in magnitude) they do not exceed values which would not have quantized to zero. If the quantizer Q is not given, then the global QMAT is used.

The procedure is applied to the matrix Ty which (at least conceptually) is the result of

$$x \xrightarrow{transform} Tx \xrightarrow{quantize} QTx \xrightarrow{dequantize} Ty,$$

Bugs It's slow.

deblockc [Maple]

Purpose Calculate AC coefficients in a smoothing scheme.

Synopsis deblockc()
deblockc(k)
deblockc(k, N)

Description deblockc is a Maple procedure to find the AC coefficients (in terms of the DC coefficients) which could be used as part of a smoothing procedure to reduce blocking artifacts (the telltale discontinuities between blocks which often follow aggressive quantizing) in JPEG.

This procedure may be of interest since it symbolically solves for the coefficients; however, no routines depend on deblockc. See the text for examples and more information. The procedure is described in Pennebaker and Mitchell [57].

Example

$$AC := \text{deblockc}(k, N) :$$

will display k (default value is 5) of the terms in the zigzag sequence $\{AC_{0,1}, AC_{1,0}, AC_{2,0}, \dots\}$ (first symbolically, then numerically) and the results are stored in AC. N is the blocksize used by the transform (default is 8).

dequant

> **Purpose** Unapply quantizing matrix.
>
> **Synopsis** dequant
> dequant(X)
> dequant(X, Q)
>
> **Description** dequant cuts an $m \times n$ matrix X into $N \times N$ matrices and dequantizes using the matrix Q. If X is not given, then $X =$ ans. If Q is not given, then $Q =$ QMAT.

do_dct

> **Purpose** Utility routine to do forward or inverse transform.
>
> **Synopsis** do_dct(X, N, inv)
>
> **Description** do_dct cuts an $m \times n$ image X into $N \times N$ subimages and transforms these images. It is designed to be called from higher-level routines such as dct and invdct. N is the size of the transform, and *inv* is a flag indicating forward (FALSE) or inverse (TRUE).
>
> **Bugs** It's slow.
>
> **See also** dct, invdct

getint

> **Purpose** Retrieve an integer from a stream.
>
> **Synopsis** getint(*fid*)
>
> **Description** getint reads the next integer from *fid*. All characters from '#' to the end of a line are ignored. Intended to be used by other scripts, such as getpgm.

getpgm

> **Purpose** Read a graymap file.
>
> **Synopsis** getpgm(*filename*)
> [$x, maxgray$] = getpgm(*filename*)
>
> **Description** getpgm reads a pgm file (in either raw P5 or ascii P2 format) and returns a matrix suitable for display with the image function. The *maxgray* return value is the maximum gray value (see pgm(5)).
>
> **Bugs** Under Octave-1.1.1, only raw P5 format can be used.

invdct

> **Purpose** Perform inverse cosine transform on an image.
>
> **Synopsis** invdct
> invdct(X)
> invdct(X, N)

Description invdct cuts an $m \times n$ image X into $N \times N$ subimages and performs an inverse cosine transform. If X is not given, then $X =$ ans. If N is not given, then N is the size of the global quantizing matrix QMAT, or, if QMAT is undefined or of size 0, then $N = 8$.

See also dct

jpeg

Purpose Converts image via transform $\rightarrow$ quantize $\rightarrow$ invert.

Synopsis jpeg(X)
jpeg(X, Q)
$[Y, r] =$ jpeg(X)
$[Y, r] =$ jpeg(X, Q)

Description jpeg takes the original image X (which may be a matrix or a filename) and uses the cosine transform and the specified quantizing matrix Q to generate a new image Y. The process is lossy at the quantizing stage, and Y will usually differ from X.

If the quantizer is not given, then the global quantizer QMAT will be used. Initially, QMAT is the 8×8 JPEG luminance matrix. The quantizer becomes the new value for QMAT.

If the ratio r is requested, then a calculation of the "lossy compression" is performed. This is returned as a percentage and measures the amount of savings obtained at the quantizing stage.

Bugs The ratio measures only the savings obtained by removing "trailing zeros" in the submatrices. This often gives useful information about the choice of quantizer, but it is not the whole story of compression.

psum

Purpose Calculate partial sums.

Synopsis psum(X, n)

Description The nth partial sum in the cosine series for X is returned.

psumgrid

Purpose Calculate partial sum grid.

Synopsis psumgrid(X)

Description The partial sums for X are collected (in zigzag order) in a matrix.

See also basisgrid, psum

quant

> **Purpose** Apply quantizing matrix.
>
> **Synopsis** quant
> quant(X)
> quant(X, Q)
> $[Y, r] =$ quant
> $[Y, r] =$ quant(X)
> $[Y, r] =$ quant(X, Q)
>
> **Description** quant cuts an $m \times n$ matrix X into $N \times N$ submatrices and quantizes
> using the $N \times N$ matrix Q. If X is not given, then $X =$ ans. If Q is not
> given, then $Q =$ QMAT, or, if QMAT is undefined or of size 0, then Q is
> chosen to be the 8×8 JPEG luminance matrix. Q becomes the new value
> for QMAT.
>
> If the ratio r is requested, then a calculation of the "lossy compression"
> is performed. This is returned as a percentage and measures the amount
> of savings obtained at the quantizing stage.
>
> **Bugs** See the Bugs under jpeg for the interpretation of the ratio.

setdefaults

> **Purpose** Set defaults for jpegtool session.
>
> **Synopsis** setdefaults
>
> **Description** Sets various global defaults, such as the quantizing matrix QMAT
> and the colormap. Should be run at the start of every session.

trailnum

> **Purpose** Count trailing zeros in zigzag sequence.
>
> **Synopsis** trailnum(X)
>
> **Description** JPEG-like compression leads to matrices which usually contain
> zeros in the high-frequency entries. Counting the number of trailing zeros
> (in the zigzag pattern) gives an indication of the compression achieved at
> the quantizing stage; trailnum returns this count.
>
> **See also** jpeg, quant

zigzag

> **Purpose** Generate traversal-through-matrix used by jpeg.
>
> **Synopsis** zigzag
> zigzag(N)
>
> **Description** The skew-diagonal-traversal-pattern of jpeg for an $N \times N$ matrix
> is returned in an $N \times 2$ matrix. N defaults to 8. The $N \times 2$ matrix contains
> the appropriate row and column indices (starting at 1).

A.3 Obtaining Octave

Octave is a high-level language, primarily intended for numerical computations. It provides a convenient command line interface for solving linear and nonlinear problems numerically. Currently, Octave runs on Unix-like platforms, OS/2, and MS-Windows (via Cygwin).[10]

Octave is free software; you can redistribute it and/or modify it under the terms of the GNU General Public License as published by the Free Software Foundation (FSF).[11] You can get Octave from a friend who has a copy, by anonymous ftp, or by ordering a tape or CD-ROM from the FSF.

Free Software Foundation	Voice: +1-617-542-5942
59 Temple Place - Suite 330	Fax: +1-617-542-2652
Boston, MA 02111-1307, USA	E-Mail: gnu@gnu.org

Octave is developed by John W. Eaton, with contributions from many folks. Complete sources, documentatation, and ready-to-run executables for several popular systems are available via www.octave.org. The *GNU Octave Manual* by John W. Eaton "is also now available and may be ordered from http://www.network-theory.co.uk/octave/manual/. Any money raised from the sale of this book will support the development of free software. For each copy sold, $1 will be donated to the GNU Octave Development Fund."

Support Programs

Octave relies on an external program to view images. The default is John Bradley's *xv*, but *xloadimage* or *xli* can also be used (OS/2 uses *ghostview*).

xv has a generous license, and use of *xv* generally requires registration. Complete details are available with the source distribution. The latest version of *xv* (or at least a pointer to it) is available via anonymous ftp on ftp.cis.upenn.edu, in the directory pub/xv; the official site is now http://www.trilon.com.

xloadimage was written by Jim Frost and may be obtained from ftp.x.org under R5contrib. The *xli* viewer was written by Graeme Gill and is based on *xloadimage*. It may be obtained from ftp.x.org in contrib/applications.

Information on *GSview*, *ghostview*, and *ghostscript* may be obtained from http://www.cs.wisc.edu/~ghost, which is maintained by the author of *GSview*, Russell Lang.

---

[10]The authors have used Octave under GNU/Linux i486 and on Sun SPARCs running Solaris. The OS/2 port was done by Klaus Gebhardt and is available from http://hobbes.nmsu.edu in os2/apps/math/. An article by Isaac Leung on the OS/2 version appears in OS/2 eZine magazine, 16 July 2002, www.os2ezine.com.

[11]The FSF is a nonprofit organization that promotes the development and use of free software. (The word "free" refers to freedom, not price.) The GNU Project was launched in 1984 to develop a complete Unix-like operating system (the *Hurd*). Variants of the GNU operating system, which use the kernel Linux, are now widely used. GNU (guh-NEW) is a recursive acronym for "GNU's Not Unix."

Appendix B

Source Listing for LZRW1-A

This appendix contains the complete listing of the LZRW1-A dictionary scheme discussed in Chapter 9, along with some additional notes on the algorithm. The authors of this book are grateful to Dr. Ross N. Williams for permission to include the sources.[1]

Williams describes LZRW1-A as "a direct descendant of LZRW1 [83]" with optimizations. These algorithms illustrate design decisions favoring speed and low resource requirements over compression. The expensive search through the history for a match has been almost completely eliminated by the use of a hash function, illustrated in Figure B.1. At each stage, a hash of the first three characters of the lookahead gives an index into the hash table. The current value in the hash table is used for attempting a match, the hash table is updated to point at the first character of the lookahead, and then the window is moved. In the case of a match, the window will be moved by at least 3 characters, and no additional updating of the hash table occurs until the next match attempt.

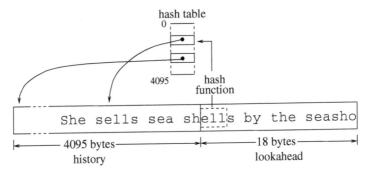

Figure B.1: *Hashing in LZRW1-A.*

In short, at each stage the hash table contains pointers to the most recent occurrence of a 3-character sequence with the same hash, and which was obtained at some previous matching attempt. From this, the dictionary can be identified as all the 3–18 character sequences which start at offsets from the hash table

---

[1]Permission obtained via email 6 August 1996. Williams can be reached electronically through http://www.rocksoft.com/ross.

(and which start in the history), together with the characters from the symbol set.

As an example, suppose the fragment '□sea' is added to the end of the "She sells..." text, giving

```
She sells sea shells by the seashore sea
        (6,3)
              (11,5)
                    (24,5)
```

where the underlines indicate characters matched against the dictionary, and the vertical line indicates the current position in the scan. At this stage, the "preferred" match is to 'e□sea' in the history. However, this fragment is not in the dictionary, and the match will actually be to the letters 'e□se' in the first two words of the example.

Definitions and documentation

The C sources for LZRW1-A are listed on the next few pages. Some changes for portability have been made; in particular, it was necessary to change the interface slightly. Williams wrote a family of LZRW algorithms,[2] and his basic framework has been retained in these sources. In addition, minor reformatting was done, and the note on patents was deleted since it was discovered that the algorithm may be covered by patent (see Appendix C). The first portion presented below consists of definitions and documentation.

```
/**********************************************************************/
/*                                                                    */
/*                             LZRW1-A.C                              */
/*                                                                    */
/**********************************************************************/
/*                                                                    */
/* Author  : Ross Williams.                                           */
/* Date     : 25 June 1991.                                           */
/* Release : 1.                                                       */
/*                                                                    */
/**********************************************************************/
/*                                                                    */
/* This file contains an implementation of the LZRW1-A data compression */
/* algorithm in C.                                                    */
/*                                                                    */
/* The algorithm is a general purpose compression algorithm that runs */
/* fast and gives reasonable compression. The algorithm is a member of */
/* the Lempel-Ziv family of algorithms and bases its compression on the */
/* presence in the data of repeated substrings.                       */
/*                                                                    */
/* The algorithm/code is based on the LZRW1 algorithm/code. Changes are: */
/*    1) The copy length range is now 3..18 instead of 3..16 as in LZRW1. */
/*    2) The code for both the compressor and decompressor has been   */
```

---

[2]LZRW 1–3 are described briefly in [26]. Complete sources are available on Williams' site (see Appendix C).

```
/*       optimized and made a little more portable.              */
/*                                                               */
/* WARNING: This algorithm is non-deterministic. Its compression */
/* performance  may vary slightly from run to run.               */
/*                                                               */
/****************************************************************/

                        /* INCLUDE FILES                        */
                        /* =============                        */
#include "port.h"       /* Defines symbols for the non portable stuff. */
#include "compress.h"   /* Defines single exported function "compress". */

/****************************************************************/

/* The following structure is returned by the "compress" function below */
/* when the user asks the function to return identifying information.    */
/* The most important field in the record is the working memory field   */
/* which tells the calling program how much working memory should be     */
/* passed to "compress" when it is called to perform a compression or    */
/* decompression.  For more information on this structure see compress.h. */

static struct compress_identity identity =
  {
  0x4B3E387B,                   /* Algorithm identification number. */
  sizeof(UBYTE**)*4096,         /* Working memory (bytes) to alg.   */
  "LZRW1-A",                    /* Name of algorithm.               */
  "1.0",                        /* Version number of algorithm.     */
  "22-Jun-1991",                /* Date of algorithm.               */
  "Public Domain",              /* Copyright notice.                */
  "Ross N. Williams",           /* Author of algorithm.             */
  "Renaissance Software",       /* Affiliation of author.           */
  "Public Domain"               /* Vendor of algorithm.             */
  };

LOCAL void compress_compress  (void *,UBYTE *,ULONG,UBYTE *,ULONG *);
LOCAL void compress_decompress(UBYTE *,ULONG,UBYTE *,ULONG *);

/****************************************************************/

/* This function is the only function exported by this module.  Depending */
/* on its first parameter, the function can be requested to compress a    */
/* block of memory, decompress a block of memory, or to identify itself.  */
/* For more information, see the specification file "compress.h".         */

EXPORT void compress(action,wrk_mem,src_adr,src_len,dst_adr,p_dst_len)
UWORD    action;         /* Action to be performed.                    */
void    *wrk_mem;        /* Address of working memory we can use.      */
UBYTE   *src_adr;        /* Address of input data.                     */
ULONG    src_len;        /* Length  of input data.                     */
UBYTE   *dst_adr;        /* Address to put output data.                */
ULONG *p_dst_len;        /* Address of longword for length of output data. */
  {
  switch (action)
    {
     case COMPRESS_ACTION_IDENTITY:
       *(struct compress_identity **) wrk_mem = &identity;
```

```
          break;
      case COMPRESS_ACTION_COMPRESS:
          compress_compress(wrk_mem,src_adr,src_len,dst_adr,p_dst_len);
          break;
      case COMPRESS_ACTION_DECOMPRESS:
          compress_decompress(src_adr,src_len,dst_adr,p_dst_len);
          break;
      }
  }


/***************************************************************************/
/*                                                                         */
/* The remainder of this file contains some definitions and two more       */
/* functions, one for compression and one for decompression. This section  */
/* contains information and definitions common to both algorithms.         */
/* Most of this information relates to the compression format which is     */
/* common to both routines.                                                */
/*                                                                         */
/***************************************************************************/
/*                                                                         */
/*                   DEFINITION OF COMPRESSED FILE FORMAT                   */
/*                   ====================================                   */
/* * A compressed file consists of a COPY FLAG followed by a REMAINDER.     */
/* * The copy flag CF uses up four bytes with the first byte being the      */
/*   least significant.                                                    */
/* * If CF=1, then the compressed file represents the remainder of the      */
/*   file exactly. Otherwise CF=0 and the remainder of the file consists    */
/*   of zero or more GROUPS, each of which represents one or more bytes.     */
/* * Each group consists of two bytes of CONTROL information followed by     */
/*   sixteen ITEMs except for the last group which can contain from one      */
/*   to sixteen items.                                                     */
/* * An item can be either a LITERAL item or a COPY item.                   */
/* * Each item corresponds to a bit in the control bytes.                   */
/* * The first control byte corresponds to the first 8 items in the         */
/*   group with bit 0 corresponding to the first item in the group and      */
/*   bit 7 to the eighth item in the group.                                 */
/* * The second control byte corresponds to the second 8 items in the       */
/*   group with bit 0 corresponding to the ninth item in the group and      */
/*   bit 7 to the sixteenth item in the group.                              */
/* * A zero bit in a control word means that the corresponding item is a    */
/*   literal item. A one bit corresponds to a copy item.                    */
/* * A literal item consists of a single byte which represents itself.      */
/* * A copy item consists of two bytes that represent from 3 to 18 bytes.   */
/* * The first byte in a copy item will be denoted C1.                      */
/* * The second byte in a copy item will be denoted C2.                     */
/* * Bits will be selected using square brackets.                           */
/*   For example: C1[0..3] is the low nibble of the first control byte.     */
/*   of copy item C1.                                                       */
/* * The LENGTH of a copy item is defined to be C1[0..3]+3 which is a        */
/*   number in the range [3,18].                                            */
/* * The OFFSET of a copy item is defined to be C1[4..7]*256+C2[0..8]        */
/*   which is a number in the range [1,4095] (the value 0 is never used).   */
/* * A copy item represents the sequence of bytes                           */
/*      text[POS-OFFSET..POS-OFFSET+LENGTH-1] where "text" is the entire     */
/*   text of the uncompressed string, and POS is the index in the text      */
/*   of the character following the string represented by all the items     */
```

```
/*      preceeding the item being defined.                              */
/*                                                                      */
/**********************************************************************/

/* The following define defines the length of the copy flag that appears */
/* at the start of the compressed file. I have decided on four bytes so  */
/* as to make the source and destination longword aligned in the case    */
/* where a copy operation must be performed.                             */
/* The actual flag data appears in the first byte. The rest are zero.    */
#define FLAG_BYTES    4  /* How many bytes does the flag use up?         */

/* The following defines define the meaning of the values of the copy    */
/* flag at the start of the compressed file.                            */
#define FLAG_COMPRESS 0  /* Signals that output was result of compression.*/
#define FLAG_COPY     1  /* Signals that output was simply copied over.   */


/**********************************************************************/
```

The compress routine

The main compression routine is listed next. The hash function can be seen in the assignment of p_entry. (See exercise 9.1.5 for information on this type of hash function.) The routine returns the original string (after the header bytes) in the case that expansion occurs (this is the overrun case).

```
LOCAL void compress_compress(p_wrk_mem,p_src_first,src_len,
                             p_dst_first,p_dst_len)
/* Input  : Specify input block using p_src_first and src_len.         */
/* Input  : Point p_dst_first to the start of the output zone (OZ).    */
/* Input  : Point p_dst_len to a ULONG to receive the output length.   */
/* Input  : Input block and output zone must not overlap.              */
/* Output : Length of output block written to *p_dst_len.              */
/* Output : Output block in Mem[p_dst_first..p_dst_first+*p_dst_len-1]. */
/* Output : May write in OZ=Mem[p_dst_first..p_dst_first+src_len+288-1].*/
/* Output : Upon completion guaranteed *p_dst_len<=src_len+FLAG_BYTES.  */
UBYTE *p_src_first,*p_dst_first; ULONG src_len,*p_dst_len; void *p_wrk_mem;
#define PS *p++!=*p_src++  /* Body of inner unrolled matching loop.     */
#define ITEMMAX 18         /* Max number of bytes in an expanded item.  */
#define TOPWORD 0xFFFF0000
{
 register UBYTE *p_src=p_src_first,*p_dst=p_dst_first;
 UBYTE *p_src_post=p_src_first+src_len,*p_dst_post=p_dst_first+src_len;
 UBYTE *p_src_max1,*p_src_max16;
 register UBYTE **hash= p_wrk_mem;
 UBYTE *p_control; register ULONG control=TOPWORD;
 p_src_max1= (src_len>=ITEMMAX) ? p_src_post-ITEMMAX+1 : p_src;
 p_src_max16= (src_len>=16*ITEMMAX) ? p_src_post-16*ITEMMAX+1 : p_src;
 *p_dst=FLAG_COMPRESS; {UWORD i; for (i=1;i<FLAG_BYTES;i++) p_dst[i]=0;}
 p_dst+=FLAG_BYTES; p_control=p_dst; p_dst+=2;

 while (TRUE)
   {register UBYTE *p,**p_entry; register UWORD unroll=16;
    register ULONG offset;
    if (p_dst>p_dst_post) goto overrun;
```

```
      if (p_src>=p_src_max16)
        {unroll=1;
         if (p_src>=p_src_max1)
            {if (p_src==p_src_post) break; goto literal;}}
      begin_unrolled_loop:
         p_entry=&hash
            [(((40543*((((p_src[0]<<4)^p_src[1])<<4)^p_src[2])))>>4) & 0xFFF];
         p=*p_entry; *p_entry=p_src; offset=p_src-p;
         if (offset>4095 || p<p_src_first || offset==0 || PS || PS || PS)
            {p_src=*p_entry; literal: *p_dst++=*p_src++; control&=0xFFFEFFFF;}
         else
            {PS || PS || PS || PS || PS || PS || PS || PS ||
             PS || PS || PS || PS || PS || PS || PS || p_src++;
             *p_dst++=((offset&0xF00)>>4)|(--p_src-*p_entry-3);
             *p_dst++=offset&0xFF;}
         control>>=1;
      end_unrolled_loop: if (--unroll) goto begin_unrolled_loop;
      if ((control&TOPWORD) == 0)
        {*p_control=control&0xFF; *(p_control+1)=(control>>8)&0xFF;
         p_control=p_dst; p_dst+=2; control=TOPWORD;}
    }

 while (control&TOPWORD) control>>=1;
 *p_control++=control&0xFF; *p_control++=control>>8;
 if (p_control==p_dst) p_dst-=2;
 *p_dst_len=p_dst-p_dst_first;
 return;

 overrun: fast_copy(p_src_first,p_dst_first+FLAG_BYTES,src_len);
          *p_dst_first=FLAG_COPY; *p_dst_len=src_len+FLAG_BYTES;
}
```

The decompress routine

Like many LZ77 schemes, the decoder is especially simple and fast. The assignment of bits to the offset and length, and the separation of the control bits into a control word minimize the amount of bit-shifting which must be done.

```
LOCAL void compress_decompress(p_src_first,src_len,p_dst_first,p_dst_len)
/* Input   : Specify input block using p_src_first and src_len.     */
/* Input   : Point p_dst_first to the start of the output zone.     */
/* Input   : Point p_dst_len to a ULONG to receive the output length. */
/* Input   : Input block and output zone must not overlap. User knows */
/* Input   : upperbound on output block length from earlier compression. */
/* Input   : In any case, maximum expansion possible is nine times.  */
/* Output  : Length of output block written to *p_dst_len.          */
/* Output  : Output block in Mem[p_dst_first..p_dst_first+*p_dst_len-1]. */
/* Output  : Writes only  in Mem[p_dst_first..p_dst_first+*p_dst_len-1]. */
UBYTE *p_src_first, *p_dst_first; ULONG src_len, *p_dst_len;
{
 register UBYTE *p_src=p_src_first+FLAG_BYTES, *p_dst=p_dst_first;
 UBYTE *p_src_post=p_src_first+src_len;
 UBYTE *p_src_max16=p_src_first+src_len-(16*2);
 register ULONG control=1;
 if (*p_src_first==FLAG_COPY)
```

```
    {fast_copy(p_src_first+FLAG_BYTES,p_dst_first,src_len-FLAG_BYTES);
     *p_dst_len=src_len-FLAG_BYTES; return;}
while (p_src!=p_src_post)
   {register UWORD unroll;
    if (control==1) {control=0x10000|*p_src++; control|=(*p_src++)<<8;}
    unroll= p_src<=p_src_max16 ? 16 : 1;
    while (unroll--)
      {if (control&1)
         {register UWORD lenmt; register UBYTE *p;
          lenmt=*p_src++; p=p_dst-(((lenmt&0xF0)<<4)|*p_src++);
          *p_dst++=*p++; *p_dst++=*p++; *p_dst++=*p++;
          lenmt&=0xF; while (lenmt--) *p_dst++=*p++;}
      else
         *p_dst++=*p_src++;
      control>>=1;
      }
   }
 *p_dst_len=p_dst-p_dst_first;
}
```

The compress header

This is a generic header file used by Williams in implementing several compression schemes.

```
/*****************************************************************************/
/*                                                                         */
/*                            COMPRESS.H                                   */
/*                                                                         */
/*****************************************************************************/
/*                                                                         */
/* Author : Ross Williams.                                                 */
/* Date   : December 1989.                                                 */
/*                                                                         */
/* This header file defines the interface to a set of functions called     */
/* 'compress', each member of which implements a particular data           */
/* compression  algorithm.                                                 */
/*                                                                         */
/* Normally in C programming, for each .H file, there is a corresponding    */
/* .C file that implements the functions promised in the .H file.          */
/* Here, there are many .C files corresponding to this header file.        */
/* Each comforming implementation file contains a single function          */
/* called 'compress' that implements a single data compression algorithm   */
/* that conforms with the interface specified in this header file.         */
/* Only one algorithm can be linked in at a time in this organization.     */
/*                                                                         */
/*****************************************************************************/
/*                                                                         */
/*                   DEFINITION OF FUNCTION COMPRESS                        */
/*                   ===============================                        */
/*                                                                         */
/* Summary of Function Compress                                            */
/* ----------------------------                                            */
/* The action that 'compress' takes depends on its first argument called   */
/* 'action'.  The function provides three actions:                         */
```

```
/*
/*     - Return information about the algorithm.                          */
/*     - Compress   a block of memory.                                    */
/*     - Decompress a block of memory.                                    */
/*                                                                        */
/* Parameters                                                             */
/* ----------                                                             */
/* See the formal C definition later for a description of the parameters. */
/*                                                                        */
/* Constants                                                              */
/* ---------                                                              */
/* COMPRESS_OVERRUN: The constant defines by how many bytes an algorithm  */
/* is allowed to expand a block during a compression operation.           */
/*                                                                        */
/* Although compression algorithms usually compress data, there will      */
/* always be data that a given compressor will expand.  Fortunately, the  */
/* degree of expansion can be limited to a single bit, by copying over    */
/* the input data if the data gets bigger during compression.  To allow   */
/* for this possibility, the first bit of a compressed representation can  */
/* be used as a flag indicating whether the input data was copied over,   */
/* or truly compressed. In practice, the first byte would be used to      */
/* store this bit so as to maintain byte alignment.                       */
/*                                                                        */
/* Unfortunately, in general, the only way to tell if an algorithm will   */
/* expand a particular block of data is to run the algorithm on the data. */
/* If the algorithm does not continuously monitor how many output bytes   */
/* it has written, it might write an output block far larger than the     */
/* input block before realizing that it has done so.  On the other hand,  */
/* continuous checks on output length are inefficient.                    */
/*                                                                        */
/* To cater for all these problems, this interface definition:            */
/* > Allows a compression algorithm to return an output block that is up  */
/*   to COMPRESS_OVERRUN bytes longer than the input block.               */
/* > Allows a compression algorithm to write up to COMPRESS_OVERRUN bytes */
/*   more than the length of the input block to the memory of the output  */
/*   block regardless of the length of the output block eventually        */
/*   returned.  This allows an algorithm to overrun the length of the     */
/*   input block in the output block by up to COMPRESS_OVERRUN bytes       */
/*   between expansion checks.                                            */
/*                                                                        */
/* The problem does not arise for decompression.                          */
/*                                                                        */
/* Identity Action                                                        */
/* ---------------                                                        */
/* > action must be COMPRESS_ACTION_IDENTITY.                             */
/* > wrk_mem must point to a pointer to struct compress_identity.         */
/* > The value of the other parameters does not matter.                   */
/* > After execution, p=*((struct identity **) wrk_mem) is a pointer      */
/*   to a structure of type compress_identity.                            */
/*   Thus, for example, after the call, p->memory will return the number  */
/*   of bytes of working memory that the algorithm requires to run.       */
/* > The values of the identity structure returned are fixed constant     */
/*   attributes of the algorithm and must not vary from call to call.     */
/*                                                                        */
/* Common Requirements for Compression and Decompression Actions          */
/* ------------------------------------------------------------           */
/*                                                                        */
```

```
/* > wrk_mem must point to an unused block of memory of a length         */
/*    specified in the algorithm's identity block. The identity block can */
/*    be obtained by making a separate call to compress, specifying the   */
/*    identity action.                                                    */
/* > The INPUT BLOCK is defined to be Memory[src_addr,src_addr+src_len-1].*/
/* > dst_len will be used to denote *p_dst_len.                           */
/* > dst_len is not read by compress, only written.                      */
/* > The value of dst_len is defined only upon termination.              */
/* > OUTPUT BLOCK is defined to be Memory[dst_addr,dst_addr+dst_len-1].   */
/*                                                                        */
/* Compression Action                                                     */
/* -----------------                                                      */
/* > action must be COMPRESS_ACTION_COMPRESS.                             */
/* > src_len must be in the range [0,COMPRESS_MAX_ORG].                   */
/* > The OUTPUT ZONE is defined to be                                     */
/*       Memory[dst_addr,dst_addr+src_len-1+COMPRESS_OVERRUN].            */
/* > The function can modify any part of the output zone regardless of    */
/*    the final length of the output block.                              */
/* > The input block and the output zone must not overlap.               */
/* > dst_len will be in the range [0,src_len+COMPRESS_OVERRUN].           */
/* > dst_len will be in the range [0,COMPRESS_MAX_COM] (from prev fact).  */
/* > The output block will consist of a representation of the input block.*/
/*                                                                        */
/* Decompression Action                                                   */
/* -------------------                                                    */
/* > action must be COMPRESS_ACTION_DECOMPRESS.                           */
/* > The input block must be the result of an earlier compression op.     */
/* > If the previous fact is true, the following facts must also be true: */
/*    > src_len will be in the range [0,COMPRESS_MAX_COM].                */
/*    > dst_len will be in the range [0,COMPRESS_MAX_ORG].                */
/* > The input and output blocks must not overlap.                        */
/* > Only the output block is modified.                                   */
/* > Upon termination, the output block will consist of the bytes         */
/*    contained in the input block passed to the earlier compression op.  */
/*                                                                        */
/**************************************************************************/

#include "port.h"

#define COMPRESS_ACTION_IDENTITY   0
#define COMPRESS_ACTION_COMPRESS   1
#define COMPRESS_ACTION_DECOMPRESS 2

#define COMPRESS_OVERRUN 1024
#define COMPRESS_MAX_COM 0x70000000
#define COMPRESS_MAX_ORG (COMPRESS_MAX_COM-COMPRESS_OVERRUN)

#define COMPRESS_MAX_STRLEN 255

/* The following structure provides information about the algorithm.     */
/* > The top bit of id must be zero. The remaining bits must be chosen    */
/*    by the author of the algorithm by tossing a coin 31 times.          */
/* > The amount of memory requested by the algorithm is specified in      */
/*    bytes and must be in the range [0,0x70000000].                      */
/* > All strings s must be such that strlen(s)<=COMPRESS_MAX_STRLEN.      */
struct compress_identity
```

```
{
  ULONG id;              /* Identifying number of algorithm.          */
  ULONG memory;          /* Number of bytes of working memory required. */

  char  *name;           /* Name of algorithm.                        */
  char  *version;        /* Version number.                           */
  char  *date;           /* Date of release of this version.          */
  char  *copyright;      /* Copyright message.                        */

  char  *author;         /* Author of algorithm.                      */
  char  *affiliation;    /* Affiliation of author.                    */
  char  *vendor;         /* Where the algorithm can be obtained.      */
};

void  compress(    /* Single function interface to compression algorithm. */
UWORD     action,  /* Action to be performed.                         */
void     *wrk_mem, /* Working memory temporarily given to routine to use. */
                   /*     If action=..IDENTITY  => Adr of id structure.   */
UBYTE    *src_adr, /* Address of input  data.                         */
ULONG     src_len, /* Length  of input  data.                         */
UBYTE    *dst_adr, /* Address of output data.                         */
ULONG *p_dst_len   /* Pointer to a longword where routine will write:  */
                   /*     If action=..COMPRESS   => Length of output data. */
                   /*     If action=..DECOMPRESS => Length of output data. */
);
```

The port header

In the original version, a fast copy routine specific to the 68000 was used. Appropriate definitions for other platforms have been placed in this header file.

```
/********************************************************************/
/*                                                                  */
/*                           PORT.H                                 */
/*                                                                  */
/********************************************************************/
/*                                                                  */
/* This module contains macro definitions and types that are likely to */
/* change between computers.                                        */
/*                                                                  */
/********************************************************************/

#ifndef DONE_PORT       /* Only do this if not previously done.     */
#define UBYTE unsigned char        /* Unsigned byte                 */
#define UWORD unsigned int         /* Unsigned word (2 bytes)       */
#define ULONG unsigned long        /* Unsigned word (4 bytes)       */
#define LOCAL static               /* For non-exported routines.    */
#define EXPORT                     /* Signals exported function.    */
#ifndef TRUE
# define TRUE 1
#endif

#ifdef HAVE_FAST_COPY_H
# include "fast_copy.h"
#else
```

```
# ifdef USE_BCOPY
#   define fast_copy bcopy
# else
#   include <string.h>
#   define fast_copy(src, dst, len) memcpy(dst, src, len)
# endif
#endif

#define DONE_PORT                    /* Don't do all this again.      */
#endif
```

Testing

A very simple program using two copies of the "She sells..." fragment as the src test string will illustrate the process. The calls may look like:

```
/* Retrieve a pointer to the compress_identity structure. */
struct compress_identity *p;
compress(COMPRESS_ACTION_IDENTITY, &p, NULL, 0, NULL, NULL);

/* allocate p->memory bytes for wrk_mem, etc.  Then call the
 * compress routine.
 */
compress(COMPRESS_ACTION_COMPRESS, wrk_mem, src, src_len, dst, &dst_len);

/* The bytes in dst should be examined.  The source can be recovered
 * with the following call.
 */
compress(COMPRESS_ACTION_DECOMPRESS, wrk_mem, dst, dst_len, src, &src_len);
```

After dst is filled by the second call to compress, it is illustrative to print the bytes and verify the contents. Also, the process might be repeated using a single copy of the "She sells..." example. After the third call to compress, the original string and length should be recovered in src and src_len, respectively.

Appendix *C*

Resources, Patents, and Illusions

This appendix is divided into three sections. The first contains information on finding software and other resources. The second introduces some of the patent issues which affect developers and users of compression algorithms. The final section is presented as somewhat of an amusement, and shows what happens when lack of basic mathematical reasoning is combined with advertising. It is centered on what is known as the "WEB Compressor," but it is a story that seems to be repeated regularly.

C.1 Resources

Vast amounts of information and source code may be found via computer using standard retrieval methods. This section lists some links which the authors have found useful, along with the site for material for this book.

Documentation and scripts for this book Material directly related to this book is maintained on

> http://www.dms.auburn.edu/compression

which is also visible by anonymous ftp under pub/compression. The documentation and scripts discussed in Appendix A appear in the 'jpegtool' subdirectory.

Frequently Asked Questions The FAQ (maintained by Jean-loup Gailly) for the newsgroups comp.compression and comp.compression.research is a good source for introductory material, pointers to source code, references, and other information. It is posted regularly on the newsgroups, and may also be obtained via http://www.faqs.org in compression-faq. The site contains FAQs for many newsgroups.

Arithmetic coding The implementation from [84] may be found via

> ftp://ftp.cpsc.ucalgary.ca/projects/ar.cod/

A separate implementation of the same coding scheme can be found in the book by Nelson and Gailly [53]. The code from [48] can be found on Moffat's page http://www.cs.mu.oz.au/~alistair.

Barnsley and Hurd [5] present arithmetic coding in the language of fractals.

Fractal image encoding Yuval Fisher's page contains a wealth of information and pointers to fractal methods:

> http://inls.ucsd.edu/Research/Fisher/Fractals

His book [20] contains a nicely-done introduction to the topic. The Waterloo Fractal Compression Project at http://links.uwaterloo.ca is another large page. Included is a pointer to the "Waterloo BragZone" which introduces a test suite and includes test results from various coders.

The second edition of [53] includes a chapter on fractal compression (by Jean-loup Gailly). On-line information is available through Nelson's page and http://www.teaser.fr/~jlgailly/.

The GNU Project and the Free Software Foundation started in 1984 to develop a complete free Unix-like operating system. A number of software packages (including *Octave*) used by the authors of this book are released under the GNU General Public License. Information about the Project and the FSF is available through http://www.fsf.org.

Info-ZIP This group supports the Zip and UnZip programs, widely used on many platforms. Their page contains pointers to source code and documentation, and information about the authors: http://www.info-zip.org. The *zlib* compression library uses the same algorithm as Zip and gzip, and the documentation may be of interest: http://www.gzip.org/zlib/.

JPEG The images in Figure 10.13 were generated with release 6 software from the Independent JPEG Group (IJG). Their software, along with a revised version of [77], errata for the first printing of [57], and other information is available via ftp://ftp.uu.net/graphics/jpeg.

League for Programming Freedom The LPF is an organization that opposes software patents and user-interface copyrights (but is not opposed to copyright on individual programs), http://lpf.ai.mit.edu/.

Mark Nelson maintains a collection of his articles, source code, information on books (such as [53]), and other notes via http://marknelson.us.

US Patent and Trademark Office A searchable database of patent information is available via http://www.uspto.gov.

Portable Network Graphics PNG is designed as a GIF successor, and uses the LZ77-variant found in *gzip*: http://www.libpng.org. A short history of PNG may be found in [60].

Wavelets The Wavelet Digest at http:/www.wavelet.org may be a good starting point. Colm Mulcahy's Mathematics Magazine article [52] contains an elementary introduction. A number of his papers, along with Matlab code and images, are available from http://www.spelman.edu/˜colm.

The images in Figures 10.18 and 10.19 were generated with Geoff Davis' Wavelet Image Compression Construction Kit, available through http://www.cs.dartmouth.edu/˜gdavis.

Ross Williams opened "Dr Ross's Compression Crypt" as the first edition of this book was going to press. http://www.ross.net/compression/ contains notes and sources for his work on various compression-related topics, including the LZRW family of algorithms.

C.2 Data compression and patents

The area of patents is a minefield for those interested in data compression. In testimony prepared by the LPF for the 1994 Patent Office Hearings, Gordon Irlam and Ross Williams write:

> As a result of software patents, many areas of software development are simply becoming out of bounds. A good example is the field of text data compression. There are now so many patents in this field that it is virtually impossible to create a data compression algorithm that does not infringe at least one of the patents. It is possible that such a patent-free algorithm exists, but it would take a team of patent attorneys weeks to establish this fact, and in the end, any of the relevant patent holders would be able to launch a crippling unfair lawsuit anyway.[1]

Companies such as Oracle, Adobe, and Autodesk presented testimony against software patents; on the other side were companies such as IBM, Intel, Microsoft, and SGI. There were middle-ground positions: Sun testified that "the [patent] system is indeed broken and needs addressing," but did not call for elimination.

Donald E. Knuth, in a letter to the patent office, writes

> In the period 1945–1980, it was generally believed that patent law did not pertain to software. However, it now appears that some people have received patents for algorithms of practical importance—e.g., Lempel-Ziv compression and RSA public key encryption—and are now legally

---

[1] From "Software Patents: An Industry at Risk" by Gordon Irlam and Ross Williams.

preventing other programmers from using these algorithms...If software patents had been commonplace in 1980, I would not have been able to create [TEX], nor would I probably ever have thought of doing it, nor can I imagine anyone else doing so...The basic algorithmic ideas that people are now rushing to patent are so fundamental, the result threatens to be like what would happen if we allowed authors to have patents on individual words and concepts...There are far better ways to protect the intellectual property rights of software developers than to take away their right to use fundamental building blocks.[2]

Perhaps the best known patent problem (other than the infamous exclusive-or patent)[3] concerns an LZ78-type scheme (Lempel-Ziv-Welch). The scheme is widely used, but the general internet user probably only heard of the patent problem when Unisys pressed for royalties in late 1994 in connection with the GIF graphics format:[4]

The LZW algorithm used in *compress* is patented by IBM and Unisys. It is also used in the V.42*bis* compression standard, in Postscript Level 2, in GIF and TIFF. Unisys sells the license to modem manufacturers for a onetime fee. CompuServe is licensing the usage of LZW in GIF products for 1.5% of the product price, of which 1% goes to Unisys; usage of LZW in non-GIF products must be licensed directly from Unisys.

And, as an example of the patent mess,

The IBM patent application was first filed three weeks before that of Unisys, but the US patent office failed to recognize that they covered the same algorithm. (The IBM patent is more general, but its claim 7 is exactly LZW.)[5]

To be precise, the patent office maintains that *algorithms* are not patentable, but an algorithm used to solve some particular problem is considered patentable. Irlam and Williams write: "Thus the 'RSA algorithm' is not patentable, but 'use of the RSA algorithm to encrypt data' is patentable...For all practical purposes, such patents can be considered patents on algorithms."

The Stac–Microsoft lawsuit involved an LZ77-type scheme:

Waterworth patented[6] the algorithm now known as LZRW1 (the "RW" is because Ross Williams reinvented it later and posted it on comp.compression on April 22, 1991). The *same* algorithm has later been patented by

---

[2]Reported in *Programming Freedom*, the Newsletter of the League for Programming Freedom, February 1995.

[3]4,197,590 Method for dynamically viewing image elements stored in a random access memory array, filed Jan 19, 1978, granted Apr 8, 1980. Cadtrack has collected large sums of money and successfully defended this patent which includes claims of "XOR feature permits part of the drawing to be moved or 'dragged' into place without erasing other parts of the drawing."

[4]A short note on the Unisys action and an introduction to software patent issues can be found in the March 1995 issue of *Scientific American* [10]. A new graphics specification, Portable Network Graphics (PNG or "ping"), was developed partly in response to the Unisys action. PNG is a lossless scheme with more capabilities than GIF.

[5]The patents are 4,814,746 (IBM) and 4,558,302 (Unisys). Much of this patent information comes from the FAQ maintained by Jean-loup Gailly, and from the LPF.

[6]4,701,745 Data compression system, filed Mar 3, 1986, granted Oct 20, 1987.

Gibson & Graybill.[7] The patent office failed to recognize that the same algorithm was patented twice, even though the wording used in the two patents is very similar.

The Waterworth patent is now owned by Stac Inc., which won a lawsuit against Microsoft, concerning the compression feature of MSDOS 6.0. Damages awarded were $120 million. (Microsoft and Stac later settled out of court.)

The Gibson & Graybill patent is very general and could be interpreted as applying to any LZ algorithm using hashing (including all variants of LZ78). However, the text of the patent and the other claims make clear that the patent should cover the LZRW1 algorithm only. (In any case the Gibson & Graybill patent is likely to be invalid because of the prior art in the Waterworth patent.)

The LZRW1 scheme was presented by Williams in [83]. The original GNU zip (*gzip*) was to have used LZRW1. Patents on arithmetic coding affect the graphics compression scheme known as JPEG:

IBM holds many patents on arithmetic coding.[8] It has patented in particular the Q-coder implementation of arithmetic coding. The arithmetic coding option of the JPEG standard requires use of the patented algorithm. No JPEG-compatible method is possible without infringing the patent, because what IBM actually claims rights to is the underlying probability model (the heart of an arithmetic coder).

From the the documents in the Independent JPEG Group's source distribution:

It appears that the arithmetic coding option of the JPEG spec is covered by patents owned by IBM, AT&T, and Mitsubishi...For this reason, support for arithmetic coding has been removed from the free JPEG software. (Since arithmetic coding provides only a marginal gain over the unpatented Huffman mode, it is unlikely that very many implementations will support it.)

More information and references (on both sides of the patent issue) can be found in the LPF materials.

---

[7]5,049,881 Apparatus and method for very high data rate-compression incorporating lossless data compression and expansion utilizing a hashing technique, filed Jun 18, 1990, granted Sep 17, 1991.

[8]Here's a few from the FAQ: 4,286,256 Method and means for arithmetic coding using a reduced number of operations, granted Aug 25, 1981.

4,463,342 A method and means for carry-over control in a high order to low order combining of digits of a decodable set of relatively shifted finite number strings, granted Jul 31, 1984.

4,467,317 High-speed arithmetic compression using concurrent value updating, granted Aug 21, 1984.

4,652,856 A multiplication-free multi-alphabet arithmetic code, granted Feb 4, 1986.

4,935,882 Probability adaptation for arithmetic coders, granted Jun 19, 1990.

C.3 Illusions

Webster defines an illusion as a "mistaken idea," and the field of data compression has had a few amusing cases. The specific example given here concerns the "WEB compressor," but the claims are perhaps the classic ones presented by those who haven't done their homework.

The FAQ has additional material on this topic. Concerning the WEB fiasco, Jean-loup Gailly writes:

> Such algorithms are claimed to be applicable recursively, that is, applying the compressor to the compressed output of the previous run, possibly multiple times. Fantastic compression ratios of over 100:1 on random data are claimed to be actually obtained.

> Such claims inevitably generate a lot of activity on comp.compression, which can last for several months. The two largest bursts of activity were generated by WEB Technologies and by Jules Gilbert. Premier Research Corporation (with a compressor called MINC) made only a brief appearance.

> Other people have also claimed incredible compression ratios, but the programs (OWS, WIC) were quickly shown to be fake (not compressing at all).

The story

The claims made by WEB Technologies are not unique, and certainly illustrate that sometimes not even common-sense analysis is performed. According to BYTE:[9]

> In an announcement that has generated quite a bit of interest, and more than a healthy dose of skepticism, WEB Technologies (Smyrna, GA) says it has developed a utility that will compress files larger than 64KB to about one-sixteenth their original size. Furthermore, WEB says its DataFiles/16 program can compress files that the program has already compressed.

We might be willing to play along at this stage. Perhaps the announcement exaggerates a little, and the company meant to say that it can achieve very good compression on a large class of files. The last sentence is also cause for concern, although it is not proof that the claims are completely bogus. However, the article goes on to say:

> In fact, according to the company, virtually any amount of data can be compressed to under 1024 bytes by using DataFiles/16 to compress its own output files multiple times.

---

[9]"Instant Gigabytes?", *BYTE Magazine* 17(6):45, June 1992. © by The McGraw-Hill Companies, Inc. All rights reserved. Used by permission.

According to the FAQ, the company's promotional materials clearly indicate that this is a lossless scheme:

> DataFiles/16 will compress all types of binary files to approximately one-sixteenth of their original size...regardless of the type of file (word processing document, spreadsheet file, image file, executable file, etc.), *no data will be lost* by DataFiles/16 [for files of at least 64K].

> Performed on a 386/25 machine, the program can complete a compression/decompression cycle on one megabyte of data in less than thirty seconds.

> The compressed output file created by DataFiles/16 can be used as the input file to subsequent executions of the program. This feature of the utility is known as recursive or iterative compression, and will enable you to compress your data files to a tiny fraction of the original size. In fact, virtually any amount of computer data can be compressed to under 1024 bytes using DataFiles/16 to compress its own output files multiple times. Then, by repeating in reverse the steps taken to perform the recursive compression, all original data can be decompressed to its original form without the loss of a single bit.

The report in the FAQ goes on to say "Decompression is done by using only the data in the compressed file; there are no hidden or extra files."

The company apparently failed to make even the most basic mathematical analysis. A simple counting argument (such as that contained in the next section) would have convinced them to review their statements.

The counting argument[10]

The WEB compressor was claimed to compress without loss *all* files of greater than 64KB in size to about 1/16th their original length. A very simple counting argument shows that this is impossible, regardless of the compression method. It is even impossible to guarantee lossless compression of all files by at least 1 bit. (Many other proofs have been posted on comp.compression, please do not post yet another one.)

Assume that the program can compress without loss all files of size at least N bits. Compress with this program all the 2^N files which have exactly N bits. All compressed files have at most $N - 1$ bits, so there are at most $2^N - 1$ different compressed files (2^{N-1} files of size $N - 1$, 2^{N-2} of size $N - 2$, and so on, down to 1 file of size 0). So at least two different input files must compress to the same output file. Hence the compression program cannot be lossless. (Much stronger results about the number of incompressible files can be obtained, but the proofs are a little more complex.)

This argument applies of course to WEB's case (take $N = 64K \cdot 8$ bits). Note that no assumption is made about the compression algorithm. The proof

---

[10]Contributed by Jean-loup Gailly. Used by permission.

applies to *any* algorithm, including those using an external dictionary, or repeated application of another algorithm, or combination of different algorithms, or representation of the data as formulas, etc. All schemes are subject to the counting argument. There is no need to use information theory to provide a proof, just basic mathematics.

This assumes, of course, that the information available to the decompressor is only the bit sequence of the compressed data. If external information such as a file name, a number of iterations, or a bit length is necessary to decompress the data, the bits necessary to provide the extra information must be included in the bit count of the compressed data. Otherwise, it would be sufficient to consider any input data as a number, use this as the file name, iteration count or bit length, and pretend that the compressed size is zero. For an example of storing information in the file name, see the program 'lmfjyh' in the 1993 International Obfuscated C Code Contest, available on all comp.sources.misc archives (Volume 39, Issue 104).

A common flaw in the algorithms claimed to compress all files is to assume that arbitrary bit strings can be sent to the decompressor without actually transmitting their bit length. If the decompressor needs such bit lengths to decode the data (when the bit strings do not form a prefix code), the number of bits needed to encode those lengths must be taken into account in the total size of the compressed data.

Conclusion

To get a more complete story, we recommend reading the BYTE article and the FAQ. The folks at BYTE were clearly skeptical, and reported:

> [A beta-test version] did create archive files that were compressed to the degree that the company claimed. The beta version decompressed these files into their original names and sizes, but, unfortunately, the contents of the decompressed files bore little resemblance to that of the original files.

The FAQ reported:

> [WEB] now says that they have put off releasing a software version of the algorithm because they are close to signing a major contract with a big company to put the algorithm in silicon. He said he could not name the company due to non-disclosure agreements, but that they had run extensive independent tests of their own and verified that the algorithm works.

> He said the algorithm is so simple that he doesn't want anybody getting their hands on it and copying it even though he said they have filed a patent on it. [He] said the silicon version would hold up much better to patent enforcement and be harder to copy.

> He claimed that the algorithm takes up about 4K of code, uses only integer math, and the current software implementation only uses a 65K buffer. He said the silicon version would likely use a parallel version and work in real-time.

Our favorite statement from the BYTE article is:

> According to...WEB Technologies' vice president of sales and marketing, the compression algorithm used by DataFiles/16 is not subject to the laws of information theory.

Also, "The company's spokespersons have declined to discuss the nature of the algorithm" and, of course, there was no product. BYTE did a followup, reporting:

> WEB said it would send us a version of the program that worked, but we never received it.

> When we attempted to follow up on the story about three months later, the company's phone had been disconnected. Attempts to reach company officers were also unsuccessful. WEB appears to have compressed itself right off the computing radar screen.[11]

Concerning stories such as the WEB compressor, Gailly adds: "similar affairs tend to come up regularly on comp.compression. The advertised revolutionary methods have all in common their supposed ability to compress significantly random or already compressed data. I will keep this item in the FAQ to encourage people to take such claims with great precautions."

The US patent office apparently doesn't read the FAQ. In July 1996, they granted a patent (5,533,051) on a "Method for Data Compression" that repeats several of the mathematically impossible claims discussed in the WEB story. Gailly has an analysis on his page (see Section C.1 or the FAQ).

---

[11]"Whatever Happened To...WEB Technologies' Amazing Compression?" *BYTE Magazine* 20(11):48, November 1995. © by The McGraw-Hill Companies, Inc. All rights reserved. Used by permission.

Appendix D

Notes on and Solutions to Some Exercises

1.1 Introduction

   2. 15

1.2 Events

   1. 30%

   2. No. Let $a = P(HH)$, $b = P(HT)$, $c = P(TH)$, and $d = P(TT)$. The considerations posed are: $b = c$ and $a + b = a + c = 1/2$. In addition, $a, b, c, d \geq 0$ and $a + b + c + d = 1$. Do these requirements determine a, b, c, and d? No: for instance, $a = b = c = d = 1/4$ and $a = d = 1/8$, $b = c = 3/8$ are two different probability assignments satisfying the requirements.

     Notice that even requiring, in addition, that $a = d$ and that $b + d = c + d = 1/2$ will not suffice to determine a, b, c, and d.

   3. 65%. With $\mathcal{S}$ standing for the population and A, B, C having the obvious meanings,

$$
\begin{aligned}
P(\mathcal{S} \setminus (A \cup B \cup C)) &= 1 - P((A \cup B) \cup C) \\
&= 1 - [P(A \cup B) + P(C) - P(A \cup B) \cap C)] \\
&= 1 - [P(A) + P(B) + P(C) - P(A \cap B) - P((A \cap C) \cup (B \cap C))] \\
&= 1 - [P(A) + P(B) + P(C) - P(A \cap B) - P(A \cap C) - P(B \cap C) \\
&\quad + P(A \cap B \cap C)] \\
&= 1 - \left[\frac{30}{100} + \frac{10}{100} + \frac{12}{100} - \frac{8}{100} - \frac{7}{100} - \frac{4}{100} + \frac{2}{100}\right] = \frac{65}{100}
\end{aligned}
$$

1.3 Conditional probability

   1. $\frac{3 \cdot 2}{14 \cdot 13} + \frac{11 \cdot 3}{14 \cdot 13} + \frac{3 \cdot 11}{14 \cdot 13} = 1 - \frac{11 \cdot 10}{14 \cdot 13} = \frac{36}{91}$

   2. $1 - \frac{11 \cdot 10 \cdot 9}{14 \cdot 13 \cdot 12} = \frac{199}{364}$

   3. $1 - \left(\frac{11}{14}\right)^2 = \frac{75}{196}$

4. (a) $\frac{1}{100,000} + \frac{99,999}{100,000}\frac{1}{150,000} = \frac{249,999}{15,000,000,000}$

(b) $\frac{\frac{99,999}{100,000}\frac{1}{150,000}}{(249,999)/15\,\text{billion}} = \frac{99,999}{249,999}$

5. $P(A \mid \text{green}) = \frac{(1/3)(7/18)}{\frac{17}{3\,18}+\frac{11}{3\,5}+\frac{13}{3\,4}} = \frac{70}{241}$, $P(\text{red}) = \frac{1}{3}(\frac{11}{18} + \frac{4}{5} + \frac{1}{4}) = \frac{299}{540}$, proportion of red balls $= 17/31$

1.4 Independence

3. $S, \emptyset, \{a,c\}, \{a,d\}, \{b,c\}, \{b,d\}$

4. $S, \emptyset$

5. Urn C will contain three red and five green balls. [Let x be the number of red and y the number of green balls in urn C. The independence requirement translates into the equation $\frac{1}{3}\frac{y}{x+y} = \frac{1}{3}\frac{1}{3}(\frac{1}{4}+1+\frac{y}{x+y})$. Solve for $\frac{y}{x+y}$: $\frac{y}{x+y} = \frac{5}{8}$. The positive integers x, y satisfying this equation with $x+y$ smallest are $x = 3$, $y = 5$.]

1.5 Bernoulli trials

1. (a) $\binom{9}{6}(\frac{7}{15})^6(\frac{8}{15})^3$ (b) $1 - [(\frac{4}{5})^9 + \frac{9}{5}(\frac{4}{5})^8]$ (c) $\sum_{k=0}^{4}\binom{9}{k}\frac{2^{9-k}}{3^9}$

2. (a) $\binom{8}{3}\frac{1}{2^8} = \frac{7}{32}$ (b) $1 - \sum_{k=0}^{2}\binom{8}{k}\frac{1}{2^8} = \frac{219}{256}$ (c) $\frac{1}{2^8}\sum_{k=0}^{3}\binom{8}{k} = \frac{93}{256}$

3. $\frac{P(n+1 \text{ heads in } 2n+2 \text{ flips})}{P(n \text{ heads in } 2n \text{ flips})} = \frac{1}{2^{2n+2}}\frac{(2n+2)!}{((n+1)!)^2} / \frac{1}{2^{2n}}\frac{(2n)!}{(n!)^2} = \frac{2n+1}{2n+2} < 1$

1.6 An elementary counting principle

3. $\binom{15}{5}\binom{8}{2}$

1.7 On drawing without replacement

2. (a) $\frac{\binom{4}{3}\binom{23}{6}}{\binom{27}{9}} = \frac{28}{325}$ (b) $\frac{\binom{10}{4}\binom{13}{4}\binom{4}{1}}{\binom{27}{9}} = \frac{56}{2185}$

1.8 Random variables and expected, or average, value

4. (a) $\frac{20}{17}$ (b) $\frac{20}{17}$

5. The probability of an atom not decaying in 24 hours (four successive 6-hour periods) is $(1-p)^4$. Therefore, the probability of an atom decaying in 24 hours is $1-(1-p)^4$. Let n be the number of undecayed atoms present at the beginning of the 24-hour period: $n(1-(1-p)^4) = \frac{n}{10}$ implies $p = 1 - (\frac{9}{10})^{1/4}$.

2.1 How is information quantified?

1. (a) $I(E_a, F_g) = \frac{1}{3}\frac{7}{18} \log \frac{\frac{1}{3}\frac{7}{18}}{\frac{1}{3}[\frac{1}{3}\frac{7}{18} + \frac{1}{3}\frac{1}{5} + \frac{1}{3}\frac{3}{4}]}$

   (b) $I(F_g|F_a) = \log \frac{18}{7}$

   (c) $I(E_a|F_g) = \log \frac{18}{7}(\frac{7}{18} + \frac{1}{5} + \frac{3}{4})$

2.2 Systems of events and mutual information

1. S_2 and S_4; S_3 and S_4

3. $I(\mathcal{E}, \mathcal{F}) = \frac{1}{36}\Big[\log \frac{1/36}{(1/6)(1/36)} + \log \frac{1/36}{(1/6)(2/36)} + \log \frac{1/36}{(1/6)(3/36)} + \log \frac{1/36}{(1/6)(4/36)} + $
   $\log \frac{1/36}{(1/6)(5/36)} + \log \frac{1/36}{(1/6)(6/36)} + \log \frac{1/36}{(1/6)(2/36)} + \cdots + \log \frac{1/36}{(1/6)(6/36)} + $
   $\log \frac{1/36}{(1/6)(5/36)} + \cdots + \log \frac{1/36}{(1/6)(6/36)} + \cdots + \log \frac{1/36}{(1/6)(1/36)}\Big] = $
   $\frac{1}{36}[24\log 3 + 10\log 2 - 10\log 5]$

4. $I(\mathcal{E}, \mathcal{F}) = \frac{1}{3}\frac{11}{18} \log \frac{(1/3)(11/18)}{(1/3)(1/3)(\frac{11}{18}+\frac{4}{5}+\frac{1}{4})} + \frac{1}{3}\frac{7}{18} \log \frac{(1/3)(7/18)}{(1/3)(1/3)(\frac{7}{18}+\frac{1}{5}+\frac{3}{4})} + $
   $\frac{1}{3}\frac{4}{5} \log \frac{(1/3)(4/5)}{(1/3)(1/3)(\frac{11}{18}+\frac{4}{5}+\frac{1}{4})} + \frac{1}{3}\frac{1}{5} \log \frac{(1/3)(1/5)}{(1/3)(1/3)(\frac{7}{18}+\frac{1}{5}+\frac{3}{4})} + $
   $\frac{1}{3}\frac{1}{4} \log \frac{(1/3)(1/4)}{(1/3)(1/3)(\frac{11}{18}+\frac{4}{5}+\frac{1}{4})} + \frac{1}{3}\frac{3}{4} \log \frac{(1/3)(3/4)}{(1/3)(3/4)(\frac{7}{18}+\frac{1}{5}+\frac{3}{4})}$

   (If simplification is desired, hire a small child.)

7. $I(\mathcal{E}, \mathcal{E}) = 0$ if and only if $\mathcal{E}$ contains one event of probability one, with the other events, if any, in $\mathcal{E}$ necessarily of zero probability.

2.3 Entropy

2. $H(\mathcal{E}) = -\sum_{k=0}^{n} \binom{n}{k} p^k (1-p)^{n-k} \log \binom{n}{k} p^k (1-p)^{n-k}$

   $H(\mathcal{S}) = -\sum_{k=0}^{n} \binom{n}{k} p^k (1-p)^{n-k} \log p^k (1-p)^{n-k}$

3. $I(\mathcal{E}, \mathcal{E}) = \sum_{i \in I}\sum_{j \in I} P(E_i \cap E_j) \log \frac{P(E_i \cap E_j)}{P(E_i)P(E_j)}$

   $= \sum_{i \in I} P(E_i \cap E_i) \log \frac{P(E_i \cap E_i)}{P(E_i)^2}$ (since $i \ne j \Rightarrow P(E_i \cap E_j) = 0$)

   $= \sum_{i \in I} P(E_i) \log \frac{1}{P(E_i)} = H(\mathcal{E}).$

5. By Theorem 2.4.4, $H(\mathcal{E}|\mathcal{F}) = H(\mathcal{E}) - I(\mathcal{E}, \mathcal{F}) = H(\mathcal{E}) \Leftrightarrow I(\mathcal{E}, \mathcal{F}) = 0 \Leftrightarrow \mathcal{E}$ and $\mathcal{F}$ are statistically independent (Theorem 2.2.13).

2.4 Information and entropy

5. Necessary and sufficient: both $\mathcal{E}$ and $\mathcal{F}$ are the trivial sort of system with exactly one event of probability 1 and the others of probability zero.

6. $H(\mathcal{E} \wedge \mathcal{E}) = H(\mathcal{E})$, $H(\mathcal{E} \mid \mathcal{E}) = 0$.

7. $H(\mathcal{E}) = \log 3$,

$$H(\mathcal{F}) = -[\tfrac{1}{3}(\tfrac{3}{8}+\tfrac{1}{3}+\tfrac{7}{13})\log\tfrac{1}{3}(\tfrac{3}{8}+\tfrac{1}{3}+\tfrac{7}{13}) + \tfrac{1}{3}(\tfrac{5}{8}+\tfrac{2}{3}+\tfrac{6}{13})\log\tfrac{1}{3}(\tfrac{5}{8}+\tfrac{2}{3}+\tfrac{6}{13})],$$

$$I(\mathcal{E},\mathcal{F}) = \tfrac{1}{3}\tfrac{3}{8}\log\frac{(1/3)(3/8)}{\tfrac{1}{3}\tfrac{1}{3}(\tfrac{3}{8}+\tfrac{1}{3}+\tfrac{7}{13})} + \tfrac{1}{3}\tfrac{1}{3}\log\frac{(1/3)(1/3)}{\tfrac{1}{3}\tfrac{1}{3}(\tfrac{3}{8}+\tfrac{1}{3}+\tfrac{7}{13})} + \tfrac{1}{3}\tfrac{7}{13}\log\frac{(1/3)(7/13)}{\tfrac{1}{3}\tfrac{1}{3}(\tfrac{3}{8}+\tfrac{1}{3}+\tfrac{7}{13})}$$

$$+ \tfrac{1}{3}\tfrac{5}{8}\log\frac{(1/3)(5/8)}{\tfrac{1}{3}\tfrac{1}{3}(\tfrac{5}{8}+\tfrac{2}{3}+\tfrac{6}{13})} + \tfrac{1}{3}\tfrac{2}{3}\log\frac{(1/3)(2/3)}{\tfrac{1}{3}\tfrac{1}{3}(\tfrac{5}{8}+\tfrac{2}{3}+\tfrac{6}{13})} + \tfrac{1}{3}\tfrac{6}{13}\log\frac{(1/3)(6/13)}{\tfrac{1}{3}\tfrac{1}{3}(\tfrac{5}{8}+\tfrac{2}{3}+\tfrac{6}{13})},$$

$$H(\mathcal{E} \wedge \mathcal{F}) = H(\mathcal{E}) + H(\mathcal{F}) - I(\mathcal{E},\mathcal{F}), \ H(\mathcal{E} \mid \mathcal{F}) = H(\mathcal{E}) - I(\mathcal{E},\mathcal{F}),$$

$$H(\mathcal{F} \mid \mathcal{E}) = H(\mathcal{F}) - I(\mathcal{E},\mathcal{F}).$$

8. Necessary and sufficient: $\mathcal{E}$ is an amalgamation of $\mathcal{F}$.

9. (b) Put all the balls of one color in B, and all the balls of the other color in C; $I(\mathcal{E},\mathcal{F}) = H(\mathcal{E})$.

(c) Make the proportions of red and green balls in B and C the same. (In this case, it will be necessary to make the numbers of red and green balls equal in each urn.) $I(\mathcal{E},\mathcal{F}) = 0$.

3.1 Discrete memoryless channels

2. (a) $p^2 q (1-p)(1-q)$ (b) $pq^2(1-p)^2$

(c) $3p^2 q(1-p)(1-q) + 3pq^2(1-p)^2 + 3p^2q(1-p)^2 + p^3(1-p)(1-q)$

(d) For $n = 1$, the answer is 1. For $n \geq 2$, the answer is $p^n + p^{n-1}(1-p) + (n-1)p^{n-2}q(1-p) + (n-2)p^{n-3}q(1-p)(1-q) + \binom{n-2}{2}p^{n-4}q^2(1-p)^2 + (n-2)p^{n-3}q(1-p)^2 + p^{n-2}(1-p)(1-q)$.

3. (a) $p^3(1-p)^2$ (b) $p^3(1-p)^2$

(c) $\binom{5}{2}p^3(1-p)^2$ (d) $p^n + np^{n-1}(1-p) + \binom{n}{2}p^{n-2}(1-p)^2$

3.2 Transition probabilities and binary symmetric channels

1. (a) $Q = \begin{bmatrix} q_{00} & q_{01} & q_{0*} \\ q_{10} & q_{11} & q_{1*} \end{bmatrix} = \begin{bmatrix} p & q & r \\ q & p & r \end{bmatrix}$ (b) $\binom{n}{k}p^{n-k}(1-p)^k$

(c) $\begin{bmatrix} p+r/2 & q+r/2 \\ q+r/2 & p+r/2 \end{bmatrix}$; yes, a BSC.   (d) $\begin{bmatrix} p & q+r \\ q & p+r \end{bmatrix}$; not a BSC unless $r = 0$.

2. (a) $p \geq (.95)^{1/15}$ (b) $p^{15} + 15p^{14}(1-p) \geq .95$ (c) $p \geq \tfrac{29}{30}$

3. (a) $p_0 p_1^2 (1-p_1)^2$ (b) $p_1^3(1-p_0)(1-p_1)$

(c) $p_0^z p_1^{n-z} + z p_0^{z-1}p_1^{n-z}(1-p_0) + (n-z)p_0^z p_1^{n-z-1}(1-p_1) + z(n-z)p_0^{z-1} \times p_1^{n-z-1}(1-p_0)(1-p_1) + \binom{z}{2}p_0^{z-2}p_1^{n-z}(1-p_0)^2 + \binom{n-z}{2}p_0^z p_1^{n-z-2}(1-p_1)^2$

5. $\widehat{B} = \{0, 1\}^3$. Transition probabilities: $q_{000,000} = q_{111,111} = p^3$; $q_{000,111} = q_{111,000} = (1-p)^3$; the other six are either $p^2(1-p)$ or $p(1-p)^2$.

6. (a) In 3.2.1 (c), $\widetilde{B} = \{0, 1\}$, $U = \begin{bmatrix} 1 & 0 \\ 0 & 1 \\ 1/2 & 1/2 \end{bmatrix}$. In 3.2.1 (d) $\widetilde{B} = \{0, 1\}$, $U = \begin{bmatrix} 1 & 0 \\ 0 & 1 \\ 0 & 1 \end{bmatrix}$.

3.3 Input frequencies

1. $p_0 = 24/53$, $p_1 = 29/53$

3. Output frequency of 0: $(1+p)/3$; output frequency of 1: $(2-p)/3$.

4. $P(b_1) = .384$, $P(b_a) = .485$, $P(b_3) = .131$

5. (a) Average cost $= .38p_1 + .22p_2 + .20p_3$. When $p_1 = .4$, $p_2 = .5$, $p_3 = .1$, the average cost is $.282$. (b) Set $p_3 = 1$, $p_1 = p_2 = 0$ to minimize cost; an unwise choice, however, since it renders the channel useless.

3.4 Channel capacity

3. $Q = \begin{bmatrix} p & 1-p \\ 1-q & q \end{bmatrix}$, $I(A, B) = p_0[p \log \frac{p}{p_0 p + p_1(1-q)} + (1-p) \log \frac{1-p}{p_0(1-p)+p_1 q}]$
   $+ p_1[(1-q) \log \frac{1-q}{p_0 p + p_1(1-q)} + q \log \frac{q}{p_0(1-p)+p_1 q}]$
   Capacity equations: $p_0 + p_1 = 1$
   $$p \log \frac{p}{p_0 p + p_1(1-q)} + (1-p) \log \frac{1-p}{p_0(1-p)+p_1 q} = C$$
   $$(1-q) \log \frac{1-q}{p_0 p + p_1(1-q)} + q \log \frac{q}{p_0(1-p)+p_1 q} = C$$

4. The capacity equations are

$$p_1 + p_2 + p_3 = 1$$

$C = .94 \log \frac{.94}{.94 p_1 + .01 p_2 + .03 p_3} + .04 \log \frac{.04}{.04 p_1 + .93 p_2 + .04 p_3} + .02 \log \frac{.02}{.02 p_1 + .06 p_2 + .93 p_3}$

$C = .01 \log \frac{.01}{.94 p_1 + .01 p_2 + .03 p_3} + .93 \log \frac{.93}{.04 p_1 + .93 p_2 + .04 p_3} + .06 \log \frac{.06}{.02 p_1 + .06 p_2 + .93 p_3}$

$C = .03 \log \frac{.03}{.94 p_1 + .01 p_2 + .03 p_3} + .04 \log \frac{.04}{.04 p_1 + .93 p_2 + .04 p_3} + .93 \log \frac{.93}{.02 p_1 + .06 p_2 + .93 p_3}$

5. $p_0 = p_1 = 1/2$ are optimal, and the capacity is $p \log \frac{2p}{p+q} + q \log \frac{2q}{p+q}$.

6. For $0 < p \le 1$ (with $0^0 = 1$ in case $p = 1$), the optimal input frequencies are
   $p_a = \frac{1-(1-p)^{1/p}}{1+p(1-p)^{(1-p)/p}}$, $p_b = \frac{(1-p)^{(1-p)/p}}{1+p(1-p)^{(1-p)/p}}$, and the capacity is $\log(1 + p(1 - p)^{(1-p)/p})$. When $p = 0$, the capacity is zero, and any relative input frequencies are "optimal."

7. (a) $p_a = (2p^p(1-p)^{1-p} + 1)^{-1}$, $p_b = p_c = (1 - p_a)/2$, $C = \log(2p^p(1 - p)^{1-p} + 1)$.

(b) $p_a = p_b = 1/2$, $C = \log 2$.

(c) No; and there is equality when and only when $p = 1/2$. Explanation left to you.

8. $p_1 = \cdots = p_n = 1/n$, $C = \log n$.

10. For $1 \le j \le n-1$; $p_j = \frac{n^n-1}{n^{n+1}-n^n+1}$, and $p_n = \frac{n}{n^{n+1}-n^n+1}$, $C = \log(n-1+n^{-n})$.

11. Capacity $= p^3 \log \frac{2p^3}{p^3+(1-p)^3} + (1-p)^3 \log \frac{2(1-p)^3}{p^3+(1-p)^3} + 3p(1-p)[p \log 2p + (1-p) \log 2(1-p)]$.

13. $p_1 = p_2 = p_3 = 1/3$, $C = 2/3 \log 2$

14. $p_1 = p_3 = 1/2$, $p_2 = 0$, $C = \frac{1}{2} \log 3 - \frac{2}{3} \log 2$

[It is somewhat shocking that one of the optimal input frequencies is zero. We are indebted to Luc Teirlinck for this example.]

4.2 Prefix-condition codes and the Kraft-McMillan inequality

1. (a) $\ell = 5$ (b) $\ell = 3$ (c) $n = 3$ (d) $m = 64$

4.3 Average code word length and Huffman's algorithm

1. (a) There are various correct answers arising from choices made in running through Huffman's algorithm, but the unique sequence of code word lengths is $2, 2, 3, 3, 3, 4, 4$. One correct answer: $e \to 01$, $a \to 10$, $d \to 001$, $b \to 110$, $f \to 111$, $g \to 0000$, $c \to 0001$.

   (b) One correct answer: $e \to 1$, $a \to *0$, $d \to *1$, $b \to **$, $f \to 00$, $g \to 01$, $c \to 0*$.

4.4 Optimizing the input frequencies

1. The answers given are not unique, and in (c) and (d), they are debatable.

   (a) $e \to 01$, $a \to 10$, $d \to 00$, $b \to 110$, $c \to 111$.

   (b) $e \to 00$, $a \to 01$, $d \to 11$, $b \to 100$, $c \to 101$.

   (c) $e \to 0$, $a \to 1$, $d \to *1$, $b \to *0$, $c \to **$.

   (d) $e \to 0$, $a \to *$, $d \to 10$, $b \to 11$, $c \to 1*$.

2. Again, the following are not unique,

   (a) $e \to 001$, $a \to 110$, $d \to 101$, $b \to 011$, $c \to 000$.

   (b) $e \to 000$, $a \to 001$, $d \to 010$, $b \to 111$, $c \to 100$.

   (c) $e \to 01*$, $a \to 0*1$, $d \to *01$, $b \to *10$, $c \to 1*0$.

   (d) $e \to 01$, $a \to 00$, $d \to 11$, $b \to **$, $c \to 1*$.

4.5 Error correction and reliability

1. (a) (i) Receive w :　00　01　0∗　10　11　1∗　∗0　∗1　∗∗

　　　　　Decode s :　a　c　a　a　b　b　a　b　a

　　(ii) $R = .79194$　(iii) $\widehat{E} = .312$

4.6 Shannon's Noisy Channel Theorem

1. $C = .95\log_2 1.9 + (.05)\log_2(.1) \approx .7136$; $H = (.5)\log_2 2 + .3\log_2(.3)^{-1} + .2\log_2 5 \approx 1.4855$; $p = 100$, if the unit of time is one second. Therefore, the upper limit on the number of source letters per second that the channel can handle, with vanishingly small error probability, is $\frac{100C}{H} \approx 48.0385$.

5.1 Replacement via encoding scheme

4. 5/3 [With $s_1 = 000, \ldots, s_8 = 111$, the original file parses into the source text $s_8 s_7 s_8 s_8 s_6 s_7 s_8 s_6 s_7 s_7$ which is encoded 010001101001101010, 18 bits compared to 30 in the original file.]

5.2 Review of the prefix condition

1. (a) Add 11; (b) add 010 and 111; (c) add 1111.

5.3 Choosing an encoding scheme

1. (a) 2/1.95　(b) 2.7/2.55
2. (a) Shannon: 2/2.2 (less than one!), Fano: 2/1.95

   (b) Shannon: 2.7/2.95 (again!), Fano: 2.7/2.55

5.4 The Noiseless Coding Theorem and Shannon's bound

1. (a) $H = .4\log_2(.4)^{-1} + 2(.25\log_2(.25)^{-1}) + .1\log_2 10 \approx 1.8610$
   $\bar{L}/H \approx 2/1.861 \approx 1.0747$

   (b) $H \approx 2.5037$, $\bar{L} = 2.7$, so $\bar{L}/H \approx 1.0784$.

2. Compute $\bar{L} = 2.6$.

   (a) $\bar{\ell} = 1.9$, so $\bar{L}/\bar{\ell} \approx 1.3684$.

   (b) It is a struggle computing $\bar{\ell}(S^2)$, but here is a tip: it is not necessary to write out the full encoding scheme; the code word lengths can be found by counting edges along the paths from the terminal (leaf) nodes of the Huffman tree to the root node. $\bar{\ell}(S^2) = 3.73$, so the compression ratio is $2\bar{L}/\bar{\ell}(S^2) = 5.2/3.73 \approx 1.3941$, assuming those "digram" frequencies are correct.

Another labor-saving remark: if the average length of a source letter, in its original incarnation as a binary word, is $\bar{L}$, then the average length of two of them together will be $2\bar{L}$, and this holds with no assumptions on the relation between the digram and the single letter frequencies.

(c) Because the relative frequency of the digram $s_i s_j$ is $f_i f_j$, by assumption, for each i and j, we have $H(S^2) = 2H(S)$, so the Shannon bound on the compression ratio is the same in both cases: $2\bar{L}/H(S^2) = 2\bar{L}/2H(S) = \bar{L}/H \approx \frac{2.6}{1.8464} \approx 1.4081$.

3. $H = L$ so the Shannon bound is $L/L = 1$. Shannon's method will give an encoding scheme with every word of length L. (In fact, if $S = \{0, 1\}^L$ is ordered correctly, the scheme will be $w \to w$ for all $w \in \{0, 1\}^L$.) Thus the compression ratio achieved by Shannon's method is 1. Huffman's algorithm cannot do worse than Shannon's method, nor better than the Shannon bound, so the compression ratio will again be 1 (and, in fact, all code words will be of length L). The same holds for Fano's method; an easy induction on L shows that the code words in the resulting scheme will all have length L.

6.1 Pure zeroth-order arithmetic coding: dfwld

1. (a) $bbbb \to 1, abcd \to 00110111, dcba \to 111110011, badd \to 0111010001$

   (b) $11 \to cbab, 010001 \to acba, 10101 \to caaa, 0101 \to acdc$

2. (a) $bbbb$ is encoded 1000. The other three words are encoded as in Exercise 6.1.1(a).

   (b) $baaca$

3. $acdcaca$

6.4 Implementing arithmetic coding

1. (a)

Next letter or rescale or underflow	L	H	New code	Underflow count
	0	16		0
d	13	16		0
$x \to 2x - 16$	10	16	1	0
$x \to 2x - 16$	4	16	1	0
b	9	10		0
$x \to 2x - 16$	2	4	1	0
$x \to 2x$	4	8	0	0
$x \to 2x$	8	16	0	0
$x \to 2x - 16$	0	16	1	0
EOF	11	13		0
$x \to 2x - 16$	6	10	1	0
$x \to 2x - 8$	4	12		1
$x \to 2x - 8$	0	16		2

The code: 11100110111. Notice that in this problem we have violated the policy guideline that EOF should be last in the ordering of the source letters.

(c) In $[0, C)$, the subintervals $[0, 3)$, $[3, 4)$, $[4, 5)$, $[5, 6)$, and $[6, 7)$ correspond to a, b, c, EOF, and d, resp. As in the text, we scale back to these subintervals to determine the current symbol (and then (6.1) is used to obtain the new current interval).

value v	current interval	interval calculation	output symbol
$1110_2 = 14$	$[0, 16)$	$w = \lfloor \frac{(14-0+1)7-1}{16-0} \rfloor = \lfloor \frac{97}{16} \rfloor = 6$	d
	$[13, 16)$	expand $x \mapsto 2(x - M/2)$	
$1100_2 = 12$	$[10, 16)$	expand $x \mapsto 2(x - M/2)$	
$\vdots$	$\vdots$	$\vdots$	$\vdots$
$1011_2 = 11$	$[0, 16)$	$w = \lfloor \frac{(11-0+1)7-1}{16-0} \rfloor = \lfloor \frac{83}{16} \rfloor = 5$	EOF

Of course, the decoding could also be done without the calculation of w. As in the decoding examples in the text, that would involve the calculation of the subintervals of $[L, H)$ corresponding to a, b, c, EOF, and d each time decoding is about to take place.

3. Here, $16 = M = 2^m$, so condition (6.2) requires $\log_2 5 = \log_2 |S| \le c$ and $c \le m - 2 = 2$, which is not possible. Although the condition in 6.4.5 is not satisfied, direct calculation in the critical case when $[L, H) = [3.9)$ or $[7, 13)$, i.e., when $H - L = 6$, shows that the last statement in the exercise holds.

4. In the expansion tagged '$x \mapsto 2x$', the values L and $H - 1$ should be replaced by $2L$ and $2H - 1$, respectively. Shifting left gives the same result as multiplying by 2, and the bitwise '$| 1$' adds 1 to an even number. Hence, $L \ll 1 = 2L$ and $(H - 1) \ll 1 | 1 = (H - 1)2 + 1 = 2H - 1$, as desired.

6. (a) The corresponding lines in the table are

string s	input	$P(1 \mid s)$	$C(s)$	$A(s)$	$A(s0)$	$A(s1)$
$\vdots$	$\vdots$	$\vdots$	$\vdots$	$\vdots$	$\vdots$	$\vdots$
0100010	1	$1/2^2$	01110.1100	.1100	.1001	.0011
01000101			01111.0101	.0011	shift 2	
01000101			0111101.0100	.1100		

(c) The handling of the stuffed bit by the decoder leads to a starting codeword of '1000'.

(d) The increased code length can be estimated. If a limit of k consecutive 1s is imposed, then a stuffed bit will be inserted every 2^k output bits on average.

7.1 Higher-order Huffman encoding

1. (a) $f_1 = .35$, $f_2 = .32$, $f_3 = .23$, $f_4 = .10$; $\bar{L} = 2.57$ and the unique code word lengths in the scheme from Huffman's algorithm are $1, 2, 3, 3$, so $\bar{\ell} = 1.98$; the compression ratio is $\bar{L}/\bar{\ell} = 2.57/1.98 \approx 1.298$.

(b) $\bar{\ell}(S^2) = 3.64$; the compression ratio is $2(2.57)/3.64 \approx 1.4121$. Review the remarks in the answer to Exercise 5.4.2(b).

(c) The following schemes are not unique, of course, but the code word lengths are, except in context s_3.

Starter scheme :	$s_1 \rightarrow 0$	Context s_1 :	$s_1 \rightarrow 0$
	$s_2 \rightarrow 10$		$s_2 \rightarrow 111$
	$s_3 \rightarrow 110$		$s_3 \rightarrow 10$
	$s_4 \rightarrow 111$		$s_4 \rightarrow 110$

Context s_2 :	$s_1 \rightarrow 110$	Context s_3 :	$s_1 \rightarrow 10$		00
	$s_2 \rightarrow 0$		$s_2 \rightarrow 0$	or	01
	$s_3 \rightarrow 10$		$s_3 \rightarrow 110$		10
	$s_4 \rightarrow 111$		$s_4 \rightarrow 111$		11

Context s_4 :	$s_1 \rightarrow 00$
	$s_2 \rightarrow 01$
	$s_3 \rightarrow 10$
	$s_4 \rightarrow 11$

Encoding	s_2	s_2	s_1	s_3	s_1	s_1	s_1	s_3	s_2	s_3	s_3	s_1	s_4
$\rightarrow$	10	0	110	10	10	0	0	10	0	10	110	10	110

(d) $[\ell_{ij}] = \begin{bmatrix} 1 & 3 & 2 & 3 \\ 3 & 1 & 2 & 3 \\ 2 & 1 & 3 & 3 \\ 2 & 2 & 2 & 2 \end{bmatrix}$; $\bar{\ell}^{(1)} = \sum_i \sum_j f_{ij} \ell_{ij} = 1.79$. Compression ratio $= \frac{2.57}{1.79} \approx 1.4358$.

2. (a) Letting $\bar{\ell}(S^2)$ denote the average length of a code word replacing a digram, using the erroneous digram frequencies to produce a scheme via Huffman's algorithm *and* to calculate the average, you obtain $\bar{\ell}(S^2) = 3.7741$, for a supposed compression ratio of $2(2.57)/3.7741 \approx 1.3619$.

(b) Using the code word lengths from the scheme alluded to in (a), but using the true digram frequencies to compute the average code word length, you obtain $\bar{\ell}(S^2) = 3.77$, for a compression ratio of $5.14/3.77 \approx 1.3634$. Notice that the encoder using the erroneous digram frequencies has done better than he thinks, but not as well as the encoder in Exercise 7.1.1(b), using the true digram frequencies.

3. In both (a) and (b) the average code word length is 1.98, the same as for zeroth-order Huffman encoding, and so the compression ratio is the same as in Exercise 7.1.1(a).

This is no accident. A little reflection reveals a moral here, that there is no point in attempting higher order encoding with a zeroth-order source.

4. $\bar{\ell}^{(0)} = 1.2$, $\bar{\ell}^{(1)} = 1.18$, $\bar{\ell}(S^2) = 1.83$.

7.2 The Shannon bound for higher-order encoding

1. $H(S) = H^{(0)}(S) =$
   $.35\log_2(.35)^{-1} + .32\log_2(.32)^{-1} + .23\log_2(.23)^{-1} + .1\log_2 10 \approx 1.8760,$
   $H(S^2) = \sum_i \sum_j f_{ij}\log_2 f_{ij} \approx 3.603,\ H^{(1)}(S) = H(S^2) - H(S) \approx 1.7271,$
   $\bar{L}/H^{(0)} \approx 2.57/1.876 \approx 1.3699,\ \bar{L}/H^{(1)} \approx 2.57/1.7271 \approx 1.4880.$
2. $H^{(0)} \approx .9219,\ H(S^2) \approx 1.7012,\ H^{(1)}(S) = H(S^2) - H^{(0)} \approx .7793.$

7.3 Higher-order arithmetic coding

1. (a) $s_2s_2s_2s_2 \to 1001,\ s_1s_2s_3s_4 \to 010000011,$
   $s_4s_3s_2s_1 \to 1111100111,\ s_2s_1s_4s_4 \to 0111101011$
   (b) $11 \to s_3s_1s_1s_3,\ 01001 \to s_1s_3s_1s_1,\ 10101 \to s_2s_3s_1s_1,\ 0101 \to s_1s_3s_1s_3$

8.1 Adaptive Huffman encoding

1. 100001101110100101101001111010100110011100100101 1
2. 1001101101111001001111001100111101111001011010
3. $s_2s_1s_2s_5s_3s_5s_1s_1s_1$
4. $s_2s_2s_2s_2s_3s_1s_2s_6s_2s_2$

8.3 Adaptive arithmetic coding

1. Note: this exercise was not done using the algorithm of Section 6.4, but rather the ivory-tower, "pure" dfwld method of Section 6.1. If you did it via Section 6.4, your answers will be different.

   (a) $bbbb \to 01011,\ abcd \to 001001,\ dcba \to 110111,\ badd \to 010011001$
   (b) $11 \to daaa,\ 010001 \to baad,\ 10101 \to cccc,\ 0101 \to bbac$

8.4 Interval and recency rank encoding

1. (a) 0010010011100100000101010010000010100010000010011001000011000101000100111
   (b) 111001111011011111001101101111011110111100110111011110111000
2. (a) $s_2s_6s_2s_2s_5s_4s_5s_5$
   (b) $s_3s_6s_4s_4s_2s_6s_2s_3s_3s_3s_5$

9.1 LZ77 (sliding window) schemes

1. Only three pairs are produced, due to the minimum on match lengths.
2. Lazy evaluation will send 'a' as a literal and (*offset*, *length*)-pairs for 'bcde' and 'fg'.

3. Both conventions favor small match distances, hopefully improving compression obtained by the Huffman back-end. Note that ignoring length-3 matches which are too distant expands the output at the dictionary stage of the scheme by 3 bits in the case of no matches starting at any of the 3 characters.

5. The material surrounding the discussion of Fibonacci hashing (with $w = 2^{16}$ and $M = 2^{12}$) in [39] applies. However, the constant $A = 40543$ does not satisfy all of the recommendations, and other choices are possible. In experiments performed by Williams during the development of LZRW1, the choice did as well as 40507 (which corresponds to a value closer to the "golden ratio recommendation") and better than 40637. The tests were not exhaustive, and Williams cautions that there may exist formulae of this basic type that do better.[1]

9.2 The LZ78 approach

1. The dictionary is

Entry	Phrase	Entry	Phrase
#0	null	#4	si
#1	M	#5	ss
#2	i	#6	ip
#3	s	#7	p

2. #1, #2, #4, #4, #6, #8, #3, #3, #2. The final trie is

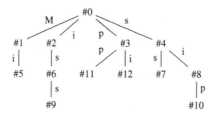

3. This is an example of the "exceptional case" in LZW; the string 'abbb' is obtained.

4. An 'a' following 'b' is allocated $1/3$ of the code space or $-\log_2 1/3 \approx 1.58$ bits.

10.2 Periodic signals and the Fourier transform

5. (a) $(z_k)^N = (e^{2\pi i k/N})^N = e^{2k\pi i} = 1$.

---

[1] Notes written during the development of LZRW1 on the experimental results were provided by Williams.

(b)

$N = 2$ $N = 3$ $N = 4$ $N = 8$

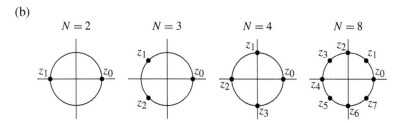

10. Take $\alpha = \langle u, w \rangle / |\langle u, w \rangle|$. *Hint:* $\|u - v\|^2 = 2$.

10.3 The cosine and sine transforms

1. $\mathcal{C}\mathbf{y} = [4.74, 12.01, 31.98, -5.27, 1.77, -2.30, 0.15, -1.67]$
   $\mathcal{S}\mathbf{y} = [-5.87, 13.24, 29.97, -0.62, 10.09, -1.39, 3.72, -1.61]$

5. $\|\hat{v}\|^2 = \langle \hat{v}, \hat{v} \rangle = \langle Ev, Ev \rangle = \langle v, \overline{E}^t Ev \rangle = \langle v, v \rangle = \|v\|^2$

10.5 An application: JPEG image compression

1. (a) $Tx = \begin{bmatrix} 640 & 0 & 0 & 0 \\ 0 & 0 & 0 & 0 \\ 0 & 0 & 0 & 0 \\ 0 & 0 & 0 & 0 \end{bmatrix}$ $QTx = \begin{bmatrix} 213 & 0 & 0 & 0 \\ 0 & 0 & 0 & 0 \\ 0 & 0 & 0 & 0 \\ 0 & 0 & 0 & 0 \end{bmatrix}$

   (b) $Tx = \begin{bmatrix} 645.00 & -3.35 & 0.50 & -0.24 \\ -3.35 & 1.21 & 0.46 & 0.00 \\ 0.50 & 0.46 & 0.00 & -0.19 \\ -0.24 & 0.00 & -0.19 & -0.21 \end{bmatrix}$ $QTx = \begin{bmatrix} 215 & -1 & 0 & 0 \\ -1 & 0 & 0 & 0 \\ 0 & 0 & 0 & 0 \\ 0 & 0 & 0 & 0 \end{bmatrix}$

   (c) $Tx = \begin{bmatrix} 280.00 & 230.70 & 40.00 & 34.33 \\ -126.17 & -11.72 & 126.17 & 28.28 \\ 40.00 & -21.65 & -40.00 & 52.26 \\ 8.97 & 28.28 & -8.97 & -68.28 \end{bmatrix}$ $QTx = \begin{bmatrix} 93 & 46 & 6 & 4 \\ -25 & -2 & 14 & 3 \\ 6 & -2 & -4 & 4 \\ 1 & 3 & -1 & -5 \end{bmatrix}$

   Dequantize: $T\tilde{x} = \begin{bmatrix} 279 & 230 & 42 & 36 \\ -125 & -14 & 126 & 33 \\ 42 & -18 & -44 & 52 \\ 9 & 33 & -13 & -75 \end{bmatrix}$ $\tilde{x} = \begin{bmatrix} 161 & 1 & 3 & -2 \\ 160 & -9 & 3 & 4 \\ 159 & 162 & -5 & 0 \\ 161 & 161 & 159 & -1 \end{bmatrix}$

   (d) $Tx = \begin{bmatrix} 379.75 & -118.54 & 23.25 & 82.93 \\ 127.60 & -22.91 & -22.43 & 85.83 \\ -129.25 & 50.24 & -77.75 & 4.36 \\ -55.83 & 4.33 & -61.34 & -125.09 \end{bmatrix}$ $QTx = \begin{bmatrix} 127 & -24 & 3 & 9 \\ 26 & -3 & -2 & 8 \\ -18 & 6 & -7 & 0 \\ -6 & 0 & -5 & -8 \end{bmatrix}$

10.6 A brief introduction to wavelets

1. A threshold value of 10 was chosen, giving the "clipped" matrices below.

   (a) $Hx = \begin{bmatrix} 640 & 0 & 0 & 0 \\ 0 & 0 & 0 & 0 \\ 0 & 0 & 0 & 0 \\ 0 & 0 & 0 & 0 \end{bmatrix}$ clipped $Hx = \begin{bmatrix} 640 & 0 & 0 & 0 \\ 0 & 0 & 0 & 0 \\ 0 & 0 & 0 & 0 \\ 0 & 0 & 0 & 0 \end{bmatrix}$

   $\tilde{x} = \begin{bmatrix} 160 & 160 & 160 & 160 \\ 160 & 160 & 160 & 160 \\ 160 & 160 & 160 & 160 \\ 160 & 160 & 160 & 160 \end{bmatrix}$

(b) $Hx = \begin{bmatrix} 645.00 & -3.00 & -0.71 & -1.41 \\ -3.00 & 1.00 & 0.71 & 0 \\ -0.71 & 0.71 & 0 & 0 \\ -1.41 & 0 & 0 & 0 \end{bmatrix}$ clipped $Hx = \begin{bmatrix} 645 & 0 & 0 & 0 \\ 0 & 0 & 0 & 0 \\ 0 & 0 & 0 & 0 \\ 0 & 0 & 0 & 0 \end{bmatrix}$

$\tilde{x} = \begin{bmatrix} 161 & 161 & 161 & 161 \\ 161 & 161 & 161 & 161 \\ 161 & 161 & 161 & 161 \\ 161 & 161 & 161 & 161 \end{bmatrix}$

(c) $Hx = \begin{bmatrix} 280.00 & 200.00 & 113.14 & 56.57 \\ -120.00 & -40.00 & 113.14 & -56.57 \\ 0 & 0 & 0 & 0 \\ -56.57 & 56.57 & 0 & -80.00 \end{bmatrix}$

clipped $Hx = \begin{bmatrix} 280.00 & 200.00 & 113.14 & 56.57 \\ -120.00 & -40.00 & 113.14 & -56.57 \\ 0 & 0 & 0 & 0 \\ -56.57 & 56.57 & 0 & -80.00 \end{bmatrix}$ $\tilde{x} = \begin{bmatrix} 160 & 0 & 0 & 0 \\ 160 & 0 & 0 & 0 \\ 160 & 160 & 0 & 0 \\ 160 & 160 & 160 & 0 \end{bmatrix}$

(d) $Hx = \begin{bmatrix} 379.75 & -141.25 & 38.54 & 5.66 \\ 139.25 & -69.75 & 78.84 & 74.95 \\ -93.34 & 50.91 & -99.00 & 44.00 \\ 89.45 & -12.37 & -44.50 & -57.00 \end{bmatrix}$

clipped $Hx = \begin{bmatrix} 379.75 & -141.25 & 38.54 & 0 \\ 139.25 & -69.75 & 78.84 & 74.95 \\ -93.34 & 50.91 & -99.00 & 44.00 \\ 89.45 & -12.37 & -44.50 & -57.00 \end{bmatrix}$ $\tilde{x} = \begin{bmatrix} 54 & 70 & 180 & 83 \\ 183 & 1 & 238 & 229 \\ 33 & 106 & 59 & 169 \\ 23 & 7 & 44 & 40 \end{bmatrix}$

2. $c_0 = c_1 = 1$, $c_j = 0$ otherwise.

Bibliography

[1] J. Aczél and Z. Daróczy. *On Measures of Information and Their Characterizations*. Academic Press, New York, 1975.

[2] N. Ahmed, T. Natarajan, and K. R. Rao. Discrete cosine transform. *IEEE Transactions on Computers*, 23(1):90–93, January 1974.

[3] M. Antonini, M. Barlaud, P. Mathieu, and I. Daubechies. Image coding using wavelet transform. *IEEE Transactions on Image Processing*, 1(2):205–220, April 1992.

[4] Robert B. Ash. *Information Theory*. Dover, New York, 1990.

[5] M. F. Barnsley and L. P. Hurd. *Fractal Image Compression*. AK Peters, Ltd., Wellesley, Massachusetts, 1992.

[6] Henry Beker and Fred Piper. *Cipher Systems: The Protection of Communication*. John Wiley & Sons, New York, 1982.

[7] Timothy C. Bell. Better OPM/L text compression. *IEEE Transactions on Communications*, 34(12):1176–1182, December 1986.

[8] Timothy C. Bell, John G. Cleary, and Ian H. Witten. *Text Compression*. Prentice-Hall, New York, 1990.

[9] Timothy C. Bell and Ian H. Witten. The relationship between greedy parsing and symbolwise text compression. *Journal of the ACM*, 41(4):708–724, July 1994.

[10] John Browning. GIF us a break. *Scientific American*, page 40, March 1995.

[11] Martin Cohn. Ziv-Lempel compressors with deferred-innovation. In Storer [70], pages 145–157.

[12] Ingrid Daubechies. Orthonormal bases of compactly supported wavelets. *Communications on Pure and Applied Mathematics*, 41(7):909–996, October 1988.

[13] ———. *Ten Lectures on Wavelets*. SIAM, Philadelphia, 1992.

[14] Lokenath Debnath. *Wavelet Transforms and their Applications*. Birkhauser, Boston, 2002.

[15] R. DeVore, B. Jawerth, and B. J. Lucier. Image compression through wavelet transform coding. *IEEE Transactions on Information Theory*, 38(2):719–146, March 1992.

[16] Peter Elias. Internal and recovery rank source coding: two on-line adaptive variable-length schemes. *IEEE Transactions on Information Theory*, 33(1):3–10, January 1987.

[17] A. Feinstein. *Foundations of Information Theory*. McGraw-Hill, New York, 1958.

[18] William Feller. *An Introduction to Probability Theory and its Applications*, volume 1. John Wiley & Sons, New York, 2nd edition, 1957.

[19] E. R. Fiala and D. H. Greene. Data compression with finite windows. *Communications of the ACM*, 32(4):490–505, 1989.

[20] Yuval Fisher, editor. *Fractal Image Compression*. Springer-Verlag, New York, 1995.

[21] Michelle J. Foster. *Operations on probabilistic finite state source automata*. PhD thesis, Auburn University, 2000.

[22] R. G. Gallager. *Information Theory and Reliable Communication*. John Wiley & Sons, New York, 1968.

[23] ———. Variations on a theme by Huffman. *IEEE Transactions on Information Theory*, 24(6):668–674, November 1978.

[24] Solomon W. Golomb. Claude Elwood Shannon (1916-2001). *Notices of the American Mathematical Society*, 49(1):8–16, January 2002.

[25] Independent JPEG Group. JPEG software release 6. Available electronically from ftp://ftp.uu.net/graphics/jpeg, Aug 1995. Email contact: jpeg-info@uunet.uu.net.

[26] Peter C. Gutmann and Timothy C. Bell. A hybrid approach to text compression. In Storer and Cohn [72], pages 225–233.

[27] R. W. Hamming. *Coding and Information Theory*. Prentice Hall, New York, 2nd edition, 1986.

[28] G. H. Hardy, John E. Littlewood, and George Polya. *Inequalities*. Cambridge University Press, London, 1934.

[29] R. V. L. Hartley. Transmission of information. *Bell System Technical Journal*, page 535, July 1928.

[30] D. G. Hoffman, D. A. Leonard, C. C. Lindner, K. T. Phelps, C. A. Rodger, and J. R. Wall. *Coding Theory: The Essentials*. Marcel Dekker, New York, 1991.

[31] R. Nigel Horspool. Improving LZW. In Storer and Reif [74], pages 332–341.

[32] ———. The effect of non-greedy parsing in Ziv-Lempel compression methods. In Storer and Cohn [73], pages 303–311.

[33] Paul G. Howard and Jeffrey Scott Vitter. Analysis of arithmetic coding for data compression. In Storer and Reif [74], pages 3–12.

[34] ———. Practical implementations of arithmetic coding. In Storer [70], pages 85–112.

[35] ———. Design and analysis of fast text compression based on quasi-arithmetic coding. In Storer and Cohn [71], pages 98–107.

[36] D. A. Huffman. A method for the construction of minimum redundancy codes. *Proceedings of the IRE*, 40(9):1098–1101, September 1952.

[37] D. S. Jones. *Elementary Information Theory*. Clarendon Press, Oxford, 1979.

[38] J. Karush. A simple proof of an inequality of McMillan. *IRE Transactions on Information Theory*, 7(2):118, April 1961.

[39] Donald E. Knuth. *The Art of Computer Programming*, volume 3. Addison-Wesley, Reading, Massachusetts, 1973.

[40] ———. Dynamic Huffman coding. *Journal of Algorithms*, 6:163–180, 1985.

[41] Neal Koblitz. *A Course in Number Theory and Cryptography*. Springer Verlag, New York, 2nd edition, 1994.

[42] Glen G. Langdon, Jr. A note on the Ziv-Lempel model for compressing individual sequences. *IEEE Transactions on Information Theory*, 29(2):284–287, March 1983.

[43] ———. An introduction to arithmetic coding. *IBM Journal of Research and Development*, 28(2):135–149, March 1984.

[44] Glen G. Langdon, Jr. and Jorma Rissanen. Compression of black-white images with arithmetic coding. *IEEE Transactions on Communications*, 29(6):858–867, June 1981.

[45] Stephane Mallat. *Multiresolution representation and wavelets*. PhD thesis, University of Pennsylvania, 1988.

[46] Michael W. Marcellin and David S. Taubman. *JPEG 2000: Image Compression, Fundamentals, Standards, and Practice*. Kluwer Academic, Boston, 2002.

[47] B. McMillan. The basic theorems of information theory. *Annals of Mathematical Statistics*, 24(2):196–219, 1953.

[48] Alistair Moffat, Radford Neal, and Ian H. Witten. Arithmetic coding revisited. Preprint. Revised from [49].

[49] ———. Arithmetic coding revisited. In Storer and Cohn [73], pages 202–211. Revised and extended in [48].

[50] Alistair Moffat, Neil Sharman, Ian H. Witten, and Timothy C. Bell. An empirical evaluation of coding methods for multi-symbol alphabets. In Storer and Cohn [71], pages 108–117. Revised and expanded in [51].

[51] ———. An empirical evaluation of coding methods for multi-symbol alphabets. *Information Processing & Management*, 30(6):791–804, November 1994.

[52] Colm Mulcahy. Plotting and scheming with wavelets. *Mathematics Magazine*, 69(5):323–343, December 1996.

[53] Mark Nelson and Jean-loup Gailly. *The Data Compression Book*. M&T Books, New York, 2nd edition, 1996.

[54] Truong Nguyen and Gilbert Strang. *Wavelets and Filter Banks*. Wellesley-Cambridge Press, Wellesley, Massachusetts, 1996.

[55] H. Nyquist. Certain factors affecting telegraph speed. *Bell System Technical Journal*, page 324, April 1924.

[56] ———. Certain topics in telegraph transmission theory. *A. I. E. E. Trans.*, 47:617, April 1928.

[57] W. B. Pennebaker and J. L. Mitchell. *JPEG Still Image Data Compression Standard*. Van Nostrand Reinhold, New York, 1992.

[58] Arthur Petrosian and Francois Meyer, editors. *Wavelets in Signal and Image Analysis: From Theory to Practice*. Kluwer Academic, Dordrecht, 2001.

[59] Majid Rabbani and Paul. W. Jones. *Digital Image Compression Techniques*. SPIE Press, Bellingham, Washington, 1991.

[60] Greg Roelofs. History of the Portable Network Graphics (PNG) format. *Linux Journal*, pages 34–40, April 1997.

[61] Walter Rudin. *Principles of Mathematical Analysis*. McGraw-Hill, New York, third edition, 1976.

[62] Khalid Sayood. *Introduction to Data Compression*. Morgan Kaufmann Publishers, San Francisco, 1995.

[63] C. E. Shannon. A mathematical theory of communication. *Bell System Technical Journal*, 27:379–423 and 623–56, 1948.

[64] ———. Certain results in coding theory for noisy channels. *Information and Control*, 1(1):6–25, 1957.

[65] C. E. Shannon and W. Weaver. *The Mathematical Theory of Communication*. University of Illinois Press, Urbana, Illinois, 1949.

[66] Paul C. Shields. The ergodic theory of discrete sample paths. *Graduate Studies in Mathematics, The American Mathematical Society*, 13, 1996.

[67] Lawrence Sirovich. *Introduction to Applied Mathematics*. Springer-Verlag, New York, 1988.

[68] Eric Stollnitz, Tony DeRose, and David Salesin. *Wavelets for Computer Graphics*. Morgan Kaufmann, San Francisco, 1996.

[69] James A. Storer. *Data Compression: Methods and Theory*. Computer Science Press, Rockville, Maryland, 1988.

[70] ———, editor. *Image and Text Compression*. Kluwer Academic Publishers, Dordrecht, 1992.

[71] James A. Storer and Martin Cohn, editors. *Proceedings, Data Compression Conference*. IEEE Computer Society Press, Los Alamitos, California, 1993.

[72] ———, editors. *Proceedings, Data Compression Conference*. IEEE Computer Society Press, Los Alamitos, California, 1994.

[73] ———, editors. *Proceedings, Data Compression Conference*. IEEE Computer Society Press, Los Alamitos, California, 1995.

[74] James A. Storer and John H. Reif, editors. *Proceedings, Data Compression Conference*. IEEE Computer Society Press, Los Alamitos, California, 1991.

[75] James A. Storer and Thomas G. Szymanski. Data compression via textual substitution. *Journal of the ACM*, 29(4):928–951, October 1982.

[76] M. Mitchell Waldrop. Claude Shannon, reluctant father of the digital age. *Technology Review*, 104(6):64–71, July/August 2001.

[77] Gregory K. Wallace. The JPEG still picture compression standard. *Communications of the ACM*, 34(4):30–44, April 1991. A revised version is available with the IJG's sources [25].

[78] David F. Walnut. *An Introduction to Wavelet Analysis*. Birkhauser, Boston, 2002.

[79] Gilbert G. Walter. *Wavelets and Other Orthogonal Systems with Applications*. CRC Press, Boca Raton, Florida, 1994.

[80] Terry A. Welch. A technique for high-performance data compression. *IEEE Computer*, 17(6):8–19, 1984.

[81] Dominic Welsh. *Codes and Cryptography*. Oxford University Press, Oxford, 1988.

[82] Ross N. Williams. *Adaptive Data Compression*. Kluwer Academic Publishers, Dordrecht, 1991.

[83] ———. An extremely fast Ziv-Lempel data compression algorithm. In Storer and Reif [74], pages 362–371.

[84] Ian H. Witten, Radford M. Neal, and John G. Cleary. Arithmetic coding for data compression. *Communications of the ACM*, 30(6):520–540, June 1987.

[85] Jacob Ziv and Abraham Lempel. A universal algorithm for seqential data compression. *IEEE Transactions on Information Theory*, 23(3):337–343, May 1977.

[86] ———. Compression of individual sequences via variable-rate encoding. *IEEE Transactions on Information Theory*, 24(5):530–536, September 1978.

Index

A

alphabet
    binary, 47
    definition, 13
amalgamation, 34, 35, 38–41, 45
approximation by partial sums, 256, 277, 278, 304–306
arithmetic coding
    underflow count, 149
arithmetic coding, 141–179, 333
    adaptive, 205, 219–221, 223, 224
    and patents, 176, 337
    bit stuffing, 177–178
    higher-order, 191–193, 219
    higher-order adaptive, 219
    implementation, 167–178
    shift condition, 170, 173
    underflow bit, 175
    underflow condition, 170, 173
    underflow count, 149–150, 175
Ash, Robert, 135
average code word length, 91, 131, 134, 135, 137, 138
    definition, 80

B

Barnsley, Michael F., 334
Bell, Timothy C., 157, 244
Bernoulli trial, 13–15, 19–22
    definition, 14
binary tree, 213, 218
binomial coefficient, 13
binomial theorem, 13
bit stuffing, 177–178

C

Calgary corpus, 234
capacity
    equations, 59

    of a channel, 57, 106–108
channel
    binary symmetric, 50, 91, 99, 111, 113, 114
    capacity, 57, 106–108
    definition, 47
    discrete, 47
    memoryless, 48
Cleary, John G., 157, 179
code, 71, 74, 75, 79, 83, 111
    ambiguous, 71, 74
    binary, 74
    block, 74, 93
    fixed-length, 74
    instantaneous, 76
    prefix-condition, 75–78, 80, 91, 123, 131
    unambiguous, 71
    uniquely decodable, 71, 74–79, 93, 95, 105, 123
    variable-length, 74
code alphabet, 71, 76, 84, 86, 90, 95, 103, 120
code text, 90
Cohn, Martin, 244
compress, 229, 234, 239, 240, 242–243
    and patents, 336
    and random data, 242
    performance comparisons, 234, 243
compression, *see also* arithmetic coding, dictionary methods, Fano's method, Huffman's algorithm, Shannon's method, JPEG, PNG
    Fourier transform, 256
    fractal methods, 301
    impossible claims, 338–341
    lossless, 119
    lossy, 119, 246
    ratio, 119, 126, 130, 133, 136–138

wavelet methods, 297
concatenation, 73
concave, 67, 189
conditional probability, 7–11
    definition, 7
cosine transform, *see* transform

D

D'Alembert-Laplace controversy, 2–4, 10
Davis, Geoff, 297, 335
deferred innovation, 240, 244
dfwld, *see* dyadic fraction with least
    denominator
dictionary guaranteed progress, 240
dictionary methods, 229–244
    decomposition, 243, 244
    LZ77, 230–237
    LZ78, 237–244
    LZC, 242
    LZFG, 244
    LZRW1, 233–234
    LZSS, 233
    LZW, 240–243
    performance comparisons among,
        234, 243
digram frequency, 182
distance of the code, 112
dyadic fraction, 127–128, 146, 162, 163,
    165
dyadic fraction with least denominator
    (dfwld), 158, 219
    definition, 142
    finding, 144–146

E

Eaton, John W., 319
Elias, Peter, 221, 222
encoding function, 71
encoding scheme, 73–75, 77, 78, 81, 82, 85,
    90, 91, 96, 103, 105, 111, 120,
    121, 123, 129, 131, 133, 138
    definition, 72
    fixed-length, 74, 75, 78, 85, 95, 96,
        105
    prefix-condition, 77, 79, 81, 84, 86,
        92, 95, 124, 134, 136, 137, 163
    uniquely decodable, 124
entropy, 30, 31, 40–45, 138, 225
    and information, 43–45

conditional, 40
definition, 40
joint, 40
of a source, 83, 134, 157–159, 162,
    189, 192, 225
of a system, 106
EOF, 159
error correction, 80, 95–106
error pattern, 112, 113
error probability
    average, 104, 106–108, 111
    maximum, 103, 107, 109, 111, 113
event, 3–7
    definition, 3
    independent, 11
    mutually exclusive, 5
    probability of, 3
    system of, 33
expected value, 18
extension
    even, 267
    odd, 269

F

Fano, R. M., 127
Fano's method, 130–131, 138
fast Fourier transform (FFT), 265
Fisher, Yuval, 334
Fourier transform, *see* transform
fractal image encoding, 334
Free Software Foundation (FSF), 319, 334
frequency
    input, *see* input frequencies
    relative input, 52
    relative source, *see* source frequencies
fundamental signal, 250

G

Gailly, Jean-loup, 179, 233, 235, 333, 338,
    339, 341
Gallager, Robert G., 209, 212, 213
Gallager order, 214–218
Gallager's method, 213, 215–217, 220
GIF, *see* Graphics Interchange Format
GNU Project, 235, 334
GNU zip (*gzip*), 229, 234–236
    hashing, 235–237
    Huffman back-end, 236, 237
    lazy evaluation, 236, 237

parameter values, 236
performance comparisons, 234, 243
Golay code, 114
Graphics Interchange Format (GIF), 229,
    240, 245, 281
    and patents, 240, 336
greedy parsing, 231, 234, 237
Gutmann, Peter C., 244
gzip, see GNU zip

H
Haar
    2D basis elements, 296
    basis elements, 295
    scaling function, 293
Hamming distance, 97, 98
Hamming weight, 111
Horspool, R. Nigel, 244
Howard, Paul G., 179
Huffman, David, 86, 127
Huffman tree, 132, 206–208, 210–215, 217,
    218
Huffman's algorithm, Huffman encoding,
    79–90, 92, 111, 122, 127,
    131–133, 136, 138, 159, 209,
    212, 213, 218
    adaptive, 205–210, 212, 213, 220–224
    higher-order, 182–186
    higher-order adaptive, 210–212
Hurd, Lyman P., 334

I
independence, statistical, *see* statistically
    independent
independent events, 11–13
Independent JPEG Group (IJG), 176, 288,
    301, 334
Info-ZIP, 229, 334
information, 25–31, 36, 37, 106, 158
    and entropy, 43–45
    rate, 57, 115–118
innovation, 240
input alphabet, 47, 81, 90, 91, 95, 102, 103
input frequencies, 52, 90, 91
input system, 56
instance, 240
instantaneous decoding, 124, 163
interval encoding, 221–223
Irlam, Gordon, 335

J
Joint Photographic Experts Group (JPEG),
    281
joint system of events, 36
Jones, Paul W., 179
JPEG, 246, 273, 281–292, 297, 301
    AC coefficient, 297, 306
    adjusting the quantizer, 307–309
    adjusting the transform size, 309
    and patents, 176, 337
    basis elements, 304
    blocking artifacts, 291, 310
    DC coefficient, 297, 306
    entropy coder, 176, 284
    luminance matrix, 285, 308
    quantizing, 285, 301
    smoothing procedure, 291, 309–313
    zigzag sequence, 305
    see also Joint Photographic Experts
        Group

K
Karush, J., 78
Knuth, Donald E., 206, 212–214, 220, 237,
    335
Knuth's algorithm, 213, 216–218
Knuth-Gallager method, 206, 212–218
Koblitz, Neal, 128
Kraft's inequality, Kraft's Theorem, 76–78,
    125
    for binary codes, 124
kth-order context, 182

L
Lagrange multiplier method, 86
Lane, Tom, 301
Langdon, Jr., Glen G., 157, 177, 229
Law of Large Numbers, 22–23
lazy evaluation, 232, 235–237
League for Programming Freedom (LPF),
    334
    patent office testimony, 335
Lempel, A., 229
Leonard, Douglas, 131, 145
LHarc, 229
LPF, *see* League for Programming Freedom
LZ77, 229–237
    sliding window, 230
LZ78, 229, 237–244

slow growth of dictionary, 230, 244
LZC, 242
LZFG, 244
LZRW1, 233–234, 321
    and patents, 235, 336
    hash function used in, 233, 237, 325
    performance comparisons, 234
    source listing, 322–331
LZSS, 233, 234, 240
    performance comparisons, 234
LZW, 240–243
    and patents, 336

M

matrix
    orthogonal, 268, 279
    unitary, 279
maximum likelihood decoding (MLD),
        97–100, 102–105, 110, 111
    incomplete, 97
    table, 97, 100, 101, 103, 105
McMillan's inequality, McMillan's
        Theorem, 77, 78, 126, 135, 139
    for binary codes, 124
MLD, *see* maximum likelihood decoding
Moffat, Alistair, 175, 179, 334
Mulcahy, Colm, 335
mutually exclusive, 5

N

n-ary Huffman sequence, 81, 86–89, 92
    definition, 79
NCWD, *see* nearest code word decoding
Neal, Radford M., 157, 175, 179
nearest code word decoding (NCWD),
        98–100, 102–105, 111–115
Nelson, Mark, 179, 233, 334
node, 132
Noiseless Coding Theorem, 83, 135, 136,
        139, 158
Noisy Channel Theorem, 106–111

O

optimal input frequencies, 57, 80, 90, 91,
        93, 95, 102
orthogonal matrix, 268, 279
output alphabet, 47, 95, 102, 103
output frequency, 54, 56
output system, 56

P

parsing, 120
patents, 240, 335–337
    compress, 336
    GIF, 240, 336
    Knuth's letter to patent office, 335
    LPF patent office testimony, 335
    LZRW1, 235, 336
    LZW, 336
    Postscript, 336
    Scientific American article, 336
    US Patent and Trademark Office, 334,
        341
    V.42*bis*, 336
PKZIP, 229, 240
PNG, *see* Portable Network Graphics
pointer guaranteed progress, 240
Poisson distribution, 194
Portable Network Graphics (PNG), 229,
        240, 245, 281, 335
prefix, 75, 76, 123, 125, 134
prefix code, *see* code
prefix condition, 75, 76, 78, 79, 90, 93, 121,
        123–126
probabilistic
    finite state automaton, 157, 188,
        198–200
    finite state source automaton, 197–204
    state (in an automaton), 188
Project Gutenberg, 196

Q

QM-coder, 176, 179

R

Rabbani, Majid, 179
random variable
    definition, 18
    expected value of, 18
reading-left-to-right, 74–76, 79, 123, 124
    definition, 73
recency rank encoding, 224–225
relative source frequencies, *see* source
        frequencies
reliability
    of a binary symmetric channel, 50,
        111
    of a channel, 102–103

of a code-and-channel system,
100–104, 114
rescaling, 146–147
Rissanen, Jorma, 157, 177, 229
Roelofs, Greg, 240

S

scaling function, 293
Shannon, C. E., 41, 57, 84, 106, 127, 136,
157, 158, 188, 189
Shannon bound, 136, 138
definition, 136
for higher-order encoding, 186–191
Shannon's method, 127–131, 134, 135,
137, 138
Shannon's sampling theorem, 264
signal, fundamental, 250
simplified nearest code word decoding
(SNCWD), 113–115
sine transform, *see* transform
SNCWD, *see* simplified nearest code word
decoding
source, 156–158, 188–191, 221, 222
model of, 188, 194–197
source alphabet, 71, 76, 77, 86, 90, 103,
108, 110, 111, 122, 123, 126,
131, 136, 138, 141, 225
source characters, 90
source entropy, 83, 134, 157–159, 162, 189,
192, 225
kth-order, 192
source frequencies, 81, 86, 87, 90, 91, 95,
96, 99, 103, 104, 106, 108, 111,
131, 156–158, 225
definition, 80
source letter, 80, 97, 105, 109, 158
source text, 80, 90, 95, 157
source word, 120
SPP, *see* strong parsing property
square wave, 251
state diagram, 195, 196
statistical independence, *see* statistically
independent
strong parsing property (SPP), 121–126,
136
definition, 121
subdividing, 143–144
system of events, 25–33
and mutual information, 33–40

definition, 33
join, 36
joint, 36
statistically independent, 36, 37, 39,
44, 45

T

Teirlinck, Luc, 66, 86
thin set, 260
transform, 247
2D cosine, 273, 277, 280
2D Fourier, 273, 276, 280
2D Haar, 296
2D orthogonal, 273, 275, 279
2D sine, 273, 280
cosine, 267, 268
fast Fourier (FFT), 265
Fourier, 254, 256
Fourier (discrete), 254
Haar (wavelet), 296
orthogonal, 259, 270
sine, 267, 269
unitary, 259
transition probability, 96, 99, 102, 103
definition, 50
trie, 238
trigram frequency, 182

U

underflow, *see* arithmetic coding
uniquely parsable, 120
unit circle, 249
unitary matrix, 271, 279

V

V.42*bis*, 229, 239, 240
and patents, 336
valid decoder-recognizer (VDR), 71–76,
79, 123, 124
VDR, *see* valid decoder-recognizer
Vitter, Jeffrey Scott, 179

W

Waterloo BragZone, 334
wavelet, 291
Haar, 293
Wavelet Image Compression Construction
Kit, 297, 335
wavelet transform, *see* transform

WEB compressor, 333, 338–341
Welsh, Dominic, 86, 135
Williams, Ross, 179, 229, 233, 235, 237,
        321, 335
Witten, Ian H., 157, 175, 179, 244

Z
zeroth-order, 120, 122, 156, 158
        replacement, 120, 122
Ziv, J., 229

9 780367 395438